Proceedings of the 15th International Symposium on Experimental and Computational Aerothermodynamics of Internal Flows

G. Rajesh · A. Sameen · C. Anbu Serene Raj
Editors

Proceedings of the 15th International Symposium on Experimental and Computational Aerothermodynamics of Internal Flows

ISAIF-15, 24–27 October 2023, Chennai, India

Springer

Editors
G. Rajesh
Department of Aerospace Engineering
Indian Institute of Technology Madras
Chennai, Tamil Nadu, India

A. Sameen
Department of Aerospace Engineering
Indian Institute of Technology Madras
Chennai, Tamil Nadu, India

C. Anbu Serene Raj
Department of Aerospace Engineering
Indian Institute of Technology Madras
Chennai, Tamil Nadu, India

ISBN 978-981-96-4081-2   ISBN 978-981-96-4082-9 (eBook)
https://doi.org/10.1007/978-981-96-4082-9

This Springer imprint is published by the registered company Springer Nature Singapore Pte Ltd.
The registered company address is: 152 Beach Road, #21-01/04 Gateway East, Singapore 189721, Singapore

Contents

Investigation of Reynolds Number Effect on Shock Wave Boundary Layer Interaction on Compressor Stator Profile

Arun Joseph, Pawel Flaszynski, Michal Piotrowicz, and Piotr Doerffer

Abstract The aerodynamic performance of a compressor significantly depends on the Reynolds number (Re) condition at which the aircraft operates. In transonic regime Shock Wave Boundary Layer Interaction (SBLI) plays an important role on aerodynamic performance and its effects are detrimental when flying at low Reynolds conditions. To investigate SBLI effects numerically and experimentally a rectilinear 3-blade test section has been designed based on the streamlines extracted from cascade simulations. The main design criterion as an inlet Mach number of 0.9 has been defined by Rolls Royce Deutschland (RRD). Based on the requirements numerical investigations have been carried out on the test section with suction slots using the EARSM turbulence model with transition features included. The advantages of using the active flow control method in the test section have been defined in this paper. A detailed comparison of flow structure at different Reynolds conditions has been performed on the suction side of the middle profile in the test section.

Keywords Internal flows · Compressors · Transonic flows · Reynolds number · SBLI · Boundary layer

1 Introduction

Modern environmental challenges regarding emission control demand a novel technology development in designing efficient engines. Designers of fans and compressors are constantly pushed to achieve higher stage pressure ratios and lower weights (Foret et al. 2021). Higher stage loading results in higher adverse pressure gradients causing complex flow patterns with chaotic separation and vortices (Hu et al. 2017). High-loading conditions generated by high-stage pressure ratios are

A. Joseph (✉) · P. Flaszynski · M. Piotrowicz
Institute of Fluid Flow Machinery Polish Academy of Sciences (IMP-PAN), Gdańsk, Poland
e-mail: ajoseph@imp.gda.pl

P. Doerffer
Gdansk University of Technology, Gdańsk, Poland

G. Rajesh et al. (eds.), *Proceedings of the 15th International Symposium on Experimental and Computational Aerothermodynamics of Internal Flows*,
https://doi.org/10.1007/978-981-96-4082-9_1

vulnerable to losses, whereas compressor blades are more sensitive to the adverse pressure gradients caused by Shock Wave Boundary Layer Interaction. The nature of SBLI critically depends on the boundary layer state (laminar or turbulent) in front of shock (Hergt et al. 2021). In transonic flows, shock waves interact with the boundary layer resulting in flow separation. The state of the boundary layer could be laminar or turbulent depending on the Reynolds number at which the engine operates. At higher altitudes, the Reynolds number drops to 4 times lower than at sea level conditions. Significant efforts were made with numerous applications to maintain the laminar boundary layer as far upstream of the typical shock as possible to reduce drag generated by skin friction (Arnal and Délery 2004). Laminar boundary layers are preferred for their ability to operate at low Reynolds numbers at higher altitudes in internal flows and as a drag-reducing design solution for external flows due to this benefit. Although laminar boundary layers benefit from less drag, they are more prone to separation in adverse pressure gradients. This may increase internal flow turbulence and total pressure losses (Davidson and Babinsky 2016).

On shock wave interaction, the laminar boundary layer is more prone to separate than the turbulent boundary layer. The engine's efficiency and aerodynamic stability are negatively impacted by rotating stalls in the blade passage if laminar separation reaches an extreme level (Tang et al. 2013). Due to the shear layer's enhanced receptivity to freestream turbulence at low Re number conditions, which causes the shear layer to quickly transition and reattach as a turbulent boundary layer, the shear layer is unconditionally unstable at these conditions (Katiyar and Sarkar 2022). Although SBLI effects in turbomachinery applications at various Reynolds numbers have been studied for decades, their adverse impacts on aerodynamic performance remain not well understood. The change from a laminar to a turbulent boundary layer exerts a significant impact on the blade's overall drag and performance at low Reynolds numbers (Becker et al. 2007). To take benefit of the laminar boundary layer and to induce transition, the transition is therefore delayed as much as possible. Flow control methods are also used upstream of the shock interaction to prevent laminar SBLI. The boundary layer transition upstream of the shock wave has a positive effect on SBLI due to the interaction being turbulent SBLI resulting reduction of separation size, losses in the wake, and shock wave unsteadiness (Grothe et al. 2021). Not much research could be found in the literature on stator cascades comparing the flow structure at different Reynolds number conditions for rather high Mach number conditions.

It is still unclear at what Reynolds conditions or altitude conditions the low Reynolds effects become more pronounced. Many investigations have been carried out on optimizing the compressor profiles for low Reynolds conditions to attain improved aerodynamic performance (Schreiber et al. 2002, 2004; Sonoda et al. 2004). It is still an open question about the 3D structure of separation bubbles and secondary flows caused by the end-wall effects at different Reynolds conditions. This paper gives an insight into the comparison of flow structure for a highly loaded compressor stator cascade at three different Reynolds conditions. A new test section has been designed to reproduce a similar flow structure as in cascade simulations. The investigation presented in this paper includes the importance of active flow control (suction slots) to reduce the end-wall effect on the flow structure in the test section. This paper focuses on the numerical investigation performed at the representative test section model and the experimental data are unavailable at the time of the submission.

2 Geometrical Description

To investigate the SBLI effects numerically and experimentally, a new compressor linear stator cascade for low Reynolds conditions has been delivered by the TEA-MAero project partner RRD (Kahraman et al. 2023). The initial chord length of the delivered cascade was 29.03 mm and for better optical access, the chord length has been up-scaled to 50 mm with a (scale factor of 1.72). Based on the existing transonic wind tunnel configuration the span-wise length has been defined as 100 mm (aspect ratio 2). The geometrical parameters of the stator cascade have been defined in Table 1.

To carry out the experimental investigation of SBLI effects on the delivered linear cascade, a new three-profile rectilinear test section has been designed for a transonic blow-down wind tunnel as shown in Fig. 1.

The design methodology of the new test section was inspired by the previous EU project TFAST (Doerffer et al. 2020) where a single passage test section was designed for the same wind tunnel configuration. The initial shape of the upper and lower walls have been defined based on the streamlines extracted from the 2D cascade simulations. The final end-wall shape optimization has been carried out based on the boundary layer developed and blockage effects developed at the upper and lower walls. The major challenge and objective of designing a test section is to reproduce a similar cascade flow structure in the test section. Obtaining the required inlet Mach number and inflow uniformity by design parameters is one of the biggest challenges in test section design (Piotrowicz et al. 2014). There are no incident effects applied to the test section, and the relative positioning of the profiles in the test section is defined as similar to a cascade configuration. The convergent nozzle present at the inlet helps in achieving uniform flow distribution and the required inlet Mach

Table 1 Geometrical parameter of stator cascade defined by RRD

Parameter	Units	Values
Chord length	mm	50
Inlet Mach number	–	0.90
Pitch to chord ratio	–	0.60
Thickness to chord ratio	–	0.05
Blade inlet angle	°	48.14
Blade exit angle	°	16.53
Turbulent intensity	%	4

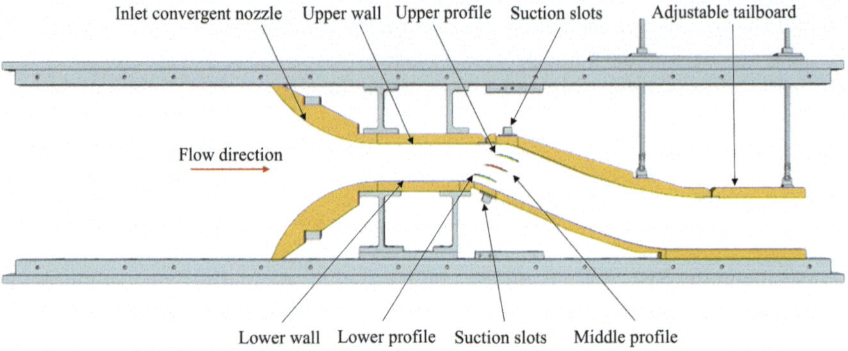

Fig. 1 The representative CAD model of the designed test section with suction slots

number upstream of the profiles (Piotrowicz et al. 2015). The outlet pressure in the test section is adjusted using an adjustable tailboard to obtain the required shock position on the suction side of the profiles. In the test section, the middle profile is mounted on the Plexiglas sidewall, while the top and lower profiles are mounted on the supporting framework. Three blades were chosen for the test section based on a number of factors such as the test section and the feasibility of the experiment. The flow structure on the suction side of the middle profile is the main focus of the investigation discussed in this paper.

3 Numerical Model Description

The commercial CFD tool from Numeca/Cadence FINE®/Turbo specially developed for turbomachinery applications has been used to perform numerical computations. On a generic structured multi-block grid configuration, the numerical algorithms from FINE®/Turbo solve steady-state 3D Reynolds Averaged Navier-Stokes equations (RANS). The structured grid for the computation was generated using Numeca/Cadence IGG (Interactive Geometry Modeler and Multi Block Structured Grid Generator). The test section mesh domain consists of 128 blocks with 25×10^6 cells as shown in Fig. 2. To achieve y^+ of 1, the mesh resolution near the walls is adapted accordingly. A second-order central difference scheme with scalar artificial dissipation formulated by Jameson (1981) was used for the spatial discretization approach. The Baseline Explicit Algebraic Reynolds Stress Model (BSL−EASRM), which is an extension of the k−ω SST turbulence model as shown in (Eqs. 1 and 2), is a two-equation nonlinear eddy viscosity turbulence model proposed by Menter (1994) has been chosen.

$$\frac{D\rho k}{Dt} = \tau_{ij}\frac{\partial u_i}{\partial x_j} - \beta^* \rho\omega k + \frac{\partial}{\partial x_j}\left[(\mu + \sigma_k\mu_t)\frac{\partial k}{\partial x_j}\right] \quad (1)$$

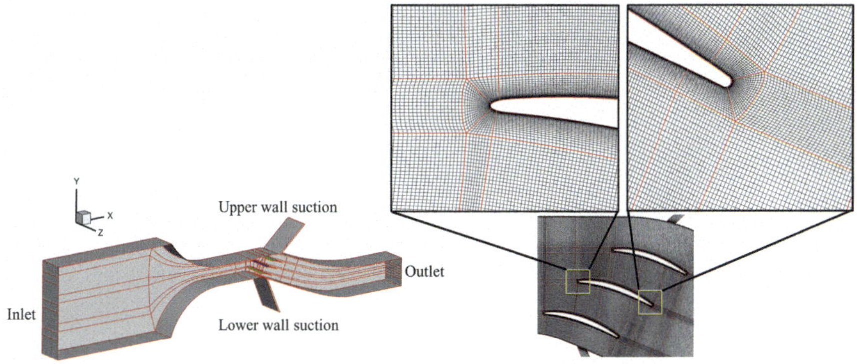

Fig. 2 Structured grid topology and mesh domain of the test section created using Numeca/IGG

$$\frac{\mathrm{D}\rho\omega}{\mathrm{D}t} = \frac{\gamma}{\nu_t}\tau_{ij}\frac{\partial u_i}{\partial x_j} - \beta\rho\omega^2 + \frac{\partial}{\partial x_j}\left[(\mu + \sigma_\omega\mu_t)\frac{\partial\omega}{\partial x_j}\right] + 2\rho(1 - F_1)\sigma_{\omega 2}\frac{1}{\omega}\frac{\partial k}{\partial x_j}\frac{\partial\omega}{\partial x_j} \tag{2}$$

To investigate the transitional effects in the flow a generalized transition model proposed by Menter and Langtry (2006) has been used for computation. This transition model is defined based on two transport equations for the intermittency (γ) and for the transition momentum thickness Reynolds number ($\tilde{Re}_{\theta t}$) as shown in (Eqs. 3 and 4).

$$\frac{\partial\rho\gamma}{\partial t} + \frac{\partial\rho u_j\gamma}{\partial x_j} = P_\gamma - E_\gamma + \frac{\partial}{\partial x_j}\left(\left(\mu + \frac{\mu_t}{\sigma_f}\right)\frac{\partial\gamma}{\partial x_j}\right) \tag{3}$$

$$\frac{\partial\rho\tilde{Re}_{\theta t}}{\partial \mathrm{t}} + \frac{\partial\rho \mathrm{u_j}\tilde{Re}_{\theta \mathrm{t}}}{\partial \mathrm{x_j}} = \mathrm{P}_{\theta\mathrm{t}} + \frac{\partial}{\partial \mathrm{x_j}}\left(\sigma_{\theta\mathrm{t}}(\mu + \mu_\mathrm{t})\frac{\partial\tilde{Re}_{\theta\mathrm{t}}}{\partial \mathrm{x_j}}\right) \tag{4}$$

The set of equations is completed with the ideal gas equation and Sutherland's law for viscosity. Numerous studies and research have demonstrated that the EARSM turbulence model is capable of accurately predicting the boundary layer separation at an adverse pressure gradient as well as secondary flows caused by end-wall interaction (Flaszynski et al. 2018; Mohammed-Taifour et al. 2021; Hellsten and Wallin 2009). The boundary conditions for computation have been defined based on the requirements specified by RRD. The inlet boundary has been defined with total quantities whereas the outlet has been defined with static pressure. The turbulent intensity has been defined as 4% and the turbulent viscosity ratio has been defined as 10 at inlet. The outlet of suction slots is defined with slightly reduced static pressure close to the end-wall to have the suction of the boundary layer from the upper and lower walls. To attain faster convergence Multi-Grid approach with 3 levels has been

Table 2 Mesh sensitivity study for test section model

Parameter	Units	Values
N_1, N_2, N_3	–	48140, 28280, 16660
h_1, h_2, h_3	mm	0.330, 0.430, 0.560
ϕ_1, ϕ_2, ϕ_3	Pa	71.380, 70.170, 69.920
r^{21}, r^{32}	–	1.304, 1.302
GCI^{21}, GCI^{32}	–	0.556, 0.118

used along with Courant-Friedrichs-Lewy (CFL) number defined as 1. In order to affirm the accuracy of the computational results, it is necessary to perform a grid convergence/Mesh sensitivity study.

3.1 Mesh Sensitivity Study

The mesh sensitivity study has been performed based on the procedure defined by Celik et al. (2008). The representative cell size (h) has been measured by creating three different mesh resolutions roughly (332×145, 280×101 and 196×85) cells (N) on the suction side of the middle blade. The y^+ in all three cases have defined to be 1. The RANS simulation has been performed for all three cases and local averaged wall shear stress on the suction side of the middle profile was chosen as a critical variable (ϕ). Based on the critical parameter the discretization error was calculated. The calculated results of the mesh sensitivity study have been shown in Table 2. Based on the calculation the mid-refinement configuration with GCI of 0.118 has been chosen for numerical investigation. This choice gives an optimal balance between accuracy and computational cost.

4 Numerical Results

The numerical results presented in this paper focuses on the investigation of flow structure in test section configuration with suction slots. To reduce the adverse effect on flow structure caused by the development of boundary layers at end-walls, a pair of suction slots have been designed on the upper and lower wall close to the blade passage. Suction slots help in removing the excess viscous layer developed on the end-walls causing flow separation and unsteadiness in blade passages. The ideal position of the suction slots at upper and lower walls have been determined based on flow uniformity achieved in each blade passage. The suction slots are connected to the outlet vacuum tank. The suction slots help in achieving required inflow conditions in the test section. Figure 3 illustrates the difference of the Mach number contour and

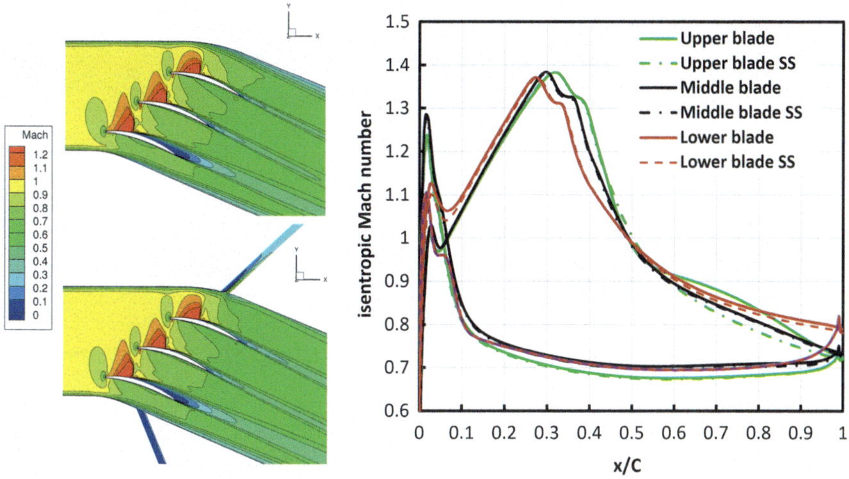

Fig. 3 Comparison of test section with and without suction slots (left) Mach number contour plot close to passage location (right) isentropic Mach number plotted on all three blades

Table 3 Boundary conditions at inlet defined in test section based on Reynolds number conditions

Cases	Reynolds number	Inlet total pressure (Pa)
High Reynolds	7.3×10^5	100,000
Mid Reynolds	5.0×10^5	67,000
Low Reynolds	2.3×10^5	31,200

the isentropic Mach number plotted at all three blades in the test section with and without suction slots. According to the isentropic Mach number plot, the suction side of the upper and lower profile benefit the most from suction slots, which also aid in producing a similar shock distribution over all three blades in comparison to the test section without suction slots. A detailed investigation of the flow structure in test sections with suction slots has been carried out to compare the effects of the Reynolds number. Based on preliminary measurements made in the IMP-PAN transonic wind tunnel, the inlet conditions in the test section based on the Reynolds number defined in Table 3 have been estimated.

The Reynolds number cases mentioned in Table 3 corresponds to different altitude conditions based on the chord length of the up-scaled cascade profile. As a design criterion, the inlet Mach number has been defined as 0.9 for all the comparisons presented in this paper as shown in Fig. 5a. The flow structure on the suction side of the middle profile has been compared using the numerical schilieren method as shown in Fig. 4.

The Numerical schilieren method is defined based on the density gradient magnitude. This shadow pattern corresponds to the expansion waves (low-density regions) and compression waves (high-density regions) that define the flow. Isentropic Mach

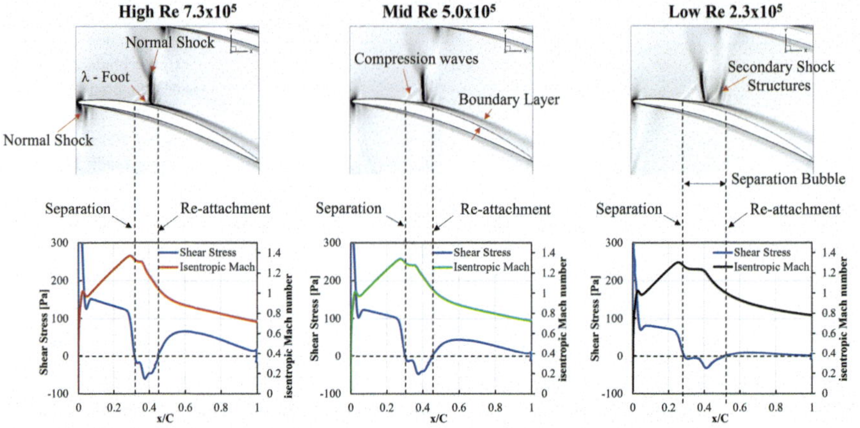

Fig. 4 Comparison of (above) numerical schilieren, (below) isentropic Mach number and wall shear stress plotted on the suction side of middle profile for three different Reynolds conditions

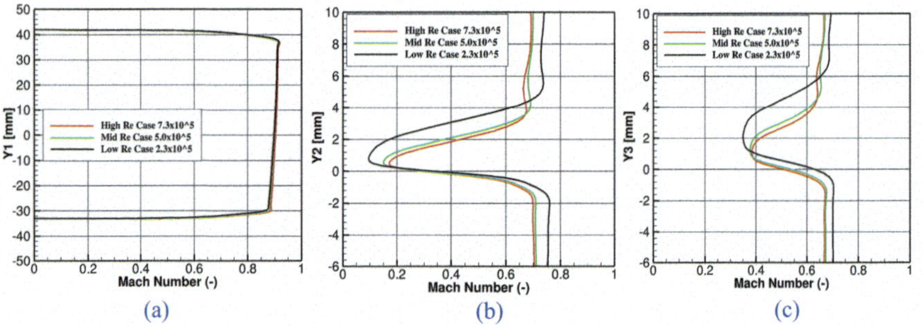

Fig. 5 Comparison of traverse at **a** 50 mm upstream the leading edge of middle profile, **b** 5 mm downstream the trailing edge of middle profile, **c** 25 mm downstream the trailing edge of middle profile

number and wall shear stress have been plotted in order to determine the distribution and position of the shock wave on the suction side of the middle profile Fig. 4. When the Reynolds number reduces, the strength of the shock wave decreases and the shock wave moves upstream, as illustrated in the numerical schlieren. The main reason is due to the location of separation bubble generated under the shock foot caused by the shock wave interaction with the boundary layer. This can be inferred from the wall shear stress plotted on the suction side of the middle profile in Fig. 4.

For high Reynolds number 7.3×10^5, a laminar separation bubble with transition and turbulent reattachment is observed between 32 and 45%. For the Mid Reynolds number 5.0×10^5 case it is observed between 30 and 47% whereas in low Reynolds condition 2.3×10^5 the separation starts at 28% and reattaches at 52% of the chord length. When Reynolds number is reduced significantly the viscous forces dominate the flow resulting in reduced shear stress which indicates higher chances of flow separation. A larger separation bubble at low Reynolds 2.3×10^5 conditions results in a larger λ-foot size and larger propagation of compression waves upstream of the normal shock wave compared to increased Reynolds conditions. It is also visible that the separation due to SBLI is more pronounced when the Reynolds number is reduced. To understand the losses due to boundary layer separation it is necessary to investigate the wake downstream of the trailing edge of the middle profile for different Reynolds conditions. From Fig. 5b and c, it is evident that the flow separation is larger when the Reynolds number is decreased from high to low conditions. It is known that high Reynolds condition leads to a thinner boundary layer compared to low Reynolds conditions (Wassell 1968).

When shock waves interact with thicker boundary layers at low Reynolds conditions, the separation increases, which is reflected in the wake as higher losses. Another aspect is visible in Fig. 5b and c, the Mach number is slightly higher for low Reynolds 2.3×10^5 case above the middle blade wake which indicates higher acceleration due to contraction of flow passage due to increased flow separation.

An extensive boundary layer analysis has been carried out in order to better understand the impacts of the Reynolds number on the suction side of the middle profile. The comparison of the velocity profile at the selected traverse has been plotted with respect to axial chord length has been shown in Fig. 6. The velocity profile appears to be laminar at traverse x/C = 0.1. For both the high 7.3×10^5 and mid 5.0×10^5 Reynolds cases, the boundary layer thickness is too thin and comparable. For the low Reynolds case, the traverse is already in a separation bubble at x/C = 0.3, indicated by the negative value of velocity, implying reversed flow. The separation bubble in the other two Reynolds cases, however, begins further downstream and is where the

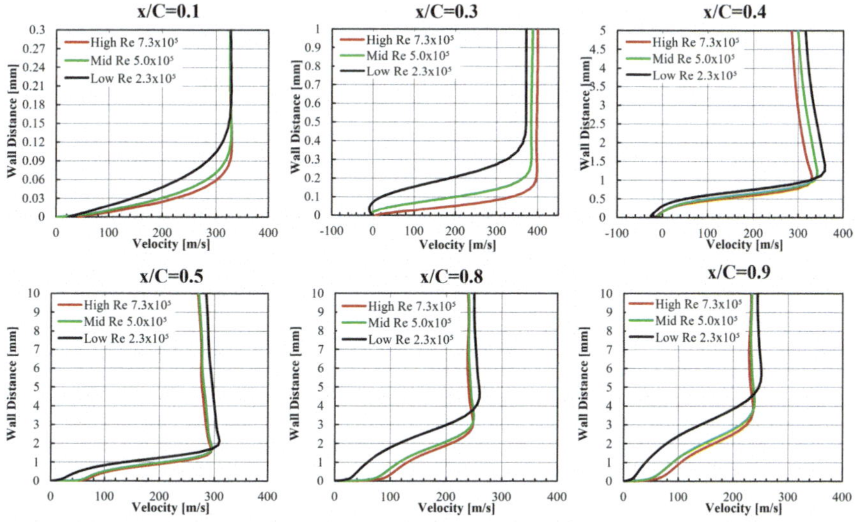

Fig. 6 Comparison velocity profiles at selected traverse location on the suction side of middle profile

shock appears. The traverse position for all three cases lies within the separation bubble at x/C = 0.4. From x/C = 0.5 it is evident that the flow is separated due to shock interaction with the boundary layer. A distinct boundary layer with a turbulent shape is visible closer to the wall in the traverses x/C = 0.8 and x/C = 0.9. These findings make it clear that the boundary layer behavior for high and mid-Reynolds conditions on the suction side is comparable.

Based on the traverse plotted at 9 locations across the suction side of the middle profile a detailed boundary layer thickness and integral parameters of the boundary layer have been plotted in Fig. 7. In the boundary layer thickness plot, it is evident that the thickness of the boundary layer increases when the Reynolds number is reduced. In the high and mid-Reynolds cases the thickness of the boundary layer is comparable till shock interaction and differences are visible downstream of the reattachment zone. It is clear from the thickness of the boundary layer that laminar SBLI is unfavorable for conditions with low Reynolds numbers because of its enhanced susceptibility for separation. Based on the displacement thickness plot, the spike in the increase of displacement thickness shows the location of separation. The displacement thickness increases significantly downstream of the shock when the Reynolds number is reduced. From Fig. 7 the differences in the momen-

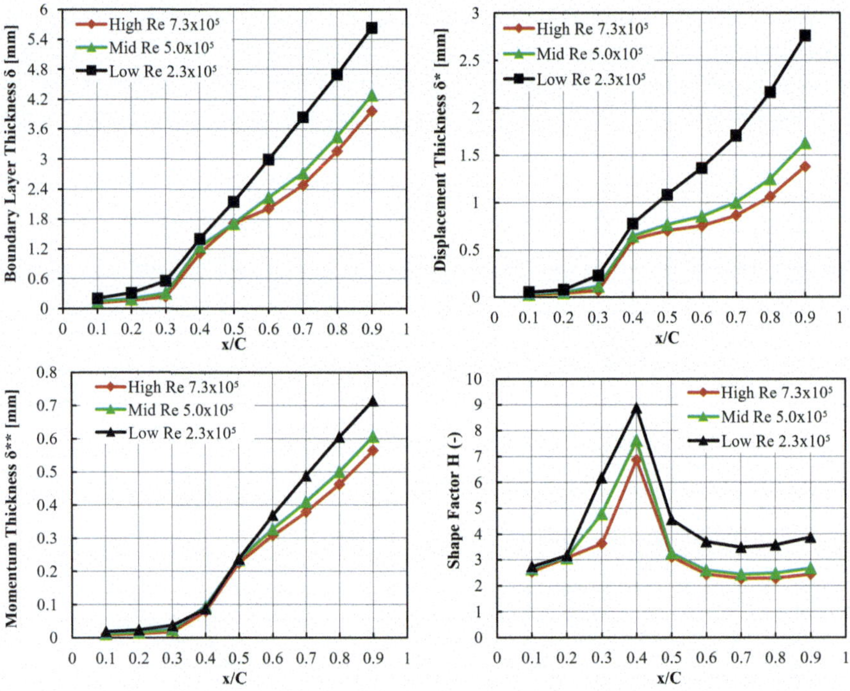

Fig. 7 A detailed comparison of boundary layer thickness and integral parameters plotted on the middle profile

tum thickness plot are comparable for all cases till x/C = 0.5 and the differences are more pronounced downstream x/C = 0.5. Based on the displacement thickness and momentum thickness, the shape factor has been plotted along the axial chord as shown in Fig. 7. Theoretically, H = 2.59 for laminar flows Blasius boundary layer (Blasius 1950) and H = 1.3 to 1.4 for turbulent flows. From shape factor plot it is evident that in all three cases, the flow upstream of the shock wave has a laminar boundary layer. Strong shock-induced boundary layer separation at x/C = 0.2 with reattachment close to x/C = 0.5 is distinguished by a steep increase in shape factor in the interaction zone and a steep decrease further downstream. The separation happens earlier and attaches downstream when the Reynolds number is decreased from high to low Reynolds number. Downstream the reattachment zone one could say the boundary layer is not any more laminar and has a transition to turbulent behavior due to SBLI. The investigations presented above are at the mid-span location in the test section.

To understand the 3D effect of SBLI, secondary flows are compared in these cases. The secondary flows developed due to the interaction of flow at viscous side walls result in flow separation and passage contraction above the suction side. The reversed flow region on suction and pressure side have been visualised using iso-surfaces of negative value of axial velocity (in blue color) as shown in Fig. 8a. The streamlines also have been plotted close to the surface to visualise the flow behaviour above the suction and pressure side of middle profile. To quantify the secondary flows developed at different Reynolds conditions Axial Velocity Density Ratio (AVDR) has been calculated using Eq. 5.

$$AVDR = (\rho_2 v_2 \sin\beta_2) / (\rho_1 v_1 \sin\beta_1) \quad (5)$$

To investigate the AVDR at middle blade passage the traverse have been extracted 10% upstream and downstream the middle profile as shown in Fig. 8b. The AVDR has been estimated using mass averaged axial velocity component and density at the traverse location shown in Fig. 8b. From the comparison of AVDR plotted in Table 4,

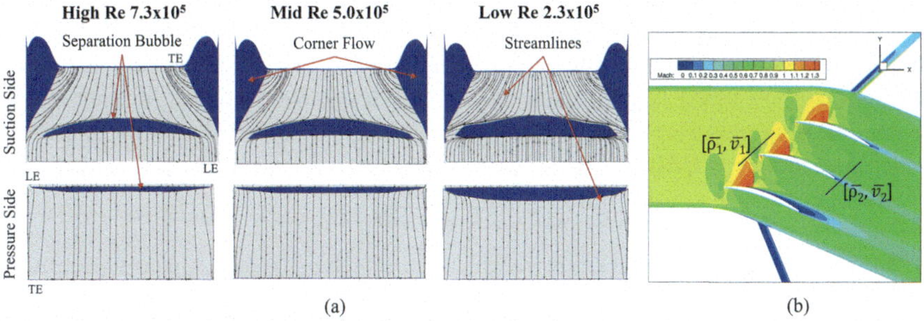

Fig. 8 Comparison of **a** Flow structure on suction and pressure side of middle profile, **b** Traverse location for calculating AVDR for different Reynolds number conditions

Table 4 Comparison of AVDR in test section

Cases	AVDR
High Reynolds (7.3×10^5)	1.152
Mid Reynolds (5.0×10^5)	1.147
Low Reynolds (2.3×10^5)	1.119

the secondary flows decrease when the Reynolds number is reduced in the middle blade passage.

5 Conclusion

The investigations presented in this paper have been performed on a stator cascade defined by Rolls Royce Deutschland (RRD). The main design objective was accomplished by designing a new rectilinear three-blade test section. To have a similar flow structure in the test section as in cascade flows, the initial shape of the upper and lower walls was designed from streamlines extracted from cascade simulations and final optimization was defined based on the boundary layer developed on the upper and lower walls. The upper and lower wall effects like flow blockage due to boundary layer separation, the unsteady effect caused by the boundary layer interaction in blade passage have been significantly reduced using suction slots at the end-walls. The main advantage of suction slots is the ability to achieve flow uniformity in all blade passages. Comparison of with and without suction slots in test section configuration shows that suction slots have a positive impact on flow structure. The numerical investigation of Reynolds number effects has been performed using the test section configuration with suction slots. Based on the isentropic Mach number and shear stress plots, the separation bubble starts upstream and reattaches further downstream when the Reynolds number is reduced. Due to increasing viscous effects at low Reynolds condition thick boundary layer interact with strong shock resulting in large separation. And the flow losses are increased when the Reynolds number is reduced. The boundary layer thickness indicates that the laminar separation is more pronounced, and has a huge impact on the loss levels when Reynolds number is decreased. As a next step, the designed test section would be mounted at the transonic wind tunnel of IMP-PAN to carry out the experimental investigation of SBLI effects at different Reynolds number conditions.

Acknowledgements This project has received funding from the European Union Horizon 2020 research and innovation programme under the Marie Sklodowska-Curie grant agreement 860909 TEAMAero (Towards Effective Flow Control and Mitigation of Shock Effects in Aeronautical Applications). Numerical simulations were performed in the Computational Centre of TASK (Trojmiejska Akademicka Siec Komputerowa) and supported by PL-Grid Infrastructure.

References

Arnal, Daniel, and Jean Délery. "Laminar-turbulent transition and shock wave/boundary layer interaction." OFFICE NATIONAL D ETUDES ET DE RECHERCHES AEROSPATIALES ONERA-PUBLICATIONS-TP 109 (2004): 69.

Becker, B., M. Reyer, and M. Swoboda. "Steady and unsteady numerical investigation of transitional shock-boundary-layer-interactions on a fan blade." Aerospace Science and Technology 11.7-8 (2007): 507–517.

Blasius, H. "The boundary layers in fluids with little friction." Zeitschrift fur mathematik und physik 56.NACA-TM-1256 (1950).

Celik, Ishmail B., Urmila Ghia, Patrick J. Roache, and Christopher J. Freitas. "Procedure for estimation and reporting of uncertainty due to discretization in CFD applications." Journal of fluids Engineering-Transactions of the ASME 130.7 (2008).

Davidson, Todd S., and Holger Babinsky. "Influence of transition on the flow downstream of normal shock wave–boundary layer interactions." 54th AIAA Aerospace Sciences Meeting. 2016.

Doerffer, Piotr, Pawel Flaszynski, Jean-Paul Dussauge, Holger Babinsky, Patrick Grothe, Anna Petersen, and Flavien Billard, eds. Transition Location Effect on Shock Wave Boundary Layer Interaction: Experimental and Numerical Findings from the TFAST Project. Vol. 144. Springer Nature, 2020

Flaszynski, P., Doerffer, P., Szwaba, R., Piotrowicz, M., & Kaczynski, P. (2018). Laminar-turbulent transition tripped by step on transonic compressor profile. Journal of Thermal Science, 27(1), 1–7.

Foret, Jonas, et al. "Experimental Aerodynamic and Aeroelastic Investigation of a Highly-Loaded 1.5-Stage Transonic Compressor with Tandem Stator." International Journal of Turbomachinery, Propulsion and Power 6.3 (2021): 21

Grothe, Patrick, Pawel Flaszynski, Ryszard Szwaba, Michal Piotrowicz, Piotr Kaczynski, Benoit Tartinville, Charles Hirsch, and Alexander Hergt. "WP3 Internal Flows Compressors." Transition Location Effect on Shock Wave Boundary Layer Interaction: Experimental and Numerical Findings from the TFAST Project (2021): 229–296

Hellsten, Antti, and Stefan Wallin. "Explicit algebraic Reynolds stress and non-linear eddy-viscosity models." International Journal of Computational Fluid Dynamics 23.4 (2009): 349–361.

Hergt, Alexander, et al. "On the Importance of Transition Control at Transonic Compressor Blades." Journal of Turbomachinery 143.3 (2021).

Hu, Jiaguo, et al. "Effects of slot jet and its improved approach in a high-load compressor cascade." Experiments in Fluids 58.11 (2017): 1–13.

Jameson, Antony, Wolfgang Schmidt, and Eli Turkel. "Numerical solution of the Euler equations by finite volume methods using Runge Kutta time stepping schemes." 14th fluid and plasma dynamics conference. 1981.

Kahraman, Selin, Jason A. Bourgeois, Christian Janke, Marius Swoboda. "Numerical investigations of transitional SBLI on a highly loaded transonic compressor passage" Proceedings of ASME Turbo Expo 2023, GT2023-102435.

Katiyar, S., and S. Sarkar. "Flow transition on the suction surface of a controlled-diffusion compressor blade using a large-eddy simulation." Physics of Fluids 34.9 (2022): 094108.

Menter, Florian R., Robin Blair Langtry, S. R. Likki, Y. B. Suzen, P. G. Huang, and S. Volker. "A correlation-based transition model using local variables part I: model formulation." Journal of turbomachinery 128.3 (2006): 413–422.

Menter, Florian R. "Two-equation eddy-viscosity turbulence models for engineering applications." AIAA journal 32.8 (1994): 1598–1605.

Mohammed-Taifour, Abdelouahab, Julien Weiss, and Louis Dufresne. "Explicit algebraic reynolds-stress modeling of pressure-induced separating flows in the presence of sidewalls." Journal of Fluids Engineering 143.10 (2021).

Piotrowicz, Michal, Pawel Flaszynski, and Ryszard Szwaba. "Influence of limiting walls on shock wave structure in single passage test section with compressor profile." TASK Quarterly 19.2 (2015): 141–152.

Piotrowicz, M., P. Flaszyński, and P. Doerffer. "Investigations of shock wave boundary layer interaction on suction side of compressor profile." Journal of Physics: Conference Series. Vol. 530. No. 1. IOP Publishing, 2014.

Schreiber, Heinz-Adolf, Wolfgang Steinert, and Bernhard Kusters. "Effects of Reynolds number and free-stream turbulence on boundary layer transition in a compressor cascade." J. Turbomach. 124.1 (2002): 1–9.

Schreiber, H. A., Steinert, W., Sonoda, T., and Arima, T. (2004). Advanced High-Turning Compressor Airfoils for Low Reynolds Number Condition-Part II: Experimental and Numerical Analysis. J. Turbomach., 126(4), 482–492.

Sonoda, T., Yamaguchi, Y., Arima, T., Olhofer, M., Sendhoff, B., and Schreiber, H. A. (2004). Advanced High Turning Compressor Airfoils for Low Reynolds Number Condition-Part I: Design and Optimization. J. Turbomach., 126(3), 350–359.

Tang, Hai-Long, et al. "High altitude low Reynolds number effect on the matching performance of a turbofan engine." Proceedings of the Institution of Mechanical Engineers, Part G: Journal of Aerospace Engineering 227.3 (2013): 455–466.

Wassell, A. B. "Reynolds number effects in axial compressors." (1968): 149–156.

Experimental Investigation on the Influence of Mass Addition on the Dynamic Characteristics of an Aluminium Tube Conveying Pulsatile Fluid

K. Sayooj, R. Kamal Krishna, and M. Unnikrishnan

Abstract Fluid-conveying tubes are frequently employed as component parts in a variety of systems, such as air conditioning, heat exchangers, solar energy, refrigeration, etc. Aluminium tubes are widely used in fluid conveying applications because of their lightweight, low density, and corrosion resistance. For the dynamic control of those systems, the vibrational characteristics of such tubes must be examined. This work focuses on the influence of discrete mass addition and its location on the dynamic properties of a horizontally positioned clamped–clamped aluminium tube transporting a pulsatile fluid. By conducting experimental modal analysis, the natural frequency of tube with and without discrete mass additions are identified. Operational modal analysis is conducted to extract the influence of internal excitation due to flow pulsation. The present study identifies a stiffness addition phenomenon due to the location of the added mass on the tube. The finding of this study points at a possible vibration control method where desired natural frequency and amplitude of vibration can be achieved by adding point masses at suitable positions at controlled flow velocities.

Keywords Experimental modal analysis · Operational modal analysis · Flow induced vibration · Mass addition

K. Sayooj (✉)
APJ Abdul Kalam Technological University, Thiruvananthapuram, India
e-mail: lncs@springer.com; d-tve20jan039@cet.ac.in

R. K. Krishna
Sree Chitra Thirunal College of Engineering, Thiruvananthapuram, India

M. Unnikrishnan
Advanced Dynamics and Control Lab, Mechanical Engineering, College of Engineering Trivandrum, Thiruvananthapuram, India
e-mail: munnikrishnan@cet.ac.in

G. Rajesh et al. (eds.), *Proceedings of the 15th International Symposium on Experimental and Computational Aerothermodynamics of Internal Flows*,
https://doi.org/10.1007/978-981-96-4082-9_2

1 Introduction

Vibration can happen in fluid conveying tubes due to the continuous transfer of energy between the fluid and the tube wall. Flow fluctuations and flaws in pipe geometry can cause vibration even in the absence of any external excitation. This flow induced vibration could be detrimental when these tubes are used to transport fluid as it may cause structural damage, reduce operational capability and affect the life span of the system. Fluid-conveying tubes are frequently employed as component parts in a variety of systems, such as air conditioning, heat exchangers, solar energy, refrigeration, etc. Hence, researchers have long been interested in the dynamics of fluid conveying tubes, and this study is still important today in determining the tube's safe operating conditions. The dynamic control of such systems requires thorough investigation on the dynamic characteristics of tubes that convey fluid such as fundamental frequency, mode shape, damping etc.

Early studies by a few researchers in the late twentieth century discovered the influence of fluid flow on the dynamics of fluid conveyance tubes. Gregory et al. (1966) analysed the unstable oscillation of tubular cantilevers conveying fluid and observed that when fluid flow is increased beyond a certain critical value, the system loses stability and lateral oscillations increases their amplitude. Naguleswaran et al. (1968) studied lateral vibrations of a pipe conveying a fluid and analysed the influence of flow speed on lateral vibrations theoretically and experimentally. Jendrzejczyk et al. (1985) conducted experiments on tubes that convey fluid for different support conditions in the subcritical and critical flow velocity region and made significant observations regarding the stabilization of the tube. A new method of vibration reduction for cantilevered flexible structures employing a controlled-speed internal fluid flow was developed and experimentally demonstrated by Y Sugiyama et al. (1996). They proposed that controlling the subcritical internal flow velocity can be a practical method for stabilising engineering systems. Recently, on an attempt to develop a testing method for investigating the dynamic characteristics of a pre-stretched horizontal fluid conveying tube, Unnikrishnan et al. (2017) estimated the factors that affect the modal parameter estimation and suggested methods to optimize them.

Influence of pulsatile fluid flow on the dynamics of tubes that convey fluid is addressed by a few researchers. Theoretical study on flexible, thin cylinders with axial, harmonically perturbed flow and their dynamic behaviour by Paidoussis et al. (Issid and Paidoussis 1974) showed that even for small pulsation, tube may be subjected to instabilities if fluid flow velocity is sufficiently high. Investigation on flexible pipe's lateral vibration conducted by Yoshizawa et al. (Yoshizawa et al. 1986) highlights the influence of fluid flow velocity and pulsation on the nonlinear behaviour of the tube. In the subcritical flow velocity area, Bajaj (Bajaj 1987) demonstrated that tubes can only have periodic solutions at half excitation frequency and a sufficiently large disturbance can cause the tube to exhibit high steady-state periodic motions. The existence of periodic and quasiperiodic oscillations in a fluid-conveying cantilevered pipe carrying a fluid with near critical flow velocity and a harmonic component was demonstrated theoretically, numerically, and experimentally by Semler et al. (Paidoussis and Semler 1996). A recent experimental study by Krishna et al. (Krishna et al. 2020) reveals that the vibration of flexible tubes under pulse excitation always stabilizes in an arbitrary plane. Resonance was induced by matching the pulse frequency to the fundamental frequency of tube and the vibration amplitudes were extracted from operational modal analysis. According to a numerical analysis of a viscoelastic pipeline by Khudayarov et al. (Khudayarov and Turaev 2022), a drop in the critical fluid flow velocity is caused by increasing the pulse frequency and excitation coefficient values.

Many studies reveal that the addition of mass to the primary structure is one way to manage such vibrations. An early study by Hill et al. (Hill and Swanson 1970) addressed stabilization of the fluid conveying pipes by adding lumped masses and suggested the possibility of stabilizing a tube by adding lumped masses of proper size at suitable location. Elnajjar et al. (Daneshmand and ElNajjar 2020) recently looked into the possibility of increasing critical fluid flow velocity by accounting for additional masses and springs at different positions along the pipe. The results show that, adding masses and springs at specific locations can significantly increase the critical flow velocity.

Above mentioned literature brings out the significance of flow induced vibrations. According to the analytical, numerical and experimental research on tubes conveying fluid, it is evident that the support conditions, tube material, tube geometry, internal and external forces influence the tube dynamics. Studies on the internal excitation of fixed–fixed tubes by flow pulsation are limited. Stability control investigations by means of discrete mass addition and location of mass addition are to be explored more.

Aluminium tubes are widely used in aforementioned applications because of their lightweight, low density, and corrosion resistance. This work examines the influence of discrete mass addition on the dynamic characteristics of a horizontally placed clamped–clamped aluminium tubes conveying a pulsatile fluid in detail. By conducting experimental modal analysis, the natural frequency of the tube with and without discrete mass additions are identified. Operational modal analysis was conducted to extract the influence of internal excitation due to flow pulsation. The influence of amount of discrete mass addition and positioning of discrete masses on deflection amplitudes, natural frequencies and damping are investigated.

2 Theory

A mathematical explanation of a generalized fluid structure interaction problem of a slender pipe conveying a steady fluid was formulated by Paidoussis (Paidoussis 2014). The general equation of small lateral motions on the basis of beam theory can be given by,

$$(E^* \frac{\partial}{\partial t} + E)I \frac{\partial^4 w}{dx^4} + \{MU^2 - T + PA(1 - 2v\delta) \\ - [(M + m)g - M \frac{\partial U}{\partial t}](L - x)\} \frac{\partial^2 w}{\partial x^2} + 2MU \frac{\partial^2 w}{\partial x \partial t} \\ + (M + m)g \frac{\partial w}{\partial x} + c \frac{\partial w}{\partial t} + (M + m0) \frac{\partial^2 w}{\partial t^2} = 0 \quad (1)$$

If the flow velocity is constant and the influence of tensioning, gravity, dissipation, and pressurisation are neglected, the Eq. (1) reduces to,

$$EI \frac{\partial^4 w}{dx^4} + MU^2 \frac{\partial^2 w}{\partial x^2} + 2MU \frac{\partial^2 w}{\partial x \partial t} + (M + m) \frac{\partial^2 w}{\partial t^2} = 0 \quad (2)$$

The terms in the Eq. (2) represents flexural restoring force, centrifugal force, Coriolis force and inertia force respectively. Any modifications to tube's geometry or mass may result in variations in the tube's dynamic response.

3 Methodology

An experimental set up was prepared to conduct modal analysis on an aluminium tube that conveys a pulsating fluid. Using experimental modal analysis (EMA), the aluminium tube's fundamental frequency was found. Operational Modal Analysis (OMA) was conducted to find the deflection amplitudes of the tube with different mass addition conditions.

3.1 Test Facility

An experimental set up is prepared to conduct modal analysis. A schematic of the test facility is shown in Fig. 1. The test facility comprises of a fixed- fixed horizontal aluminium tube of which the dynamic properties are to be studied, a fluid circulation system, a pulse generating mechanism, an excitation mechanism, a vibration response sensor and a data processing system.

The water circulation system consists of a water pump which pumps water to the overhead tanks arranged to maintain a constant head. Pulse generating mechanism consists of a rotary valve and a VFD controlled motor. Tube is kept horizontal and fixed at both the ends. The dimensions of the tube are listed in Table 1.

Electromagnetic modal shaker, Sentek MS200 [20] is used to excite the tube. The excitation signal is produced using Keysight 33500B arbitrary function generator (AFG) [21] and amplified using Sentek LA-500 power amplifier [22]. An impedance head, Dytran 5860B [23] is linked to the shaker through a stinger rod and is attached to the aluminium tube. Polytec LS400 Laser Doppler Vibrometer (LDV) [24] is used

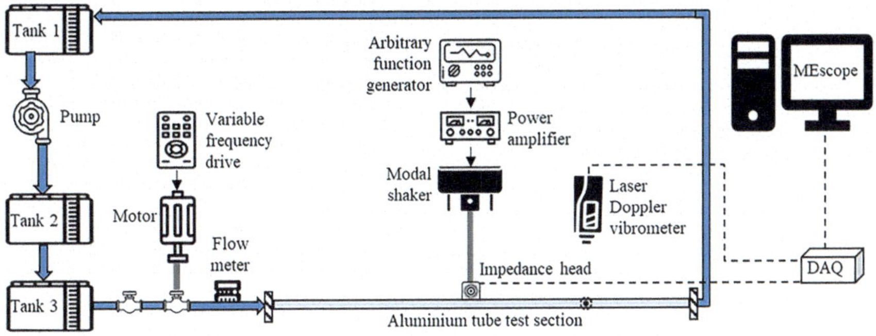

Fig. 1 Test facility for the modal testing of an aluminium tube conveying pulsatile fluid

Table 1 Dimensions of the aluminium tube

Property	Value
Length	1000 mm
Thickness	0.5 mm
Inside diameter	10.7 mm
Outside diameter	11.7 mm
Density	0.00271 g/mm^3
Young's modulus	70 GPa

to acquire the dynamic responses of the points. Signals from the impedance head and LDV are acquired through NI DAQ 4431 dynamic signal analyser [25] and further analysed using vibration analyser software MEscope Version 21.0.4.29.

3.2 *Test Procedure*

The tube is kept horizontal and clamped at both the ends. The tube is divided into equally spaced sections and reflective tapes are attached on all these points. Electromagnetic shaker is attached to the tube via a stinger rod to provide a constrained excitation. The stinger rod is carefully connected to the test section so that the excitation force is applied along the axis of the rod. The stinger rod isolates the shaker from the test section and reduce the mass loading of the shaker. Excitation is given near the center of the tube in order to extract the first fundamental mode of the tube (Fu and He 2001). A sine sweep signal with a frequency range of 0 to 150 Hz and an amplitude of 8Vpp is generated using AFG and amplified using a power amplifier. To ensure little interaction between adjacent signals, proper selection of application time and pause time is essential. Optimum application time of 7 s and a pause time of 2 s was provided. Data acquisition parameters are carefully selected by optimizing the factors affecting the modal parameter estimation (Unnikrishnan et al.

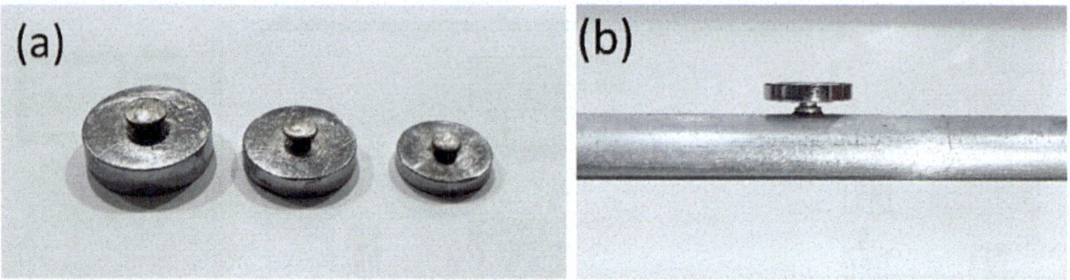

Fig. 2 **a** 20 g, 10 g and 5 g point masses used for the study **b** Point mass attached on aluminium tube

2017). 8192 samples were taken on the time domain at a time resolution of 0.0008 and 4096 samples were taken on frequency domain with a frequency resolution of 0.153. Number of averages were taken as 50.

Laser from LDV is pointed at the reflector tapes to capture the response of the tube. Point masses used for the study and the manner in which they are attached to the tube using adhesives is shown in Fig. 2.

The schematic representation of point mass addition on the tube in 8 different configurations is given in Fig. 3.

The details of positioning of point mass in each configuration is listed in Table 2. EMA is conducted and the frequency response functions are extracted. The fundamental frequency of the Aluminium tube is identified by analysing the frequency response function.

To extract the response of the tube from internal excitation alone, OMA is performed. Tube is excited by flow pulsation, rather than by using an electromagnetic shaker, with a pulse rate matching with the first fundamental frequency of the tube identified from EMA. Time response from LDV at the center of the tube is post processed. Fast Fourier Transform (FFT) of the response signal helps to identify the magnitude of vibration at the measured point at the fundamental frequency. The data

Fig. 3 Schematic representation of different configurations of point mass addition on the tube

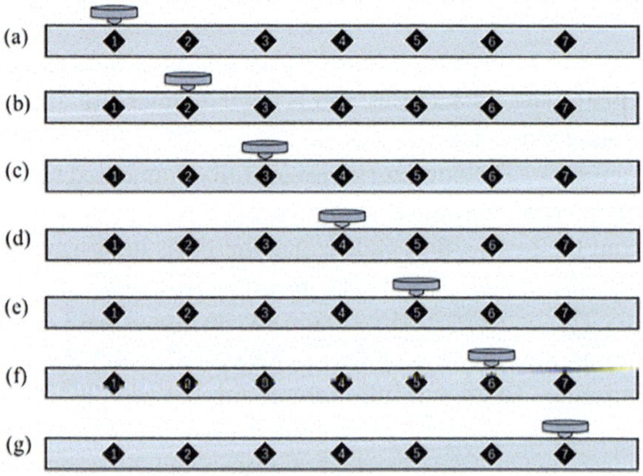

Table 2 Details of position of mass addition in each configuration

Configuration	Distance from the inlet (mm)
a	125
b	250
c	375
d	500
e	625
f	750
g	875

acquisition and signal processing methods to conduct OMA on the tube conveying a pulsatile flow were optimized (Krishna et al. 2020).

4 Results and Discussions

EMA conducted using constrained excitation revealed the dynamic characteristics of the Aluminium tube. The obtained frequency response functions from 3 different points of the tube and corresponding coherence plots are shown in Fig. 4. Unit coherence function value was achieved for the modal shaker excitation range of 0–150 Hz, indicating a good measurement. Significant drop in coherence function is only observed at resonant and anti-resonant points. The first fundamental frequency of the empty aluminium tube is found as 60.2 Hz as observed from the first peak of the frequency response function, Fig. 4(a).

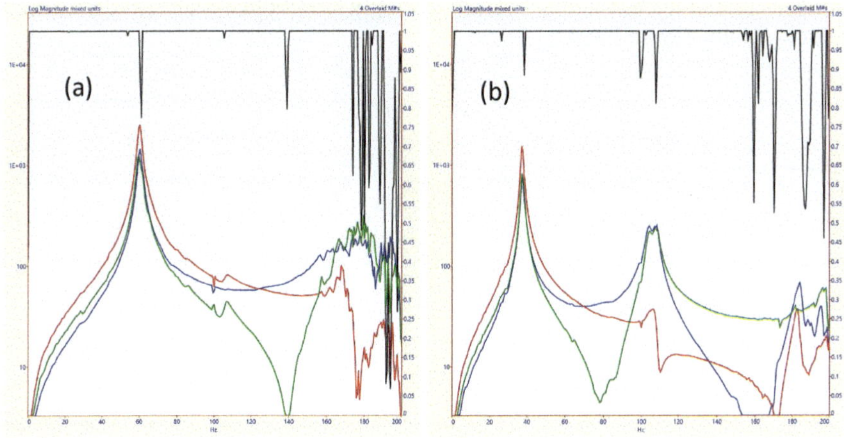

Fig. 4 Combined frequency response function and coherence plot of **a** empty tube **b** fluid filled tube

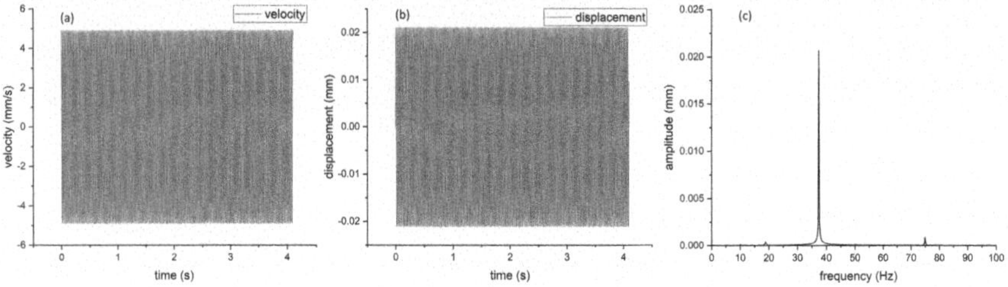

Fig. 5 **a** measured vibration response from LDV **b** integrated displacement plot **c** frequency plot from FFT

When the tube is filled with water, the frequency response function peak shifts left and the first fundamental frequency decreased to 37.4 Hz, Fig. 4(b). The reduction in size of the frequency response function peak indicates an increase in damping of the tube. These results account for the influence of the added mass of fluid.

OMA is conducted and the amplitude of vibration due to internal flow pulsation is identified. The measured vibrational response of the tube and post processed measurements are shown in Fig. 5.

Amplitude of vibration at the center of the aluminium tube conveying a 37.4 Hz pulsatile fluid without any discrete mass addition is obtained as 0.0211 mm.

4.1 *Influence of Mass Addition on Natural Frequency*

Discrete masses are added at different points of the tube to identify the influence of discrete mass addition on the dynamic characteristics of the tube. Figure 6(a) shows the first fundamental frequency of an empty tube plotted against various added masses, obtained from EMA.

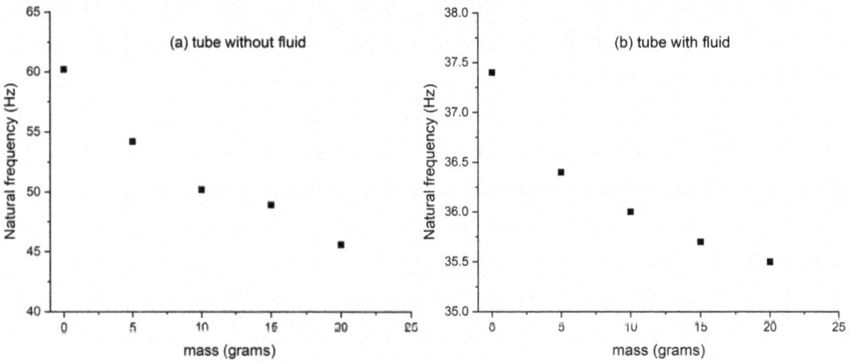

Fig. 6 Natural frequency values of **a** empty tube **b** filled tube

It can be observed from figure that the fundamental frequency of the tube decreases on discrete mass addition. As the quantity of added mass is increased the fundamental frequency is further decreased.

4.2 Influence of Location of Mass Addition on the Natural Frequency

Figure 7 shows the frequency value variations corresponding to different locations of mass addition and it is observed that the frequency reduction is more when mass is added towards the center of the tube.

Compared to the empty tube, the variations in frequency are less for fluid filled tube. In fluid filled condition, the mass of the fluid adds up with the mass of the tube, hence reducing the natural frequency compared to the empty tube condition.

Figure 8 shows the percentage reduction in natural frequency for mass addition at different position on the tube.

When a 5 g mass is added at 125 mm from the tube inlet, a frequency reduction of around 2.5% is observed. When the same mass is moved towards the other end of the tube, the frequency reduction increased to a maximum of 10% till the centre and then decreased back to 2.5% towards 875 mm position. This trend is also observed for a heavier mass of 10 g. This means that the frequency reduction is more for the heavier mass. However, in the fluid conveying tube, the percentage reduction in natural frequency upon mass addition is much less than that of the empty tube. Even though the reduction of natural frequency at different positions of mass addition is less, the same pattern of reduction in natural frequency as that of the empty tube is observed in the fluid conveying tube. And these results show that the location of mass addition has an undeniable importance on the natural frequency of fluid conveying tubes.

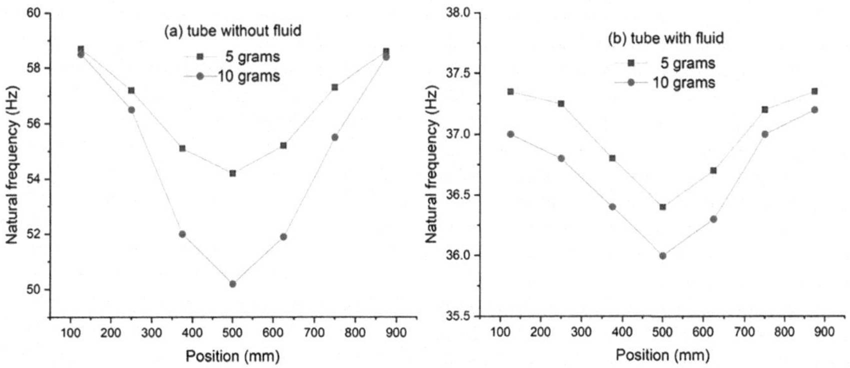

Fig. 7 Natural frequency of the aluminium tube obtained from EMA **a** without fluid **b** with fluid

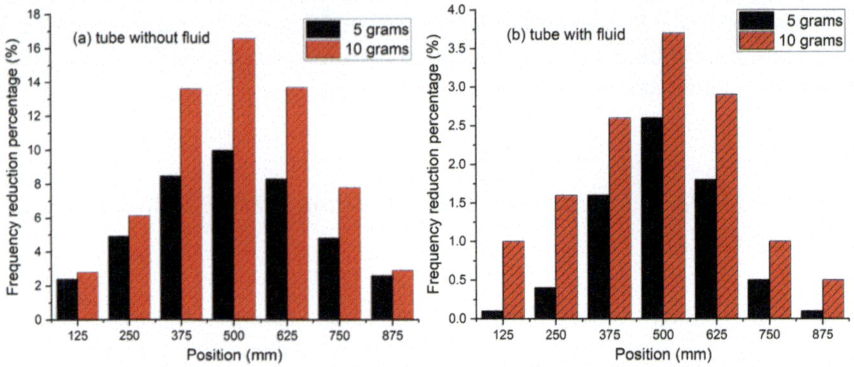

Fig. 8 Percentage reduction in natural frequency for mass addition at different positions from inlet of the tube **a** without fluid **b** with fluid

The phenomenon of stiffness addition of the tube by the addition of point mass at different positions of the tube could be observed from the above results. Stiffness of the tube is least at the center of the tube which increases towards the ends of the tube. Hence the point mass influence is observed more at the center of the tube, away from the fixtures. This idea points to a possible vibration control method within the bending limit of the tube to maintain the natural frequency of the tube at the desired level by adding pointed masses at various points of the tube.

4.3 Influence of Mass Addition on the Amplitude of Vibration

The amplitudes of vibration at the center of the tube obtained from OMA, plotted against the added point masses at the center of the tube is plotted in Fig. 9.

It can be seen from figure that the amplitude of vibration of the tube decreases on discrete mass addition. As the quantity of added mass is increased the amplitude of vibration is further decreased. Even though heavier masses results in higher static deflections, amplitude of vibration reduces for heavier masses.

4.4 Influence of Location of Mass Addition on the Amplitude of Vibration

When a tube vibrates at its first mode, maximum deflection is experienced at the center of the tube and it reduces towards the fixtures. However, addition of point masses reduces the amplitude of vibration. Figure 10(a) shows the changes in amplitude of vibration of the center of the tube corresponding to the position of mass addition along the length of the tube.

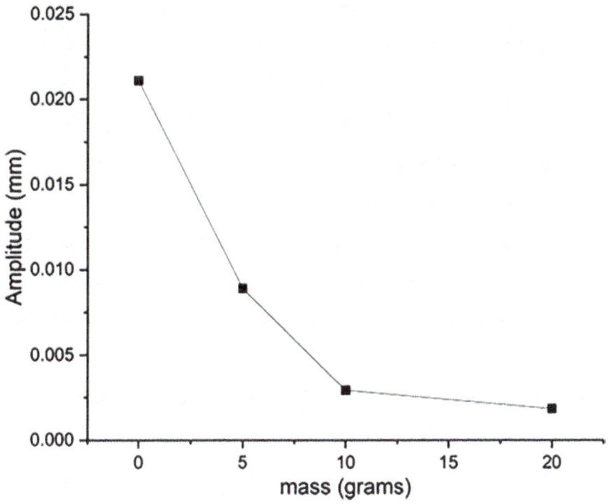

Fig. 9 Amplitude of vibration for different point masses added at the center of the tube

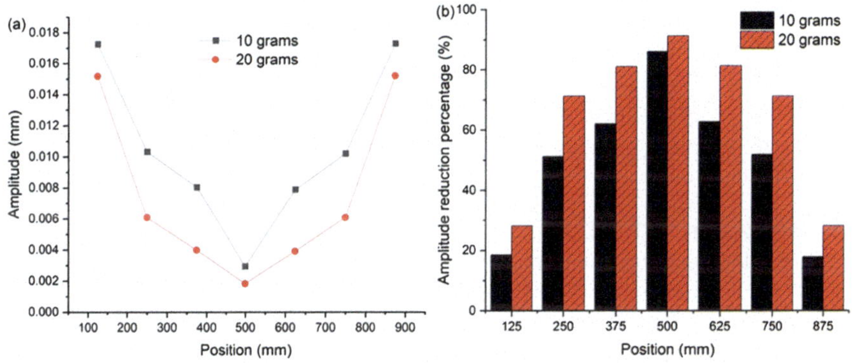

Fig. 10 **a** amplitudes of vibration and **b** percentage reduction in amplitude of vibration, for mass addition at different positions from the inlet end of the tube

The percentage reduction in amplitude of vibration corresponding to mass addition locations is shown in Fig. 10(b). When a mass of 10 g is added, a maximum amplitude reduction of 86.1% was obtained at the center of the tube and it decreased to a minimum of 17.8% towards the ends. By increasing the amount of the mass addition to 20 g, a maximum amplitude reduction of 91.3% was observed at the center of the tube and reduced to 28% towards the end of the tube, hence the trend is followed.

It is observed that the amplitude reduction is more when mass is added towards the center of the tube. Addition of masses contributes to the stabilization of vibration (Daneshmand and ElNajjar 2020). This stabilization effect is more when the mass is added at the center of the tube and it decreases towards the fixtures.

5 Conclusions

In this study, the preparation of an experimental method to analyse the vibrational behavior of fixed–fixed tubes conveying pulsatile fluids is discussed. The influence of discrete mass addition on the dynamic characteristics of an aluminium tube that convey water is captured using the developed experimental set up. External as well as internal excitation are studied in order to investigate the natural frequency and amplitude of vibration of the aluminium tube. Experimental modal analysis brought out the influence of mass addition on the natural frequency of the tube. Natural frequency of the tube is observed to reduce with mass addition. A phenomenon of stiffness addition is observed where, with the shifting of the same added mass away from the center of the fixed–fixed tube, natural frequency increased due to the increase in the stiffness of the tube. Operational modal analysis provided a clear insight on the dominant flow induced vibrations due to flow pulsation. Amplitude of vibration reduced with the increase in mass addition. Here also, with the shifting of mass addition position, the amplitude of vibration decreased towards the ends of the tube.

These finding of this study points at a possible vibration control method where desired natural frequency and amplitude of vibration can be achieved by the careful addition of the correct amount of point masses at suitable positions. This method can be further studied and applied to similar cases to control amplitudes of vibrations in different fields such as industrial piping, pressure vessels, pneumatic connectors, heat exchangers etc.

Acknowledgements The work discussed in this paper is performed in the Advanced Dynamics and Control Lab, Department of Mechanical Engineering, College of Engineering Trivandrum, affiliated to APJ Abdul Kalam Technological University, Trivandrum, Kerala, India.

References

P. Avitable, Modal testing: A practitioner Guide, Willey, 2018.
A. K. Bajaj: Nonlinear dynamics of tubes carrying a pulsatile flow. Dynamics and Stability of Systems: An International Journal 2(1), 19-41, (1987).
J. ElNajjar, F. Daneshmand: Stability of horizontal and vertical pipes conveying fluid under the effects of additional point masses and springs, Ocean Engineering 206, 106943 (2020).
R. W. Gregory, M. P. Paidoussis: Unstable oscillation of tubular cantilevers conveying fluid I. Theory. Proceedings of the Royal Society of London. Series A. Mathematical and Physical Sciences, vol. 293, pp. 528-542. The Royal Society Publishing, London (1966).
J. He, Z. Fu, Modal Analysis. Butterworth Heinemann, Oxford, UK, 2001.
J. L. Hill, C. P. Swanson: Effects of Lumped Masses on the Stability of Fluid Conveying Tubes, ASME. J. Appl. Mech. June, 37(2), 494–497 (1970).
J. A. Jendrzejczyk, S. S. Chen: Experiments on Tubes Conveying Fluid. Thin-Walled Structures 3, 109-134 (1985).
B. Khudayarov, F. Turaev: Numerical simulation of a viscoelastic pipeline vibration under pulsating fluid flow. Multidiscipline Modeling in Materials and Structures, Vol. 18(2), 219-237 (2022).

R. K. Krishna, A. G. Nath, S. A. Kumar, M. Unnikrishnan: Dynamic Behaviour of Flexible Silicone Tube Subjected to Pulsatile Flow. J. Inst. Eng. India Ser. C 101(6), 963–975 (2020).
Naguleswaran and Williams 1968 S. Naguleswaran, C. J. H. Williams: Lateral vibration of a pipe conveying a fluid. Journal of Mechanical Engineering Science 10(3), 228-238 (1968).
M. P. Paidoussis, Fluid-Structure Interaction Slender structures and axial flow. Vol.1. Second edition. Academic press, Oxford, UK (2014).
Paidoussis and Issid 1974M. P. Paidoussis, N. T. Issid. Dynamic stability of pipes conveying fluid. Journal of Sotmd and Vibration 33(3), 267-294 (1974).
C. Semler, M. P. Paidoussis: Nonlinear analysis of the parametric resonances of a planar fluid–conveying cantilevered pipe. Journal of Fluids and Structures 10, 787–825 (1996).
Y. Sugiyama, T. Katayama, E. Kanki: Stabilization of cantilevered flexible structures by means of an internal flowing fluid. Journal of Fluids and Structures 10, 653 – 661 (1996).
M. Unnikrishnan, A. Rajan, B. B. Raghunathan, J. Kochupillai: Dynamic testing of a pre-stretched flexible tube for identifying the factors affecting modal parameter estimation. J. Inst. Eng. India Ser. C, vol. 98, 421-430 (2017).
M. Yoshizawa, H. Nao, E. Hasegawa, Y. Tsujioka. Lateral vibration of a flexible pipe conveying fluid with pulsating flow. Bulletin of JSMW, Vol. 29 (253), 2243-2250 (1986).

Non-modal Local Stability Analysis of Heated Flat-Plate Boundary Layer with Different Temperature-Dependent Viscosity Models

Mayank Thummar, Ramesh Bhoraniya, and Vinod Narayanan

Abstract This article presents the local stability analysis of the boundary layer flows over a heated surface with variable viscosity using a non-modal approach. The three-different variable viscosity models have been considered (namely, linear, exponential, and inverse models) and they are a function of temperature only. The base flow profiles deflect towards the wall with increased sensitivity parameter ($\epsilon > 0$) resulting in narrow boundary layers for all viscosity models. However, the reverse effect has been observed in all models for $\epsilon < 0$. The standard procedure has been adopted to obtain the linearized governing stability equations in the normal velocity-vorticity form by eliminating the pressure term. The heat balance equation is coupled with the normal velocity-vorticity equations by temperature-dependent viscosity. The governing stability equations are spatially discretized using the Chebyshev spectral collocation approach. The discretized governing equations along with appropriate boundary conditions form an initial value problem (IVP). The transient growth in energy is mathematically computed by linear superposition of the non-orthogonal eigenvectors and large energy amplification has been achieved for a short time even though all eigenmodes lie down in the lower stable half-plane. The maximum transient energy growth ($G_{\max}$) reduces with the increased sensitivity parameter ($\epsilon > 0$) for all viscosity models resulting in an increased critical Reynolds number and flow becoming modally more stable. However, the reverse effect has been found for $\epsilon < 0$. Among all viscosity models, the linear viscosity model produced a lower energy amplification and temporal growth rate (ω_i) for a given value of ϵ.

Keywords Non-modal growth · Reynolds number · Viscosity models · Boundary layer

---

M. Thummar (✉) · R. Bhoraniya
Department of Mechanical Engineering, Marwadi University, Rajkot, Gujarat, India
e-mail: mayankthummar1993@gmail.com

V. Narayanan
Department of Mechanical Engineering, IIT Gandhinagar, Gandhinagar, Gujarat, India

G. Rajesh et al. (eds.), *Proceedings of the 15th International Symposium on Experimental and Computational Aerothermodynamics of Internal Flows*,
https://doi.org/10.1007/978-981-96-4082-9_3

1 Introduction

The stability characteristics of boundary layer flow of fluid over a heated surface with variable fluid properties are fascinating topics for enumerating experimental applications such as the airflow across warmed aeroplane wings to avoid ice development, the flow of fluid through the heat exchanger, chemical vapour decomposition (CVD) in reactors, saline jets, volcanic eruptions, etc. The stability of the laminar boundary layer over a heated or cooled surface was first numerically investigated by Wazzan et al. (1968). The O-S equation has been modified to compute an end effect of the variable viscosity in the thin boundary layer. A numerical method such as Gram-Schmidt orthogonal technique has been used to solve the modified O-S equation. The results indicate that the presence of a viscosity gradient within the boundary layer strongly destabilized the flow. It also noted that the heating effect stabilizes the flow, while the cooling effect destabilizes the flow. However, Wazzan et al. (1968) had neglected the base flow temperature quantity and temperature perturbation around base flow quantity. While Herwing and Schafer (1992) have considered base flow temperature and perturb quantities and solved the extended version of the modified O-S equation. In that study, pressure and temperature dependence fluid properties have been considered and solved the full version of the modified O-S equation using the same method as mentioned in article (Wazzan et al. 1968). The results show that a non-uniform reduction in viscosity close to the wall improves flow stability, but a uniform decrease in viscosity causes flow instability. The stability of the boundary layer over a plate under the influence of heating was experimentally studied by Strazisar et al. (1997). The findings demonstrate that the uniform heating of the boundary layers greatly enhanced the critical Reynolds number up to a certain temperature difference beyond which the opposite impact was found. Thus the overall presence of temperature gradient within the thin boundary layer increases the stability of the boundary layer and transition Reynolds number (Strazisar et al. 1997; Gazley and Wazzan 1985). The temporal stability of the boundary layer under the effect of uniform heating was also studied by Wall and Wilson (1997) for two-different viscosity models. In that study, the leading sixth-order eigenvalue problem has been solved for parallel and non-parallel (i.e., modified) problems. The results show that the parallel problem is always found to be more stable than the modified problem at a higher Reynolds number and stability results depend on the types of viscosity models. Additionally, it was discovered that a non-uniform drop in fluid viscosity inside the narrow boundary layer region promotes flow stability, but a non-uniform rise in fluid viscosity reduces flow stability. Similarly, spatial local stability of the boundary layer flow with variable fluid property (i.e., viscosity only) was also studied by Miller et al. (2018). The results indicate that the boundary layer becomes stable for liquid-type viscosity (i.e., $\epsilon \geq 0$), while unstable for gas-type viscosity (i.e., $\epsilon < 0$). The temporal stability and transient analysis of the heated boundary layer with variable viscosity were first performed by Thummar et al. (2023) using the global approach. Their results show that the boundary layer becomes more stable due to lower the temporal growth rate for $\epsilon > 0$ than the $\epsilon < 0$. The maximum transient growth in energy and wavelet structure of the optimal perturbations are also reduced with the increasing temperature-dependence parameter, ϵ.

In the present study, transient growth analysis has been performed for incompressible boundary layer flows over a heated surface with different viscosity/temperature relationships using a local approach. The three-different viscosity-temperature relations are addressed here in order to study the effects of viscosity models and temperature-dependence parameter on the transient growth. The governing stability equations are derived by a standard procedure in the form of normal velocity-vorticity form by eliminating the pressure term. The viscosity perturbation term is derived using Taylor-series expansion of base viscosity term up to first-order accuracy. The stability and transient codes are validated against the articles of Wall and Wilson (1997) and Butler and Farrell (1993) by considering the isothermal boundary layer ($\epsilon = 0$) respectively.

2 Problem Formulation

The governing stability equations are derived from the equations of motion by standard procedures. The standard form of the linearised governing stability equations becomes

$$\frac{\partial u_p}{\partial t} + U_b \frac{\partial u_p}{\partial x} + U_b{}' v_p = -\frac{\partial p_p}{\partial x} + \frac{1}{Re}[\mu_b \nabla^2 u_p + U_b{}'' \mu_p + \mu_b{}'(\frac{\partial u_p}{\partial y} + \frac{\partial v_p}{\partial x}) + U_b{}' \frac{\partial \mu_p{}'}{\partial y}] \tag{1}$$

$$\frac{\partial v_p}{\partial t} + U_b \frac{\partial v_p}{\partial x} = -\frac{\partial p_p}{\partial y} + \frac{1}{Re}[\mu_b \nabla^2 v_p + 2\mu_b{}' \frac{\partial v_p}{\partial y} + U_b{}' \frac{\partial \mu_p}{\partial x}] \tag{2}$$

$$\frac{\partial w_p}{\partial t} + U_b \frac{\partial w_p}{\partial x} = -\frac{\partial p_p}{\partial z} + \frac{1}{Re}[\mu_b \nabla^2 w_p + \mu_b{}'(\frac{\partial w_p}{\partial y} + \frac{\partial v_p}{\partial z})] \tag{3}$$

$$\frac{\partial T_p}{\partial t} + U_b \frac{\partial T_p}{\partial x} + T_b{}' v_p = \frac{1}{Pe} \nabla^2 T_p \tag{4}$$

$$\frac{\partial u_p}{\partial x} + \frac{\partial v_p}{\partial x} + \frac{\partial w_p}{\partial x} = 0 \tag{5}$$

with $\nabla^2 = \frac{\partial^2}{\partial x^2} + \frac{\partial^2}{\partial y^2} + \frac{\partial^2}{\partial z^2}$, $Re = U_\infty \delta^* / \nu$ (based on displacement thickness, δ^*), $Pr = \mu C_p / k$, $Pe = Re \times Pr$. In the above equations, the subscript 'b' and 'p' represent the base and perturb quantities respectively and superscript prime (') denotes a wall-normal derivative term.

Equations 1–5 are derived by considering following assumptions:

(i) The equations of motion are satisfied by the base flow solution, which is steady,
(ii) base flow is one-dimensional which varies only in the wall-normal direction,
(iii) parallel flow assumption is used with two-homogeneous spatial direction, and
(iv) non-linear term (i.e., $u\frac{\partial u}{\partial x}$) can be neglected due to the small value of disturbances as per linear stability theory.

The Grashof (Gr) and Generation (N_{GN}) numbers are not considered in the stability equations because the analytical value of these numbers is very small than the Reynolds (Re) and Peclet (Pe) numbers. Thus Buoyancy and viscous heat generation effects are omitted safely in the present investigation (Thummar et al. 2023; Ozgen 2004). Also, viscosity is only considered a variable property rather than other properties (i.e., density, thermal diffusivity, and thermal conductivity) because of relative change in viscosity of water is 65%, while the relative change in other properties is maximum upto 3% over a definite range of temperature (Lee et al. 2013; Incropera and Dewitt 2006).

The following three different viscosity models have been used for the non-modal linear stability analysis as

$$\text{Model} - 1: \quad \mu_b = \exp(-\epsilon T_p) \tag{6}$$

$$\text{Model} - 2: \quad \mu_b = 1 - \epsilon T_p \tag{7}$$

$$\text{Model} - 3: \quad \mu_b = \frac{1}{1 + \epsilon T_p} \tag{8}$$

where, ϵ is the dimensionless temperature dependence parameter and it indicates the sensitivity of the fluid viscosity to change in temperature. The perturb viscosity (μ_p) term present in the governing equations is derived using the series expansion of μ_b up to first-order accuracy in terms of perturb temperature (T_p) quantity.

$$\mu_p = \frac{\partial \mu_b}{\partial T_b} T_p \tag{9}$$

Equations 1–3 are converted into the normal velocity (v_p) vorticity (η_p) form by eliminating pressure term via standard procedures (Schmid and Brandt 2014) for non-modal analysis. The below sets of boundary constraints are applied at the wall (Eq. 10) and free-stream (Eq. 11) for perturb quantities.

$$v_p(0, t) = \eta_p(0, t) = T_p(0, t) = Dv_p(0, t) = 0 \tag{10}$$

$$v_p(\infty, t) = \eta_p(\infty, t) = T_p(\infty, t) = Dv_p(\infty, t) = 0 \tag{11}$$

where, $D = \partial/\partial y$.

The stability equations are discretized in the wall-normal direction using a numerical technique like the Chebyshev method. The discretized governing stability equations along with appropriate boundary conditions form an IVP as

$$\frac{\partial}{\partial t}\Phi(y,t) = -i\mathcal{L}\Phi(y,t) \tag{12}$$

with the initial condition $\Phi(y,0) = \Phi_0$. The above equation can be converted into the eigenvalue problem by considering the time-dependent Fourier transformation as

$$\Phi(y,t) = \hat{\Phi}(y)e^{i(\alpha x+\beta z-\omega t)} \tag{13}$$

where, α and β represent the wavenumbers in the streamwise (x) and spanwise (y) directions respectively, and ω denotes the complex wave frequency. The unreal part of the ω indicates the decay or growth of perturbations based on $\omega_i < 0$ or > 0. Then, Eq. 12 becomes

$$\mathcal{L}\hat{\Phi} = \omega\hat{\Phi} \tag{14}$$

with $\hat{\Phi} = [\hat{v}_p \;\; \hat{\eta}_p \;\; \hat{T}_p]^{tr}$ and $\mathcal{L} = \mathcal{B}^{-1}\mathcal{A}$ is the operator for linear evolution.

The energy growth is computed by defining the gain function as

$$G(t) = \max_{\hat{\Phi}_0} \frac{||\hat{\Phi}(t)||_E^2}{||\hat{\Phi}(0)||_E^2} \tag{15}$$

In the above equation, $G(t)$ represents the ratio of energy development of small perturbations at a given time t to the energy of perturbations at initial time $(t = 0)$, over all possible optimized initial conditions. The optimized initial conditions can be different at each instant of time as a result $G(t)$ shows the energy envelop of small disturbances corresponding to different optimal initial conditions at different times. Before proceeding to non-modal stability analysis, it is mandatory to define the overall energy of the disturbances (i.e., kinetic and potential energy). The overall energy $(E(t))$ of the disturbances is derived similarly as mentioned in articles (Butler and Farrell 1993; Jose et al. 2020; Hanifi et al. 2025).

$$E(t) = \frac{1}{8k^2}\int_0^{L_y} (k^2|\hat{v}_p|^2 + |D\hat{v}_p|^2 + |\hat{\eta}_p|^2 + k^2|\hat{T}_p|^2)\, dy \tag{16}$$

The last expression in Eq. 16 is known as potential energy, and it is influenced by the strength of the thermal stratification in the background. For a detailed description of the non-modal analysis, readers are requested to refer to articles (Schmid and Brandt 2014; Hanifi et al. 2025; Schmid and Henningson 2001). For non-modal stability analysis, we considered two different nomenclature in the present investigation as similar to those defined by Hanifi et al. (2025). First, maximum transient growth in energy over the entire time horizon $[0\ t]$ is defined as $G = G_{\max}$ and time correspondings to $G = G_{\max}$ is called $t = t_{\max}$. Second, optimal transient energy growth before exponential (or modal) growth overtakes the early non-modal behaviour is defined as $G = G_{opt}$ and time corresponds to $G = G_{opt}$ is called $t = t_{opt}$.

3 Base Flow Solution

A steady, incompressible solution of the laminar boundary layer equation is obtained by introducing the similarity variables:

$$U_b = \mathcal{F}'(\eta), \quad V_b = [\eta\mathcal{F}'(\eta) - \mathcal{F}(\eta)]/2\sqrt{Re}, \quad T_b = \mathcal{G}(\eta), \quad \text{where,} \ \eta = y\sqrt{\frac{U_\infty}{x\nu}} \tag{17}$$

Therefore, base flow equations become

$$(\mu_b\mathcal{F}'')' + 0.5\mathcal{F}\mathcal{F}'' = 0, \quad \mathcal{G}'' + 0.5Pr\mathcal{F}\mathcal{G}' = 0 \tag{18}$$

where, prime ($'$) represents the differentiation with respect to η and μ_b is defined in Eqs. 6 to 8. Equation 18 is a non-linear coupled ODE, solved by the 4th Order Runge-Kutta method along with the Newton-Raphson shooting algorithm, subjected to the following boundary constrains.

$$\mathcal{F}(0) = \mathcal{F}'(0) = \mathcal{G}(0) - 1 = 0, \quad \mathcal{F}'(\infty) - 1 = \mathcal{G}(\infty) = 0 \tag{19}$$

The numerical solution is converged up to tolerance criteria of 10^{-8} at $\eta = 20$. Thus, base flow domain height $\eta = 20$ is sufficient to apply the free-stream condition. In the present investigation, $Pr = 1$ is set for all computations of the mean flow as well as stability results. The numerical solution of the mean flow is converted from base grid points to physical grid points using the cubic spline interpolation function for non-modal analysis.

Figure 1a shows the variation of the base velocity profile for different values of ϵ corresponding to the exponential viscosity-temperature relation model (i.e., model-1). The velocity profile has a zero value at the wall, increases monotonically in the wall-normal direction (y), and finally approaches to outer edge velocity at some finite distance from the wall. The velocity profile deflects towards the wall for $\epsilon > 0$, while it deflects away from the wall for $\epsilon < 0$. Thus boundary layer becomes narrow as ϵ is increased. The impact of ϵ on the boundary layer stability is predicted by the shape of the U' and U'' profiles. For $\epsilon \geq 0$, U' profile has a maximal positive value at the wall with no point of inflection. While maximum velocity gradient (U') is exist at some finite distance from the wall in a case of $\epsilon < 0$ and it shows the 'point of inflection' as shown in Fig. 1b. At the point of inflection, U'' has a zero value as shown in Fig. 1c. For $\epsilon < 0$, U'' profile changes the sign from positive to negative in the y direction and finally vanishes at the outer edge of the boundary layer. While U'' profile does not

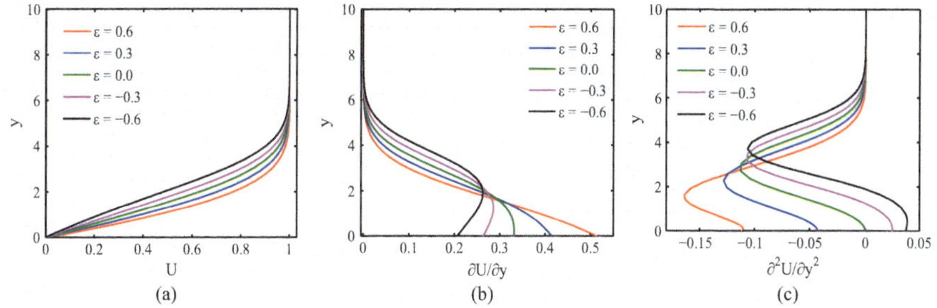

Fig. 1 The base flow **a** U, **b** U', and **c** U'' profiles for different values of ϵ for viscosity model-1

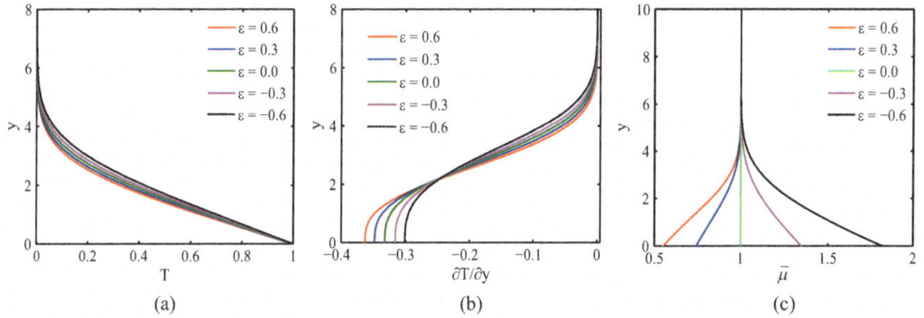

Fig. 2 The base flow **a** T, **b** T', and **c** μ_b profiles for different values of ϵ for viscosity model-1

change the sign in the y direction for $\epsilon \geq 0$. For $\epsilon = 0$, the variation of the velocity profile is linear in the y direction and U' is constant close to the wall. Thus $\epsilon = 0$ is the limiting case for the linear behaviour of the mean flow profile nearer to the boundary layer region. Figure 2a shows the temperature profile compressed towards the wall with increasing ϵ and that leads to a narrow thermal boundary layer. Due to a narrow boundary layer, temperature gradient (T') is increased at the wall with ϵ as shown in Fig. 2b. The effect of ϵ on the T profile is weak than the U profile. The variation in the viscosity of a fluid is depicted in Fig. 2c for a range of ϵ. The fluid viscosity decreases close to the wall for $\epsilon > 0$, while increases for $\epsilon < 0$. A drastic change in the viscosity of the fluid occurs for $\epsilon < 0$ than the $\epsilon \geq 0$. The same behaviour of basic state profiles are observed for other viscosity models corresponding to above mentioned range of ϵ, which are not shown here only for brevity.

4 Code Validation

For validation of the temporal stability and transient energy growth codes, we considered a case of the isothermal boundary layer. The non-isothermal boundary layer is converted into the Blasius (or Isothermal) boundary layer by selecting $\epsilon = 0$.

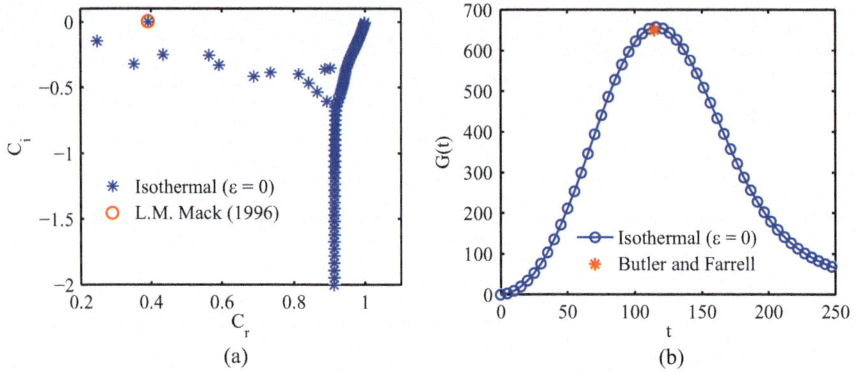

Fig. 3 Comparison of **a** eigenspectrum and **b** energy curve $G(t)$ with existing results of **a** Mack (1976) and **b** Butler and Farrell (1993) respectively for Isothermal boundary layer ($\epsilon = 0$). The results are obtained corresponding to parameters: **a** $\alpha = 0.179$, $\beta = 0$, $Re_{\delta^*} = 580$ and **b** $\alpha = 0.15$, $\beta = 0.96$, $Re_{\delta^*} = 1000$

Figure 3a demonstrates the comparison of the eigenspectrum acquired from the temporal stability analysis of the isothermal boundary layer ($\epsilon = 0$, $\mu_b = 1$) with the existing result of the Mack (1976) for $\alpha = 0.179$ and $Re_{\delta^*} = 580$. The least stable eigenmode ($C = 0.3935 + 0.0019i$) is found in a good match with the existing result of Mack (1976). Similarly, we validate the transient energy growth code for 3D disturbances corresponds to $\epsilon = 0$. Figure 3b shows the comparison of the transient energy curve ($G(t)$) with results of Butler and Farrell (1993) for $\alpha = 0.15$, $\beta = 0.96$, and $Re_{\delta^*} = 1000$. The maximum transient growth in energy ($G_{\max} = 658$) at $t_{\max} = 116$ is found in good match with the existing results of Butler and Farrell (1993). The optimal initial conditions and corresponding optimal perturbations at $t = t_{\max}$ are also found in good agreement with each other (not shown here only for brevity).

5 Results and Discussion

The non-modal analysis is performed for non-isothermal boundary layer flow with different viscosity models using a local approach. The temporal stability results are obtained by the spectral discretization method via the Chebyshev polynomial. A cubic spline interpolation function is used to transform the scale from $[-1\ 1]$ to $[0\ L]$ (i.e., from Chebyshev to real physical scale). A sufficient number of Gauss-Lobatto or collocation points (N) are considered for convergence of the stability results. The stability equations are then solved in the form of normal velocity-vorticity via the QZ algorithm with 121 collocation points. The least stable temporal eigenmode ($C = \omega/\alpha$) and maximum transient energy growth ($G = G_{\max}$) at $t = t_{\max}$ are almost identical for collocation points, $N = 111$ and $N = 121$ as shown in Table 1. Thus, all reported results in the present study are obtained using collocation points, $N = 121$.

Table 1 Grid convergence study for non-isothermal boundary layer flow with $\epsilon = 0.75$ corresponding to $\alpha = 0.15$, $\beta = 0$, and $Re_{\delta^*} = 1000$. The exponential model is used for grid convergence study

N	Eigenmode, $C = \omega/\alpha$	$G = G_{\max}$	$t = t_{\max}$
121	$0.04471 - 0.002473i$	54.41874	106.3505
111	$0.04474 - 0.002479i$	54.41048	106.3661
101	$0.04484 - 0.002487i$	54.30699	106.3860
91	$0.04498 - 0.002499i$	54.16774	106.4123

5.1 *Temporal Stability Analysis*

Figure 4 shows the neutral stability curves ($\omega_i = 0$) for a span of ϵ from -0.5 to 0.5 with an increment of 0.25 for three different viscosity/temperature relationships given by Eqs. 6–8 corresponding to $Pr = 1$. The area enclosed by the neutral stability curve represents the linearly unstable region and only small disturbances present in this region are enough for instability of the flow. From Fig. 4, it is evident that the neutral stability curves shift to the right with increasing ϵ for all viscosity models. Thus critical Reynolds number is increased with increasing ϵ and flow becomes stable even at a higher Reynolds number. It is also observed that for a given value of ϵ, neutral stability curves for the linear model (i.e., viscosity/temperature relationship given by Eq. 8) are always surrounded by corresponding curves of the exponential and inverse

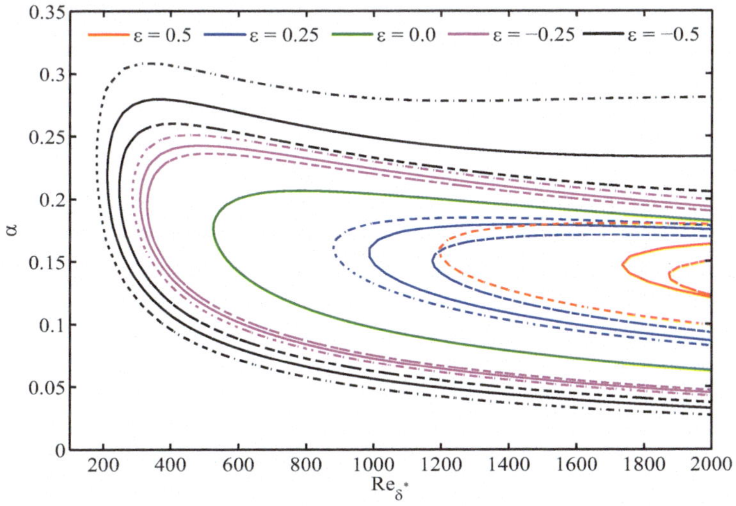

Fig. 4 Neutral stability curve for a range of ϵ from -0.5 to 0.5 with an increment of 0.25 corresponding to different viscosity models. A solid (–), dash (– – –), and dash-dot (– · –) lines represent the $\alpha - Re$ curves for viscosity model: 1, 2, and 3 respectively

models (i.e., viscosity/relationship given by Eqs. 6 and 7). Thus viscosity/relation given by Eq. 8 is less stable, while the linear viscosity/temperature relation by Eq. 7 is more stable at every disturbance wavenumber. However, the exponential viscosity/temperature relation (Eq. 6) is more stable than model-3 (i.e., Eq. 8) but less stable than model-2 (i.e., Eq. 7).

5.2 *Transient Growth in Energy*

In this subsection, we study the impact of temperature dependence parameter ϵ on the transient growth in energy and stability of the flow. Figure 5a shows the transient energy growth for different ϵ with $\alpha = 0.25$, $\beta = 0.1$, and $Re_{\delta^*} = 400$. A linear superposition of the eigenmodes depicts that the energy starts to grow algebraically with time, approaches to peak value, and then finally decays or amplifies depending on the leading stable or unstable eigenmode as per modal linear theory. From Fig. 5a, it is interpreted that the transient energy growth reduces as increases ϵ, and the time ($t_{\max}$) at which maximal transient energy growth ($G_{\max}$) is achieved also reduces with ϵ. The reason behind it, as the value of ϵ increases, the kinematic and dynamic viscosity of fluid close to the wall reduces as a result velocity profile (U and U') becomes narrow. The velocity gradient (U') at the wall is also increased with ϵ as shown in Fig. 1b which produces a higher shear gradient. This modification in the U' profile is particularly beneficial for reducing the energy of the perturbations and create more stable flows (Miller, Griffiths and Hussain 2018). Moreover, loss of perturbation energy due to viscous dissipation is also reduced with increasing the ϵ, but this energy is insufficient to defeat the stabilizing impact produced by the base flow (Ozgen 2004). Thus, inviscid reactions to modifications in the base-flow profiles have a significant impact on the stability of the boundary layer. A notable transient energy growth is obtained followed by the exponential growth of energy for $\epsilon = -0.5$, while exponential decay of energy is observed after the substantial transient growth for $\epsilon \geq 0$ (see Fig. 5a). Thus flow is found modally unstable for $\epsilon = -0.5$ (i.e., $\omega_i > 0$), modally stable for $\epsilon \geq 0$ (i.e., $\omega_i < 0$), and almost neutrally stable for $\epsilon = -0.25$ (i.e., $\omega_i \approx 0$). The transient energy growth predicted by including and excluding the unstable eigenmode is shown in Fig. 5b. A faster transient energy growth is observed within a short time followed by an exponential decay of energy by including only stable modes and it is higher than the transient growth predicted by a single unstable model. For including all modes (i.e., stable and unstable modes), substantial transient growth in energy is observed within a short time and once the transient effect subsides modal growth (or unstable mode) overtakes the early non-modal behaviour.

The impact of the parameter ϵ on the optimal perturbations (u_p, v_p, w_p, T_p) corresponding to maximum ($G_{\max}$) and optimal (G_{opt}) transient energy growth is shown in Fig. 6 for linearly stable and unstable of the flow respectively. From Fig. 6, it is noticeable that the perturbation profiles shift toward the wall as ϵ is increased. This happened due to narrowing the corresponding base flow profiles as noticed in Figs. 1 and 2. A like behaviour has been noticed by Wall and Wilson (1997) and they called

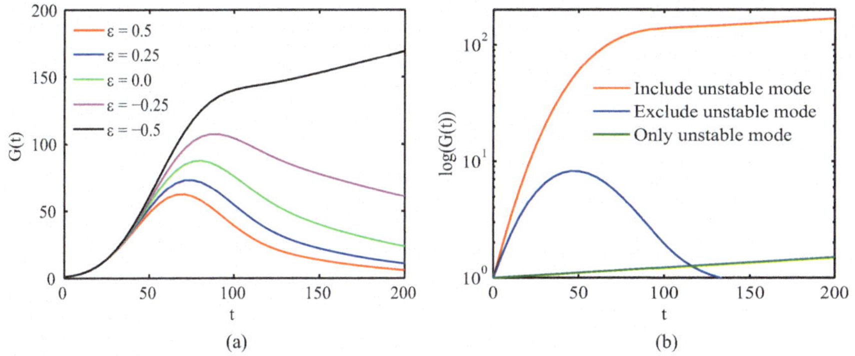

Fig. 5 Comparison of transient growth in energy, **a** $G(t)$ for values of ϵ from -0.5 to 0.5 with an increment of 0.25 for viscosity model-1 with $\alpha = 0.25$, $\beta = 0.1$, and $Re_{\delta^*} = 400$ and **b** energy growth predicted by stable and unstable eigenmodes corresponding to $\epsilon = -0.5$

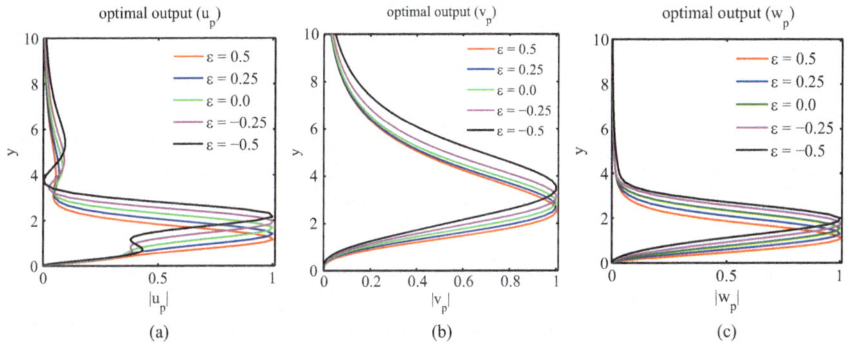

Fig. 6 Comparison of modulus of **a** u_p, **b** v_p, and **c** w_p components of optimal perturbations for different values of ϵ corresponding to model and parameters mentioned in Fig. 5. Each profile is normalized by its absolute maximum value

as a "compression effect" due to a change in the base flow profile with ϵ. It is noticed from Fig. 6a that the shape of the optimal $|u_p|$ profiles nearer to the wall differs for $\epsilon \geq 0$ than $\epsilon < 0$. The $|u_p|$ profile has two different maxima for $\epsilon < 0$, while it has only one maxima for $\epsilon \geq 0$. In Fig. 6b, we observed a drastic change in $|v_p|$ profile for $\epsilon < 0$, and the profile maximal is flatter. A similar trend is observed in w_p profile with ϵ. The variation of the temperature perturbation profile with ϵ is not shown a significant change than the velocity perturbations (not seen here only for brevity).

Next, we performed the non-modal stability analysis for three-different temperature-dependent viscosity models to investigate which viscosity/temperature relation produces large transient amplification and leads to laminar-turbulent transition. Figure 7a and b show the comparison of $G(t)$ for three-different viscosity/temperature relationships given by Eqs. 6–8 for $\epsilon = 0.3$ and -0.3 respectively. The critical parameters (such as $\alpha = 0.179$, $\beta = 0$, and $Re_{\delta^*} = 580$) associated with

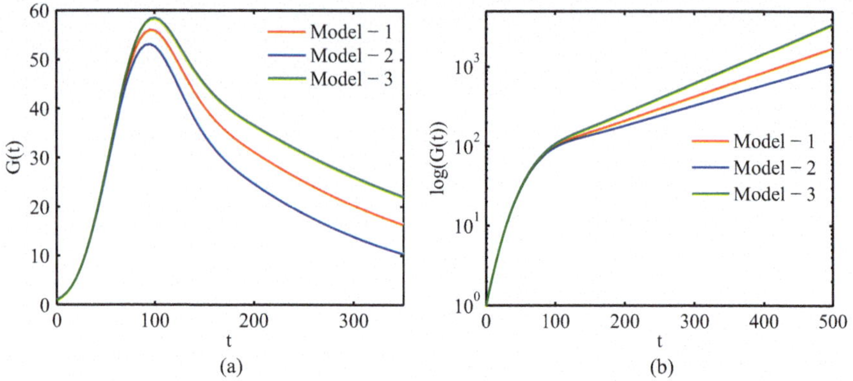

Fig. 7 Comparison of $G(t)$ for **a** $\epsilon = 0.3$ and **b** $\epsilon = -0.3$ with different temperature/viscosity models corresponding to $\alpha = 0.179$, $\beta = 0$, and $Re_{\delta^*} = 580$

the isothermal boundary layer ($\epsilon = 0$) are considered for $\epsilon = 0.3$ and -0.3. The flow is found neutrally stable ($\omega_i \approx 0$) corresponding to critical parameters for the isothermal boundary layer. In Fig. 7a, we observed a substantial transient energy growth followed by an exponential decay of energy for all viscosity models. The flow is found modally stable because of $\omega_i < 0$. All modes lie down in the stable half plane still we get notable transient growth due to the non-normal operator $\mathcal{L}$. The lower and higher transient growth in energy corresponding to optimal initial conditions at $t = t_{\max}$ is found for models 2 and 3 respectively. Thus, the exponential viscosity model is found to be more stable than the inverse model, but less stable than the linear model. However, the flow is found modally unstable for $\epsilon < 0$ because of $\omega_i > 0$. At an early time, non-modal growth is dominant, after that non-modal effect subsides with time, and finally, modal growth overtakes the early non-modal behaviour. At the large asymptotic time, modal or exponential growth represents the leading unstable eigenmode. For $\epsilon = -0.3$, optimal transient growth (G_{opt}) in energy is observed at $t = t_{opt}$ followed by the exponential growth of energy. The G_{opt} is found nearly the same for all viscosity models as shown in Fig. 7b. While higher exponential growth in energy is observed for model 3, which leads to a transition of flow from laminar to turbulent. The optimal perturbations corresponding to optimal initial conditions at $t = t_{\max}$ and $t = t_{opt}$ are also obtained for all viscosity models corresponding to $\epsilon = 0.3$ and -0.3 respectively. The optimal perturbations are found nearer to the wall for viscosity model 2 and away from the wall for viscosity model 3 (not shown here only for brevity).

5.3 Parametric Study

The non-modal stability problem depends on various parameters such as Reynolds number (Re_{δ^*}), sensitivity parameter (ϵ), streamwise (α) and spanwise (β) wavenumbers, and the time horizon t. Therefore, it is important to uncover the effects of these parameters on the short-time (non-modal) and asymptotic long-time (modal) energy growth of the fluid system. First, the impact of temperature-dependent viscosity on the transient growth mechanism is investigated by varying the value of the sensitivity parameter (ϵ) while keeping Re and Pr fixed, in order to determine the conditions that lead to maximum energy growth. For each considered value of ϵ, the optimal growth (G_{opt}) and the corresponding time (t_{opt}) are calculated for various values of α and β. For the stable fluid system, G_{opt} and t_{opt} are equivalent to maximum growth ($G_{\max}$) and time ($t_{\max}$), respectively. Figure 8 displays the contours of the optimal growth G_{opt} for different values of ϵ (−0.5, 0.0, 0.5) corresponding to viscosity model 1, $Re_{\delta^*} = 400$, and $Pr = 1$. The area enclosed by the solid grey line in the $\alpha - \beta$ plane represents the unstable region where exponential energy amplification occurs, and flow is governed solely by the modal stability theory. This indicates that the flow is modally unstable, exhibiting significant asymptotic growth for values of α and β lying within the shaded region, specifically for $\epsilon = -0.5$. A similar topology has been observed in the contours of the optimal growth for all $\epsilon < 0$, only difference is that the unstable area in the $\alpha - \beta$ plane diminishes as ϵ increases and disappears entirely for $\epsilon = 0$. Consequently, the only notable optimal growth is for $\epsilon \geq 0$, and the flow becomes modally stable for values of $0 < \alpha < 0.3$ and $0 < \beta < 1$. For the value of $\epsilon = 0$, the optimal growth is achieved at $\alpha = 0$ and $\beta = 0.38$. This is indicated by the magenta cross symbol, corresponding to $G_{opt} = 482.2$. These values are consistent with the optimal streamwise and spanwise wavenumbers discovered for the Blasius boundary layer (Schmid and Henningson 2001). In this scenario, the initial perturbation that yields the maximum energy growth of the system is a pair of counter-rotating streamwise-independent vortices.

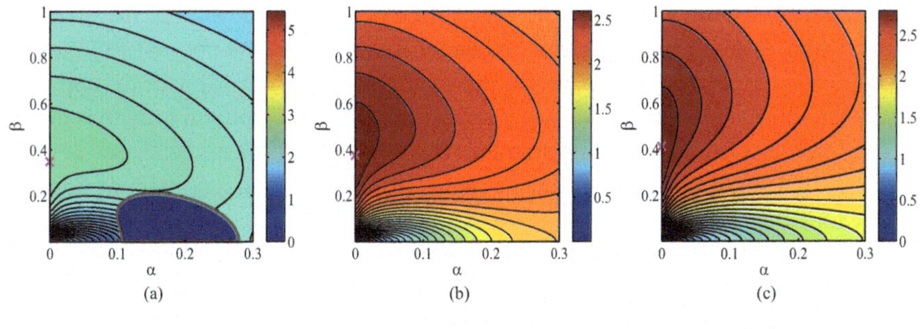

Fig. 8 Optimal growth (G_{opt}) contours as a function of streamwise (α) and spanwise (β) wavenumbers for viscosity model-1 with **a** $\epsilon = -0.5$, **b** $\epsilon = 0.0$, and **c** $\epsilon = 0.5$ at fixed $Re_{\delta^*} = 400$ and $Pr = 1$. The area enclosed by a solid grey line represents the unstable region. The magenta cross symbol represents the conditions of optimal amplification

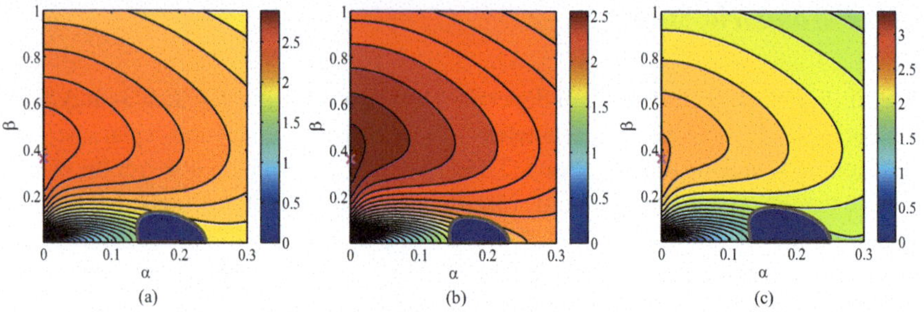

Fig. 9 Optimal growth (G_{opt}) contours as a function of streamwise (α) and spanwise (β) wavenumbers for viscosity **a** model-1, **b** model-2, and **c** model-3 with $\epsilon = -0.25$ at fixed $Re = 400$ and $Pr = 1$. The area enclosed by a solid grey line represents the unstable region. The magenta cross symbol represents the conditions of optimal amplification

These perturbations lead, at the optimal time, to elongated regions with alternating low and high momentum in the spanwise direction. This pattern resembles a pair of streaks and is attributed to the well-known lift-up mechanism (Butler and Farrell 1993). The optimal streamwise wavenumber does not change as ϵ increases, and it remains zero across the range of ϵ from -0.5 to 0.5. On the other hand, the optimal spanwise wavenumber increases slightly with ϵ and reaches its maximum value of $\beta = 0.413$ at $\epsilon = 0.5$. Thus, the thermal boundary layer with temperature-dependent viscosity yields optimal perturbation growth corresponding to $\alpha = 0$ (indicating no streamwise dependency, resulting in streamwise vortices with streaks). The value of β increases slightly with ϵ, but the optimal growth decreases as ϵ decreases, reaching approximately $G_{opt} = 316.54$ for $\epsilon = -0.5$. The influence of ϵ on the optimal time t_{opt} is also determined as a function of α and β for three distinct ϵ values. However, the outcomes are not presented here for the sake of brevity. The effect of α and β on the optimal growth is also evaluated for three different viscosity models, as illustrated in Fig. 9, with $\epsilon = -0.25$, $Re_{\delta^*} = 400$, and $Pr = 1$, aiming to assess the influence of viscosity models on the maximum growth. The same $\alpha - \beta$ structure is observed for all three models; the only difference lies in the size of the unstable region, which is smaller for model 2 and larger for model 3. The optimal streamwise wavenumber ($\alpha_{opt} = 0$) and spanwise wavenumber ($\beta_{opt} = 0.40$) are also approximately the same for all three viscosity models. However, the optimal growth is larger ($G_{opt} = 620.97$) for model 2 and smaller ($G_{opt} = 594.56$) for model 3 (see Fig. 9).

Receptivity analysis has not been conducted in this paper, but it remains a potential avenue for future researchers to explore. This analysis aims to calculate the external forcing frequency responsible for the peak response in the energy of the fluid system. A comprehensive explanation of the receptivity analysis or input-output analysis can be found in the article (Thummar et al. 2023).

6 Conclusions

The non-modal local stability analysis of the laminar boundary layer over a heated or cooled surface is performed with different variable viscosity models. The three-different viscosity/temperature relation given by Eqs. 6 (exponential model), 7 (linear model), and 8 (inverse model) are considered in the present investigation. The linearised governing stability equations are derived via standard procedure by eliminating the pressure term. The base flow solution is obtained by the 4th Order Runge-Kutta method along with the Newton-Raphson shooting algorithm. The velocity and thermal boundary layer become narrow with increasing temperature (ϵ) for all viscosity models. For $\epsilon < 0$, the viscosity of the fluid increases nearer to the wall as a result boundary layer becomes broader. While the boundary layer becomes narrow due to reducing the viscosity of fluid close to the wall for $\epsilon \geq 0$. The point of inflection is observed for $\epsilon < 0$, however, no point of inflection is observed for $\epsilon \geq 0$. The velocity profile is more significantly affected by ϵ than the temperature profile. For all viscosity models, the critical Reynolds number increases as the temperature-dependent parameter, ϵ increases. Among all viscosity models, the highest critical Reynolds number is found for a linear viscosity/temperature relation given by Eq. 7 corresponding to a given value of ϵ. The flow is found modally more stable for $\epsilon \geq 0$, while unstable for $\epsilon < 0$. The energy growth is also reduced with increasing the temperature dependence parameter ϵ. For $\epsilon > 0$, notable transient energy growth is observed followed by an exponential decay of energy even at a higher Reynolds number, while exponential growth in energy is observed within a short time after the substantial transient growth even at a lower Reynolds number. The optimal perturbations corresponding to optimal initial conditions are moved towards the wall with increasing the ϵ due to narrowing the boundary layers. Among all variable viscosity models, lower transient energy growth and temporal growth rate are found for the linear model, while higher transient energy growth and temporal growth rate are found for the inverse model.

References

Butler, K. M., Farrell, B.F.: Three-dimensional optimal perturbations in viscous shear flow. Physics of Fluids 4, 1637–1650 (1993).

Gazley, C. and Wazzan, A.R.: Control of water boundary layer stability and transition by surface temperature-distribution in Laminar-Turbulent Transition, In: Kozlov, V.V. (eds.), pp. 153–162. Berlin, Heidelberg: Springer (1985).

Hanifi, A., Schmid, P.J., Henningson, D.S.: Transient growth in compressible boundary layer. Physics of Fluids 8(3), 1637–1650 (826–837).

Herwing, H., Schafer, P.: Influence of variable properties on the stability of two-dimensional boundary layers. J. Fluid Mech. 243, 1–14 (1992).

Incropera, F.P. and Dewitt, D.P.: Fundamentals of Heat and Mass Transfer. 6th edn. Wiley, US (2006).

Jose, S., Brandt, L., Govindarajan, R.: Localisation of optimal perturbations in variable viscosity channel. International J. of Heat and Fluid Flow 85(108588) 1–9 (2020).
Lee, J., Jung, S.Y., Sung, H.J., and Zaki, T.A.: Effect of wall heating on turbulent boundary layers with temperature-dependent viscosity. J. Fluid Mech. 726, 196–225 (2013).
Mack, L.M.: A numerical study of temporal eigenvalue spectrum of the Blasius boundary layer. J. Fluid Mech. 615, 497–520 (1976).
Miller, R., Griffiths, P.T., Hussain, Z.: Stability of the Blasius boundary layer over a heated plate in a temperature-dependent viscosity flow. Phy. Rev. Fluids 3(113902), 1–22 (2018).
Ozgen S.: Effect of heat transfer on stability and transition characteristics of boundary-layers. International Journal of Heat Mass Transfer. 47(22), 4697–4712 (2004).
Schmid, P. J., Brandt, L.: Analysis of fluid systems: stability, receptivity, sensitivity. Appl. Mech. Rev. 66(2), 024803 (2014).
Schmid, P.J., Henningson, D.S..: Stability and Transition in Shear Flows : Applied Mathematical Sciences (142). Springer-Verlag, New York (2001).
Strazisar, A.J., Reshotko, E., and Prahl, J.M.: Experimental study of the stability of heated laminar boundary layers in water. J. Fluid Mech. 83(2), 225–247 (1997).
Thummar, M., Bhoraniya, R., Kant, R., Narayanan, V.: Stability and receptivity analysis of flat-plate boundary layer with suction and blowing. J Braz. Soc. Mech. Sci. Eng. 45, 415 (2023).
Thummar, M., Bhoraniya, R., Narayanan, V.: Modal and Non-Modal Stability of the Heated Flat-Plate Boundary Layer with Temperature-Dependent Viscosity. Fluid Dynamics 58, 450–475 (2023).
Wall, D.P., Wilson, S.K.: The linear stability of flat-plate boundary-layer flow of fluid with temperature-dependent viscosity. Phys. Fluids 9(10), 2885–2897 (1997).
Wazzan, A.R., Okanura, T., Smith, A.M.O.: The stability of water flow over heated and cooled flat plates. J. Heat and Mass Transfer 90, 109–114 (1968).

Suppression of Heat Transfer in Natural Convection of Air by a Permanent Magnet

Takuya Masuda, M. M. A. Alam, Toshio Tagawa, and Yasutaka Hayamizu

Abstract A magnetic field was applied with permanent magnets to the natural convection of oxygen-containing air in a cubic container consisting of a pair of vertical walls maintained at different temperatures. Two-dimensional numerical simulations were performed on this configuration to investigate the change in heat transfer due to the arrangement and shape of the prismatic magnets. The magnets were placed in all four directions of the container: top, bottom, left, and right. When a magnet is placed on the bottom or left side (hot wall side) of the container, the heat transfer is either enhanced or suppressed, though the effect is subtle. On the other hand, when a magnet is placed on the top or right side (cold wall side) of the container, the heat transfer is significantly suppressed, relying on the position of the magnet. A magnet placed near the cooling wall excludes hot oxygen, adsorbs cold oxygen, and forms vortices on the surface of the cooling wall, thereby blocking the approach of hot air. Unlike the heat transfer suppression by an adiabatic wall, this heat transfer suppression mechanism acts directly on the working fluid, keeping the surface of the cold wall at a low temperature.

Keywords Boussinesq approximation · Open FOAM · Paramagnetic fluid

Nomenclature

B	Dimensionless magnetic flux density, $= \mathbf{b}/(\mu_v h_R)$
b	Magnetic flux density, T, $b^2 = \mathbf{b} \cdot \mathbf{b}$

T. Masuda · Y. Hayamizu
National Institute of Technology, Yonago College, Tottori, Japan

M. M. A. Alam (✉)
Osaka Sangyo University, Osaka, Japan
e-mail: alam@tm.osaka-sandai.ac.jp

T. Tagawa
Tokyo Metropolitan University, Tokyo, Japan

G. Rajesh et al. (eds.), *Proceedings of the 15th International Symposium on Experimental and Computational Aerothermodynamics of Internal Flows*,
https://doi.org/10.1007/978-981-96-4082-9_4

$\mathbf{G}$	Dimensionless gravitational acceleration, $= \mathbf{g}/g$
$\mathbf{g}$	Gravitational acceleration, $(0, -g, 0)$, $\mathrm{m/s^2}$
$\mathbf{H}$	Dimensionless magnetic field intensity, $= \mathbf{h}/h_R$
$\mathbf{h}$	Magnetic field intensity, A/m
$\mathbf{U}$	Dimensionless velocity, $= l_e \mathbf{u}/\alpha$
$\mathbf{u}$	Velocity, m/s
$\mathbf{X}$	Dimensionless cartesian coordinates, $= \mathbf{x}/l_e$
$\mathbf{x}$	Cartesian coordinates, (x, y, z), m
Ma	Magnetic number, $= \chi_0 \xi (\mu_v h_R)^2/(\mu_v \mu_{ra} g l_e)$
Nu	Average nusselt number
Pr	Prandtl number, $= \nu/\alpha$
Ra	Rayleigh number, $= g\beta(\theta_h - \theta_c) l_e^3/(\alpha\nu)$
L	Dimensionless length, $= l/l_e$
l	Length, m
l_e	Length of one side of an enclosure, m
l_g	Gap between an enclosure and a permanent magnet, m
P	Dimensionless pressure, $= l_e^2(p - p_0)/(\rho_0 \alpha^2)$
p	Pressure, Pa
T	Dimensionless time, $= \alpha t/l_e^2$
t	Time, s
α	Thermal diffusivity, $\mathrm{m^2/s}$
β	Volumetric coefficient of expansion, $\mathrm{K^{-1}}$
χ	Magnetic susceptibility of oxygen, $\mathrm{m^3/kg}$
μ_r	Relative permeability, μ_{ra} in air and μ_{rm} inside a magnet
μ_v	Vacuum permeability, H/m
ν	Kinetic viscosity, $\mathrm{m^2/s}$
Φ	Dimensionless magnetical potential, $= \phi/(h_R l_e)$
ϕ	Magnetical potential, A
ρ	Density, $\mathrm{kg/m^3}$
Θ	Dimensionless temperature, $= (\theta - \theta_0)/(\theta_h - \theta_c)$
θ	Temperature, K
ξ	Mass fraction of oxygen

Subscripts

0	Reference state at $\theta_0 = (\theta_h + \theta_c)/2$
h, c	Value of a hot and a cold wall
W, D, H	Width, depth, and height of a permanent magnet
pm	Position of the center of a permanent magnet
R	Remanence

1 Introduction

Natural convection has been extensively studied from the viewpoints of natural ventilation, solar heating systems, reactor cooling, heat storage, crystal growth of molten silicon, and so on. The advantage of natural convection is that it does not require a power source to drive the fluid and heat. Therefore, in order not to impair this advantage, the device controlling natural convection should not have any mechanical moving parts. The use of magnetic forces has attracted interest as a measure for this purpose.

All substances have magnetism, and the magnitude of which is expressed by the magnetic susceptibility. The larger the magnetic susceptibility of a substance, the stronger the effect of the magnetic field. Ferromagnetic substances such as iron, cobalt, and nickel have a higher magnetic susceptibility, while many other substances have a much lower magnetic susceptibility. Among substances other than ferromagnetic substances, those with a positive magnetic susceptibility are called paramagnetic substances, while those with a negative magnetic susceptibility are called diamagnetic substances. Most of the fluids around us are diamagnetism, with the exception of oxygen, which is a paramagnetic substance.

Although the magnetic forces exerted by magnets on paramagnetic and diamagnetic substances are usually negligible, the magnetic susceptibility of oxygen is still approximately 1000 times greater than that of other elements in the air, making the effect of the magnetic forces relatively pronounced. For this reason, practical applications of introducing magnetic fields to magnetic fluids have long been limited to the applications involving oxygen, such as magnetic oxygen meters, convection of oxygen, and separation of red blood cells from the blood. However, with the advent of superconducting magnets and neodymium magnets, it became relatively easy to realize the powerful magnetic fields of several tesla. This has led to various new discoveries, such as the levitation of water droplets, nitrogen jets, carbon nanotube alignment, the magnetic Archimedes effect, crystal material processing, protein crystallization, and the effect on the evaporation rate of water.

As it is known that the magnetic susceptibility of a paramagnetic material depends on the reciprocal of the absolute temperature, and this property is called Curie's law. Therefore, the degree of influence that a paramagnetic fluid receives from a magnetic field depends on its temperature. Using this principle, heat transfer in convection can be controlled by the application of a magnetic field (Braithwaite et al. 1991). In particular, the effect of applying a magnetic field by means of a superconducting magnet or a superconducting coil on natural convection in a cubic container consisting of a pair of opposing vertical walls that are differentially heated has been variously investigated both experimentally and computationally (Tagawa et al. 2002; Shigemitsu et al. 2003; Bednarz et al. 2004; Kaneda et al. 2004).

A permanent magnet attracts a paramagnetic fluid, and the colder the fluid, the stronger the attraction. In other words, a permanent magnet excludes a hot paramagnetic fluid and attracts a cold one. By taking advantage of this characteristic, a permanent magnet can also be used to control the natural convection of a paramagnetic fluid. However, the magnetic field formed by permanent magnets is not uniform. Therefore, whether the heat transfer is enhanced or inhibited depends on the arrangement, size, and combination of permanent magnets.

With respect to Rayleigh–Benard convection, when a pair of magnets with N and S poles next to each other are placed near the heated bottom wall, the magnetic force at the junction of the pair of magnets is significantly stronger in the same direction as gravity, resulting in a local enhancement of the downward convection, and consequently, the mean Nusselt number is increased by the above-mentioned arrangement of the magnets (Wada et al. 2020). In a duct where one vertical wall is partially heated at a constant heat flux, if a permanent magnet is placed behind the heated wall, the local heat transfer coefficient is suppressed near the lower end of the magnet and enhanced near the upper end (Kaneda et al. 2019, 2020). The natural convection of paramagnetic fluid driven by a horizontal temperature gradient in a two-dimensional square container is enhanced by a square permanent magnet placed at the bottom of the container, and the degree of enhancement depends on the strength and position of the magnet (Song and Tagawa 2018). In addition, as the Curie effect is independent of the presence or absence of gravity, convection can be generated in a weightless environment when a permanent magnet is placed close to a paramagnetic fluid (Yang et al. 2003; Ashouri et al. 2010; Jiang et al. 2014, 2015; Song et al. 2019).

Although superconducting magnets are powerful, they require a cryogenic environment to maintain the superconducting state, making their running costs extremely high. Permanent magnets, on the other hand, are far less expensive because they continue to form a magnetic field without an external energy supply. In particular, neodymium magnets can generate magnetic fields on the order of tesla.

Most of the research on convection control using permanent magnets is aimed at enhancing heat transfer, and there is no research that actively suppresses heat transfer. The method using permanent magnets, however, achieves heat transfer suppression by directly controlling convection. This is a different principle from the use of adiabatic walls. Therefore, this method has the potential to be used for applications that cannot be realized with adiabatic walls.

In this study, permanent magnets were applied to the natural convection of air in a two-dimensional square container consisting of a pair of vertical walls kept at different temperatures. Although this configuration is similar to that of Song and Tagawa (Song and Tagawa 2018), the placement of the magnets was considered in all four directions (top, bottom, left, and right), not just at the bottom side of the container, and multiple aspect ratios of the rectangular bar magnet shape were also tested. Two-dimensional numerical simulations were performed for these configurations to investigate the change in the heat transfer coefficient associated with the magnet arrangement and shape.

2 Methods

2.1 System Considered

Figure 1 shows the problem structure, where an air-filled cubic container was assumed. A pair of vertical walls facing each other are isothermally and differentially heated. The hot wall on the left side facing the front is kept at θ_h and the cold wall on the right side is kept at θ_c. The other walls are adiabatic. Gravity acts in the negative direction on the y-axis.

A prism-shaped neodymium permanent magnet was placed around the container. The N-pole side faces towards the container. The size of the magnet is expressed in terms of width l_W, depth l_D, and height l_H, where l_H is the length from the N-pole to the S-pole and l_D is the spanwise (z-axis) length of the magnet. The position of the magnet is indicated by the center point coordinates (x_{pm}, y_{pm}, z_{pm}) on the N-pole side surface. The gap between the vessel wall and the magnet was set to a dimensionless length of $L_g = 0.04$. The magnetic field was assumed to be static. The properties of the vessel wall with respect to the magnetic field were assumed to be equivalent to those of air.

2.2 Derivation of Model Equation Including Magnetic Force

As known from Maxwell's equations in classical electromagnetism, the magnetic flux density is conserved:

$$\nabla \cdot \mathbf{b} = 0. \tag{1}$$

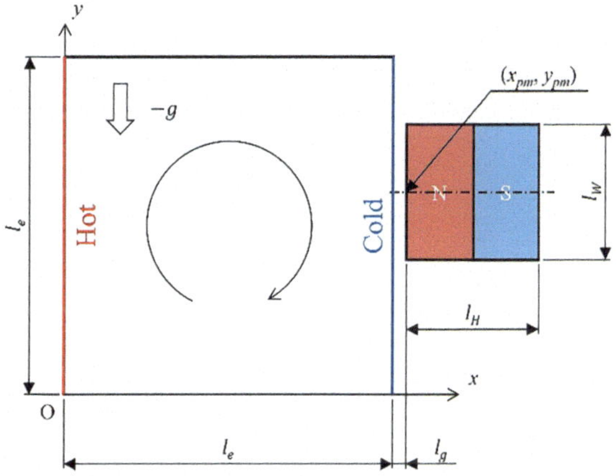

Fig. 1 Schematic illustration of the physical model

The intensity of the magnetic field was assumed to be the gradient of the magnetic potential:

$$\mathbf{h} = -\nabla\phi. \tag{2}$$

The magnetic flux density is determined from the magnetic field intensity and the remanence (also called residual magnetization):

$$\mathbf{b} = \mu_v\mu_r\mathbf{h} + \mathbf{b}_R. \tag{3}$$

This equation applies both in air and inside the permanent magnet. The relation between the remanence and the residual magnetic field intensity is. $b_R = \mu_v\mu_r h_R$. Therefore, Eq. (3) is restated as

$$\mathbf{b} = \mu_v\mu_r(\mathbf{h} + \mathbf{h}_R). \tag{4}$$

Taking the divergence of Eq. (4),

$$\nabla^2\phi = \nabla \cdot \mathbf{h}_R \tag{5}$$

is obtained using Eqs. (1) and (2).

For the relative permeability of air and neodymium magnet, $\mu_{ra} = 1$ and $\mu_{rm} = 1.05$ were adopted, respectively. The $\mathbf{b}_R$ inside the magnet was considered to be uniform. For example, $\mathbf{b}_R = (b_R, 0, 0)$ inside a magnet with the N-pole facing the positive direction on the x-axis, where b_R is the magnitude of the remanence $|\mathbf{b}_R|$. The remanence of a neodymium magnet is generally between 1 T and 1.3 T. In this study, the remanence inside the magnet was set to $b_R = 1.05$T. In this case, $\mu_v h_R = b_R/\mu_{rm} = 1$T for the representative magnetic flux density. Although the remanence varies with temperature, in this study the remanence is considered to be a constant because the permanent magnet is embedded in an isothermal or adiabatic wall. Note that in air, $b_R = 0$T.

The governing equations for the magnetic field, Eqs. (2), (4), and (5), were non-dimensionalized as follows:

$$\nabla^2\Phi = \nabla \cdot \mathbf{H}_R, \tag{6}$$

$$\mathbf{H} = -\nabla\Phi, \tag{7}$$

$$\mathbf{B} = \mu_r(\mathbf{H} + \mathbf{H}_R). \tag{8}$$

The magnetic field is obtained by solving the Poisson equation shown as Eq. (6).

The magnetic susceptibility of air can be regarded as the product of the magnetic susceptibility and mass fraction of oxygen $\chi\xi$ (Bai et al. 1999). In this way, the oxygen-containing air itself was treated as a paramagnetic fluid. The magnetic force on air was expressed as $+\rho\chi\xi\nabla b^2/(2\rho_0\mu_v\mu_{ra})$ and incorporated as a magnetic force term in the equations of motion.

The working fluid was assumed to be incompressible. However, the temperature dependence of the density and magnetic susceptibility was taken into account in the buoyancy and magnetic force terms of the equation of motion. Furthermore, the magnetic field was assumed not to change with the movement of the fluid. The governing equations for the flow field are the equation of continuity, the Navier–Stokes equation, and the energy equation, shown as Eqs. (9), (10), and (11), respectively:

$$\nabla \cdot \mathbf{u} = 0, \tag{9}$$

$$\frac{D\mathbf{u}}{Dt} = -\frac{1}{\rho_0}\nabla p + \nu\nabla^2\mathbf{u} + \frac{\rho}{\rho_0}\mathbf{g} + \frac{\rho\chi\xi}{2\rho_0\mu_v\mu_{ra}}\nabla b^2, \tag{10}$$

$$\frac{D\theta}{Dt} = \alpha\nabla^2\theta. \tag{11}$$

Note that $b^2 = \mathbf{b}\cdot\mathbf{b}$.

When there is no temperature difference between the left and right heat transfer walls, that is, $\theta_h = \theta_c = \theta_0$, the air inside is at rest. The equation of motion at that time is as follows:

$$0 = -\frac{1}{\rho_0}\nabla p_0 + \frac{\rho_0}{\rho_0}\mathbf{g} + \frac{\rho_0\chi_0\xi}{2\rho_0\mu_v\mu_{ra}}\nabla b^2. \tag{12}$$

Subtracting Eq. (12) from Eq. (10),

$$\frac{D\mathbf{u}}{Dt} = -\frac{1}{\rho_0}\nabla(p - p_0) + \nu\nabla^2\mathbf{u} + \frac{\rho - \rho_0}{\rho_0}\mathbf{g} + \frac{(\rho\chi - \rho_0\chi_0)\xi}{2\rho_0\mu_v\mu_{ra}}\nabla b^2. \tag{13}$$

If the temperature difference is small, the following approximation is valid (Tagawa et al. 2002):

$$\rho - \rho_0 = -\rho_0\beta(\theta - \theta_0), \tag{14}$$

$$\rho\chi - \rho_0\chi_0 = -2\rho_0\chi_0\beta(\theta - \theta_0). \tag{15}$$

Among them, Eq. (14) is called the Boussinesq approximation. Substituting Eqs. (14) and (15) into Eq. (13) yields the equation of motion in the following form:

$$\frac{D\mathbf{u}}{Dt} = -\frac{1}{\rho_0}\nabla(p - p_0) + \nu\nabla^2\mathbf{u} - \beta(\theta - \theta_0)\mathbf{g} - \frac{\chi_0\xi\beta(\theta - \theta_0)}{\mu_v\mu_{ra}}\nabla b^2. \tag{16}$$

The boundary conditions at the wall surface are given below. As the wall surface is no-slip, the velocity is $\mathbf{u} = (0, 0, 0)$. The pressure boundary condition is zero normal gradient. The temperature boundary conditions of $\theta = \theta_h$ and $\theta = \theta_c$ were imposed for the hot ($x = 0$) and cold ($x = l_e$) walls, respectively, and $\partial\theta/\partial y = 0$ was imposed for the adiabatic walls ($y = 0, l_e$).

The dimensionless governing equations were obtained from Eqs. (9), (16), and (11) as follows:

$$\nabla \cdot \mathbf{U} = 0, \tag{17}$$

$$\frac{D\mathbf{U}}{DT} = -\nabla P + \mathrm{Pr}\,\nabla^2\mathbf{U} - \mathrm{Pr}\,Ra\Theta\mathbf{G} - \mathrm{Pr}\,RaMa\Theta\nabla B^2, \tag{18}$$

$$\frac{D\Theta}{DT} = \nabla^2\Theta, \tag{19}$$

where the dimensionless parameter Ma is the ratio of magnetic force to gravity. In this paper, this is referred to as the magnetic number.

In this study, the conditions $Pr = 0.71$, $Ra = 10^5$, and $Ma = 1$ were considered, unless otherwise stated. The temperature difference required to realize this configuration was confirmed. Here, a neodymium magnet and air under the standard condition (101.325 kPa, 273.15 K) were assumed. Enumerating the constants, $g = 9.81\mathrm{m/s^2}$, $\beta = 1/273\mathrm{K^{-1}}$, $\nu = 1.5 \times 10^{-5}\mathrm{m^2/s}$, $b_R = 1.05\mathrm{T}$, $\mu_v = 4\pi \times 10^{-7}\mathrm{H/m}$, $\mu_{ra} = 1$, $\mu_{rm} = 1.05$, $\chi_0 = +1.34 \times 10^{-6}\mathrm{m^3/kg}$, and $\xi = 0.232$. The α is derived from Pr and ν. These constants were assumed not to vary with temperature. As most of the components of Ra and Ma are physical properties, the only element that can be changed as a practical matter is the length of one side of the container l_e. The vessel length and temperature difference at which the above conditions are realized are $l_e = 0.0252\mathrm{m}$ and $\theta_h - \theta_c = 55.0\mathrm{K}$. For reference, when $l_e = 0.01, 0.02, 0.05$, and $0.1\mathrm{m}$, in order, $\theta_h - \theta_c = 882, 110, 7.06$, and $0.882\mathrm{K}$, and $Ma = 2.52, 1.26, 0.504$, and 0.252.

Nusselt number was evaluated on the hot wall surface. The local Nusselt number Nu_{loc} is defined as

$$Nu_{loc} = -\frac{d\Theta}{dX}, \tag{20}$$

and the average Nusselt number Nu is given by

$$Nu = \int_0^1 Nu_{loc}dY \Big/ \int_0^1 dY. \tag{21}$$

Numerical Methodologies

OpenFOAM was used to numerically simulate the research subject. This is an open-source and free-licensed thermal-fluid analysis library that is widely used in the study of thermal convection. Standard solvers are available for heat transfer and buoyancy-driven flow using the Boussinesq approximation, as well as a solver for analyzing the magnetic field formed by permanent magnets (magneticFoam). However, no solver suitable for the thermo-fluid motion in this research was available, and therefore an existing buoyancy-driven flow solver (buoyantBoussinesqSimpleFoam) with an additional magnetic force term was developed on our own. The treatment of the buoyancy term in the existing solver was used as a reference. Examples of OpenFOAM customization can be seen in the analyses of the electromagnetic thermal fluid (Singh and Gohil 2019; Singh and Chandy 2020) and the magnetic fluid (Yang 2021).

The governing equations were discretized using the finite volume method. The central difference was used to discretize the advection terms. The magnetic field was obtained by an iterative solution method prior to the flow analysis. The flow field was solved by a steady-state solution method based on the SIMPLE method. This algorithm was terminated when the L1 norm of the initial residual of the discretized governing equation fell below 10^{-8}. A non-uniform mesh densely clustered near the wall was used as the mesh for the flow field, and the number of cells is 80^2. The lengths of the cells varied in isosceles, where the ratio of the length of the largest to the smallest cell was set to 4.

3 Results and Discussion

In this study, unless otherwise stated, the condition was set to $Pr = 0.71$, $Ra = 10^5$, $Ma = 1$, $L_g = 0.04$, and $L_W = L_H = 0.4$. For the above condition, we first investigated the effect on the Nusselt number of different magnet configurations (top, bottom, left, and right).

The case where the magnet was on the lower side of the container was verified. Here, the central axis of the magnet was in the range of $0 \leq X_{pm} \leq 1.5$. As shown in Fig. 2a, heat transfer is suppressed in the range of $0 \leq X_{pm} \leq 0.8$, while heat transfer is slightly enhanced in the range of $0.9 \leq X_{pm} \leq 1.5$. For the latter, heat transfer is considered to be slightly enhanced due to the accelerated settling of cold air.

As shown in Fig. 2b, heat transfer is slightly enhanced when the central axis of the magnet is in $0.6 \leq Y_{pm} \leq 1.1$, while it is suppressed in other cases of $0 \leq Y_{pm} \leq 0.5$ and $1.2 \leq Y_{pm} \leq 1.5$. The former heat transfer enhancement is attributed to the fact that the hot fluid is encouraged to disperse from the hot wall. However, at the same time, heat transfer from the hot wall to the working fluid is also hindered, thus limiting the heat transfer enhancement effect.

As shown in Fig. 2c, heat transfer is suppressed throughout the range of $0 \leq X_{pm} \leq 1.5$. In the range of $0 \leq X_{pm} \leq 0.5$, Nu is almost unchanged. When the central axis of the magnet moves from $X_{pm} = 0.5$ to 0.6, Nu decreases significantly.

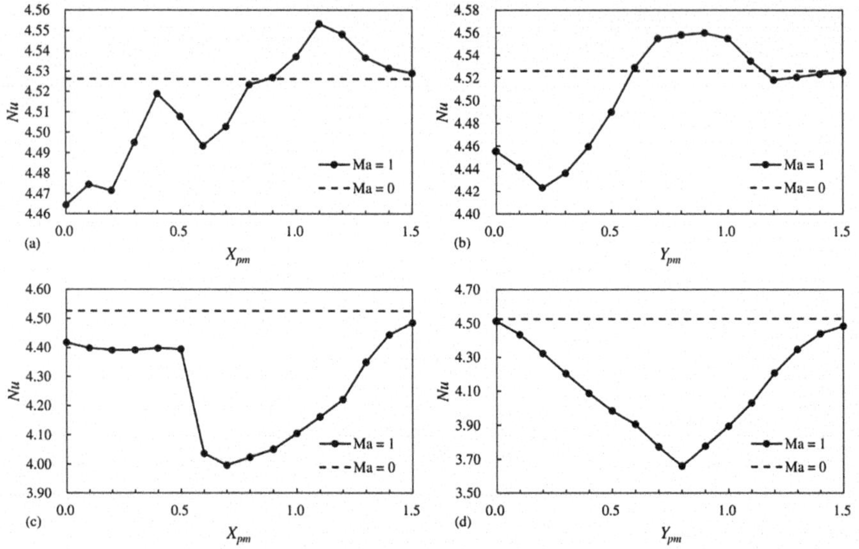

Fig. 2 Distributions of Nu over X_{pm} and Y_{pm} of $L_W = L_H = 0.4$ at $Ra = 10^5$ and $Ma = 1$. A permanent magnet is located on the **a** bottom, **b** left, **c** top, and **d** right sides

At $X_{pm} = 0.7$, Nu reaches a minimum. In this case, Nu is reduced by 11.7% compared to the case without the magnet. Finally, in the range $0.7 \le X_{pm} \le 1.5$, Nu increases as the magnet moves in the $X+$ direction. When the magnet is on the upper side of the container, the warm air is prevented from approaching the cold wall, while the cold air is kept on the upper side of the container, thereby significantly reducing heat transfer.

As shown in Fig. 2d, heat transfer is significantly suppressed throughout the range of $0 \le Y_{pm} \le 1.5$. The distribution of Nu with respect to Y_{pm} has a clean V-shape. The Nu is minimum at $Y_{pm} = 0.8$. In this case, $Nu = 3.65790$, which is 19.2% suppressed compared to no magnetic field. This is the smallest of the four directional configurations. When the magnet is on the cold wall side of the container, the warm air is prevented from approaching the cold wall, and conversely, the cool air is attracted to the cold wall, which significantly suppresses heat transfer.

Comparing the upper and right magnet arrangements, the minimum value of the Nusselt number for each is smaller in the right arrangement than in the upper arrangement. Therefore, the right-hand placement was judged to be the most suitable for suppressing heat transfer. In the following, the case where the magnet is placed on the right-hand side of the container was mainly investigated.

Figure 3a, b, c, d show the convection pattern when the magnet is placed on the right side of the vessel, where a magnet with $L_W = L_H = 0.4$ is placed at $Y_{pm} = 0.8$. Figure 3a, b, c are the temperature Θ contours, while Fig. 3b, d are the contours of the velocity magnitude |U|, and the white lines in the figures represent streamlines. As can be seen from these figures, apart from the main circulation vortex caused by

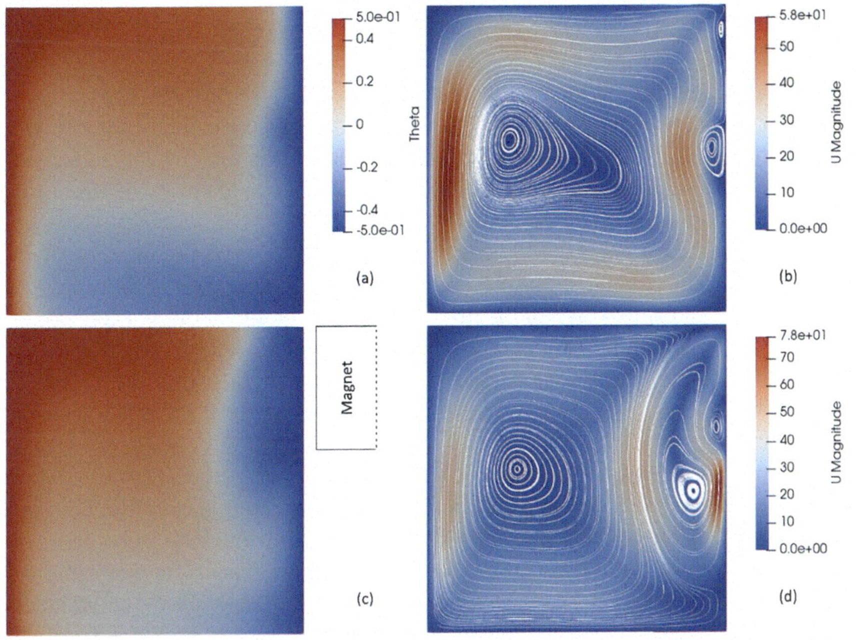

Fig. 3 Contours **a** and **c** of Θ and **b** and **d** of |U| with stream lines **a** and **b** at $Ma = 1$ and **c** and **d** at 10, where a permanent magnet with $L_W = L_H = 0.4$ is located at $Y_{pm} = 0.8$

buoyancy, additional circulation vortices are formed in the vicinity of the magnetic corners. The formation of such circulation vortices strongly inhibits convective heat transfer.

This additional vortex is due to the tangential component of the magnetic force $\partial B^2/\partial Y$. Above the top and from the central axis to the bottom of the magnet, $\partial B^2/\partial Y$ acts to move the cold air downwards. In contrast, below the bottom and from the central axis to the top, it acts to move the cold air upwards. This upward action forms an additional circulating vortex. The main circulation vortex caused by the buoyancy force is clockwise when viewed from the $Z+$ side, while the additional circulation vortices near the magnet corners are counterclockwise.

At $Ma = 1$, circulating vortices are added slightly below the top and bottom edges of the magnet, as shown in Fig. 3b. The size of the lower vortex is larger than the upper vortex, because ∇B^2 is stronger on the outside of the magnet extension than on the inside. These additional circulating vortices become larger as Ma increases. At $Ma = 10$, as shown in Fig. 3d, the vortices that were separated into an upper and a lower vortex at $Ma = 1$ merge to form an additional clockwise circulating vortex within the counterclockwise circulating vortex.

4 Conclusions

In this study, the effect of the arrangement and geometry of prismatic neodymium permanent magnets in close proximity to a cubic vessel on the natural convection of oxygen-containing air in the vessel was investigated by two-dimensional numerical simulation. Among the vertical walls of the vessel, the left side toward the front was heated, and the right side was cooled.

The length of each magnet was set to 0.4 relative to the length of one side of the cube container, and the position of the magnet was varied at the top, bottom, left, and right of the container. When the magnet is placed on the bottom or left side of the container, although heat transfer is enhanced or suppressed, the effect is slight. On the other hand, if the magnet is placed on the top or right side of the container, heat transfer is significantly suppressed. Note that the extent of the effect depends on the position of the magnet.

The principle of the cooling effect of permanent magnets placed near the cooling wall can be summarized in the following three ways. The magnetic fluid at high temperature is excluded, the magnetic fluid at low temperature is adsorbed, and vortices are formed near the cooling wall. These mechanisms suppress thermal convection. Therefore, unlike the suppression of heat conduction by an adiabatic wall, the wall surface is kept at a low temperature.

References

Ashouri, M., Ebrahimi, B., Shafii, M., Saidi, M., Saidi, M.: Correlation for Nusselt number in pure magnetic convection ferrofluid flow in a square cavity by a numerical investigation, J. Magn. Magn. Mater. 322(22), 3607–3613 (2010).

Bai, B., Yabe, A., Qi, J., Wakayama, N. I.: Quantitative analysis of air convection caused by magnetic-fluid coupling, AIAA J. 37(12), 1538–1543 (1999).

Bednarz, T., Tagawa, T., Kaneda, M., Ozoe, H., Szmyd, J. S.: Magnetic and gravitational convection of air with a coil inclined around the x axis, Numer. Heat Transf. A: Appl. 46(1), 99–113 (2004).

Braithwaite, D., Beaugnon, E., Tournier, R.: Magnetically controlled convection in a paramagnetic fluid, Nature 354(6349), 134–136 (1991).

Jiang, C., Feng, W., Zhong, H., Zeng, J., Zhu, Q.: Effects of a magnetic quadrupole field on thermomagnetic convection of air in a porous square enclosure, J. Magn. Magn. Mater. 357, 53–60 (2014).

Jiang, C., Shi, E., Hu, Z., Zhu, X., Xie, N.: Numerical simulation of thermomagnetic convection of air in a porous square enclosure under a magnetic quadrupole field using LTNE models, Int. J. Heat Mass Transf. 91, 98–109 (2015).

Kaneda, M., Noda, R., Tagawa, T., Ozoe, H., Lu, S.-S., Hua, B.: Effect of inclination on the convection of air in a cubic enclosure under both magnetic and gravitational fields with flow visualization, J. Chem. Eng. Japan 37(2), 338–346 (2004).

Kaneda, M., Nazato, K., Fujiwara, H., Wada, K., Suga, K.: Effect of magnetic field on natural convection inside a partially-heated vertical duct: Experimental study, Int. J. Heat Mass Transf. 132, 1231–1238 (2019).

Kaneda, M., Fujiwara, H., Wada, K., Suga, K.: Natural convection of paramagnetic fluid along a vertical heated wall under a magnetic field from a single permanent magnet, J. Magn. Magn. Mater. 502, 166574 (2020).
Shigemitsu, R., Tagawa, T., Ozoe, H.: Numerical computation for natural convection of air in a cubic enclosure under combination of magnetizing and gravitational forces, Numer. Heat Transf. A: Appl 43(5), 449–463 (2003).
Singh, R. J., Chandy, A. J.: Numerical investigations of the development and suppression of the natural convection flow and heat transfer in the presence of electromagnetic force, Int. J. Heat Mass Transf. 157, 119823 (2020).
Singh, R. J., Gohil, T. B.: The numerical analysis on the development of Lorentz force and its directional effect on the suppression of buoyancy-driven flow and heat transfer using OpenFOAM, Comput. Fluids 179, 476–489 (2019).
Song, K. W., Tagawa, T.: Thermomagnetic convection of oxygen in a square enclosure under non-uniform magnetic field, Int. J. Therm. Sci. 125, 52–65 (2018).
Song, K., Wu, S., Tagawa, T., Shi, W., Zhao, S.: Thermomagnetic convection of paramagnetic gas in an enclosure under no gravity condition, Fluids 4(1), 49 (2019).
Tagawa, T., Shigemitsu, R., Ozoe, H.: Magnetizing force modeled and numerically solved for natural convection of air in a cubic enclosure: effect of the direction of the magnetic field, Int. J. Heat Mass Transf. 45(2), 267–277 (2002).
Wada, K., Kaneda, M., Suga, K.: Rayleigh-Bénard convection of paramagnetic liquid under a magnetic field from permanent magnets, Symmetry 12(3), 341 (2020).
Yang, W. M.: On the boundary conditions of magnetic field in OpenFOAM and a magnetic field solver for multi-region applications, Comput. Phys. Commun. 263, 107883 (2021).
Yang, L., Ren, J., Song, Y., Guo, Z.: Free convection of a gas induced by a magnetic quadrupole field, J. Magn. Magn. Mater. 261(3), 377–384 (2003).

Numerical Investigation on Unsteady Flow Past a Hydrophobic Semi-circular Cylinder

S. Harigovind, A. Rehan, K. Nandakumar Chandran, K. K. Krishnaram, and Ranjith S. Kumar

Abstract We introduce a fully-coated slippery semi-circular cylinder in a uniform flow and studied the physics of the same. To mimic the hydrophobic condition on the semi-circular cylinder Navier-slip modeling is employed. Finite Element Analysis(FEM) has been used to solve the momentum and continuity equations. The study has been validated for a circular cylinder with slip implementation with the available studies and was found satisfactory. Four flow velocities (Re = 50, 100, 150, 200) have been employed and three slip lengths (β'= 0,0.05,0.1) have been applied to the cylinder walls to make it slippery. The hydrodynamic parameters under investigation are the coefficient of lift, coefficient of drag, and Strouhal number. As the wall slip increases, a slight decrease in lift coefficient is noted while a significant drop in drag coefficient is observed. An increase in Re caused a rise in both coefficient values. An increased shedding frequency was observed at higher slip lengths and with an increase in flow velocity and is indicated by the Strouhal number. The objectives of the study were well addressed by the designed parametric study and the performance characteristics of the fully coating had been analyzed. The numerical study performed was able to provide a reflection of the physical phenomenon in the simplest manner.

Keywords Hydrophobicity · Navier-slip · Coefficient of lift · Coefficient of drag

1 Introduction

Fluid dynamics research has many practical applications, and flow past solid bodies is a crucial area of study. A semi-circular cylinder is one of these bodies that has drawn a lot of interest because of its distinctive shape and wide range of practical uses. For the design and optimization of several engineering structures, such as airfoils,

S. Harigovind (✉) · A. Rehan · K. Nandakumar Chandran · K. K. Krishnaram · R. S. Kumar
Micro/Nano Fluidics Research Laboratory, Department of Mechanical Engineering, College of Engineering Trivandrum (Government of Kerala), Thiruvananthapuram, Kerala, India
e-mail: harigovinds2002@gmail.com

G. Rajesh et al. (eds.), *Proceedings of the 15th International Symposium on Experimental and Computational Aerothermodynamics of Internal Flows*,
https://doi.org/10.1007/978-981-96-4082-9_5

heat exchangers, and offshore constructions, it is crucial to comprehend the flow characteristics around a semi-circular cylinder. The introduction of slippery walls to the semi-circular cylinder helps in revealing whole new flow patterns and structures and their effect on the hydrodynamic parameters like the coefficient of lift and Strouhal numbers(St) have been noticeable. Due to boundary layer separation, recirculation zones, and vortices being present, the flow through a semi-circular cylinder is intrinsically complicated. Such intricate fluid dynamics processes have effects on a variety of applications, including maritime engineering, heat transport, and aerodynamics. Moreover, for the design and operation of offshore platforms, breakwaters, and coastal protection systems, it is crucial in the offshore sector to comprehend how waves interact with semi-circular structures. Predicting the performance and stability of diverse engineering systems requires an accurate characterization of the flow patterns and the related parameters, such as lift and drag forces, wake structure, and pressure distribution.

Bluff body dynamics have been prominent for several decades and extensive research on various bluff body shapes like circular, elliptical cylinders (Zdravkovich 1997, 2003; Sumer and Fredsoe 2006; Williamson 1993; Coutanceau and Defaye 1991) have been performed and the effects of various hydrodynamic parameters have also been studied. Rajani et al. (2009) studied flow past a cylinder in two-dimensional as well as three-dimensional domains and studied RMS amplitudes in unsteady periodic flows of the aerodynamic coefficients. However, the studies in the area of semi-circular cylinders have been very less. Chandra and Chhabra (2001, 2011) studied the flow over semi-circular cylinders within the range $0.01 \leq Re \leq 39.5$, where Re denotes the Reynolds number which is the ratio of inertial forces to viscous forces and Prandlt number from 0.72 and 100 and was able to find the critical Reynolds number for the formation of wakes and vortex sheddings. The present study tries to study the effect of slippery walls on the semi-circular cylinder. Various studies on slip have been performed on circular cylinders like the work by Seo et al. (2012) where slip on the walls where implemented by using Navier-slip conditions and flow was investigated for a total of 4 Re values of 20, 40, 80, and 100 and six slip lengths. The simulation results showed that as the slip length was increased, the drag coefficient decreased as the drag's frictional component was reduced, and the shear layer which was formed along the surface of the cylinder pushed the point of seperation away and showed a greater separation angle than when no slip was implemented. Various experimental (Kim et al. 2015) and numerical studies (Amal et al. 2023; Muralidhar et al. 2011; Gu et al. 2019) discussing the application of partial slip/roughness on the bluff body surfaces are discussed in previous literature, leaving us an opportunity to explore uniform coating over brand-new surfaces as in the present study.

An in-depth analysis of flow past a hydrophobic semi-circular cylinder is studied in this research. We seek to deepen our comprehension of this particular geometry, contribute to the creation of optimized engineering designs, and advance the area of fluid dynamics by investigating the flow patterns, forces, and parameters like the coefficient of lift and strouhal number.

2 Numerical Methodology

This study focuses on the study on the laminar flow past a hydrophobic semi-circular cylinder and in this section, we focus on the governing equations, followed by the validation, and finally the grid-independent study.

2.1 Governing Equations

The conservation of mass and the conservation of momentum equations have been used as the governing equations for the study. The flow was assumed to be Newtonian, and incompressible, and a 2-D domain was employed and the flow was taken as laminar. The continuity and momentum equations are given by,

$$\nabla \cdot \boldsymbol{u} = 0, \tag{1}$$

$$\frac{\partial \boldsymbol{u}}{\partial t} + \boldsymbol{u} \cdot \nabla \boldsymbol{u} = -\frac{1}{\rho} \nabla p + \nu \nabla^2 \boldsymbol{u} \tag{2}$$

where $\boldsymbol{u}$, p, ρ, ν symbols denote the inflow velocity, pressure, density, and kinematic viscosity respectively.

Figure 1 represents the computational domain used for the current study. A domain of size 20.5 D_{eq} * 12 D_{eq}, where D_{eq} represents the diameter of the semi-circular cylinder, was taken for the study, and the semi-circular cylinder close to the inlet side. The inlet of the domain was initialized as uniform velocity (U). The outlet wall was initialized as gauge pressure equal to 0. The upper and lower walls were taken as symmetry. The red outline over the semi-circular cylinder helps in portraying the hydrophobic surface. Hydrophobicity was brought to the surface of the cylinder with

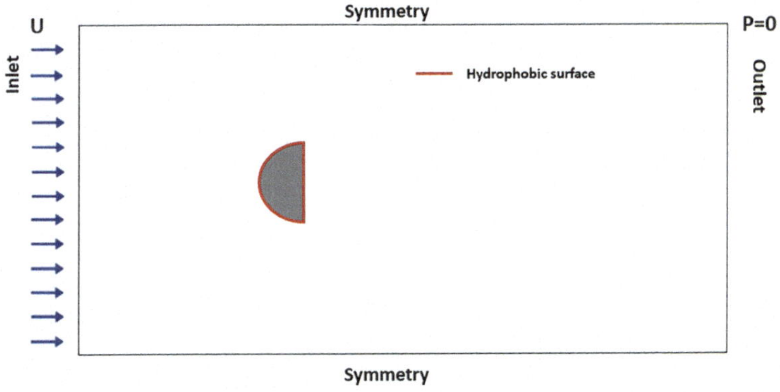

Fig. 1 Computation domain of the semi-circular cylinder used in the current study

the help of the Navier slip model which is known for its impressive results and easy execution and due to the same has been used in all types of flow models be it high velocity or very low. The Navier slip model is given by the equation,

$$\boldsymbol{u}_{\parallel} = \beta \boldsymbol{n} \cdot \left[\nabla \boldsymbol{u} + (\nabla \boldsymbol{u})^T\right] \cdot (1 - \boldsymbol{nn}),$$

where $\boldsymbol{u}_{\parallel}$ denotes the velocity which is tangential to the surface and β represents the slip length employed.

The governing equations mentioned above are resolved using the finite element technique using COMSOL Multiphysics. A fully linked strategy is used to complete the time-dependent investigation, which entails combining all of the unknowns into a single system of equations. According to recent research in the literature (Seo and Chang 2012), $\frac{\Delta t U}{D} < 1$ is used as the time step for the simulation, and here total duration ($\tau = \frac{tU}{D}$) for the investigation is chosen as 200.

The non-dimensionalized quantities used for the current study include Reynolds number (Re) and is given by the expression,

$$Re = \frac{\rho U D_{eq}}{\mu} \tag{3}$$

The hydrodynamic parameters required are the coefficient of lift (C_L) and drag (C_D) which are expressed by,

$$C_L = \frac{2F_L}{\rho U^2 D_{eq}} \tag{4}$$

$$C_d = \frac{2F_d}{\rho U^2 D_{eq}} \tag{5}$$

Slippery walls are implemented using the Navier slip model and the non-dimensionalized slip length (β') is expressed as,

$$\beta' = \frac{\beta}{D_{eq}} \tag{6}$$

It is to be noted that $\beta' = 0$, represents a no-slip surface and $\beta' \neq 0$ corresponds to a hydrophobic surface.

2.2 Validation

Validation of the study has been performed by comparing the coefficient of lift graphs plotted against non-dimensionalized time with the numerical study performed by Seo et al. (2012). The classic case of a circular cylinder has been used for the validation.

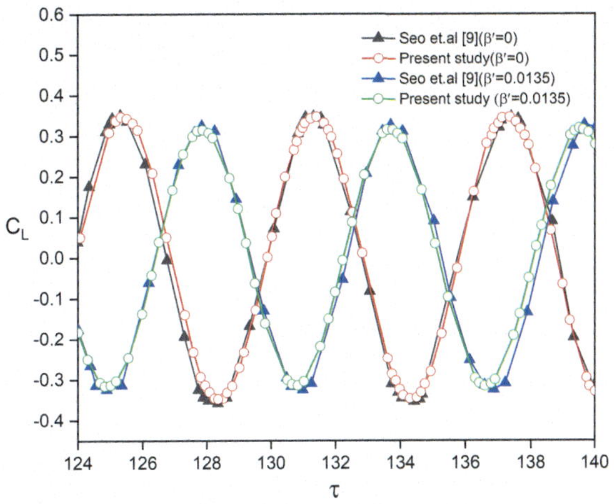

Fig. 2 Time history of coefficient of lift for a circular cylinder at $Re = 100$ and $\beta' = 0, 0.0135$. The results of the present simulation are plotted with the numerical study of Seo et al. (2012)

Simulations at Re = 100 and $\beta' = 0, 0.0135$ was performed and the errors between the lift graphs between the present study and the study by Seo et al. have been found to be very minimal and to be in agreement. Therefore, the validation graphs help in confirming the method of the study is accurate (see Fig. 2).

The Table 1 shows the comparison of the done studies in the semi-circular area and the coefficient of drag as well as the strouhal numbers of the present study are compared with available literature and the procured values have been found to be in accordance with the available literature.

2.3 *Grid Independent Study*

Grid independent study for the present study for the semi-circular cylinder was performed at Re = 100 for the parameter coefficient of drag. It was observed that at mesh elements higher than 79522, the parameter had started to stabilize and at lower values a rather significant fluctuation was observed. Therefore, a mesh element value of 79522 was taken for the simulation to save the computational time and complexity (see Table 2).

Table 1 Validation of current study with the available numerical studies at Re = 100

	St	Cd
Lange et al. (1998)	0.1656	1.327
Dalal et al. (2008)	0.1587	1.4147
Bhinder et al. (2012)	0.1697	1.46
Present	0.1697	1.46

Table 2 Grid independent study conducted for the numerical investigation at Re = 100

Mesh elements	Cd
100000	1.46
79522	1.46
40000	1.443
28000	1.378

3 Results and Discussion

3.1 *Effect of Slip*

Velocity and Pressure contours Fig. 3 represents the velocity contours for flow over the semi-circular cylinder at Re = 100 for different $\beta' = 0, 0.05$ & 0.1 at time intervals of $\tau = 50, 55, 60, 65$ and 70. Figure 4 represents the pressure contours of the same case for the specified time intervals. It is clearly visible that the wall velocity increases with the increase in slip length. The time contours also help portray the onset of vortex shedding, and the shedding phenomenon was found to increase with slip length. In this case, the semi-circular cylinder is kept upright and due to the same flow separation does not occur as there is not enough surface area for the phenomenon to happen. It is observed at $\tau = 50$ that the flow field is seen undisturbed with no shedding occurring and the flow is smooth for all the slipcases. The semi-circular cylinder was found to have two same eddies in its wake region. As the time step reaches $\tau = 55$, a disturbance is observed in the slipcases of $\beta' = 0.05$ & 0.1 as seen in Fig. 3g and h whereas the flow field is still stable at no slip ($\beta' = 0$). On further increase to $\tau = 60$, the start of the disturbance is evident in the no-slip case and the onset of vortex shedding is clearly visible for the slip-implemented cases as shown in Fig. 3h and m. The later time contours show a clear transition to the shedding phenomenon in all the cases. It has been observed that increased shedding occurs as the slip length is increased and can be clearly seen in the pressure contours Fig. 4e, j, and o which compare the pressure cases for the slip lengths $\beta' = 0, 0.05$ & 0.1. An increased oscillating pressure regions are visible in the $\tau = 70$ time stamp for the pressure distribution and it is seen to intensify with the increase in slip length from $\beta' = 0$ to $\beta' = 0.05$ and $\beta' = 0.1$.

Figure 5 portrays the effect of slip lengths on the coefficient of lift at Re = 100 and it is plotted against non-dimensionalized time (τ). It is observed that as the flow changes from no-slip ($\beta' = 0$) to slip ($\beta' = 0.05$) there is a decrease in the amplitude of the coefficient of lift but as the slip length is increased to $\beta' = 0.1$ there is no significant change in the amplitude. As the flow velocity across the walls of the semi-cylinder increases with the implementation of slip the skin friction across the walls is reduced and that affects the pressure distributions on the top and bottom part of the structure thereby decreasing the lift. But as the slip increases we see minimal change in amplitude which is because of the lack of surface area for producing the

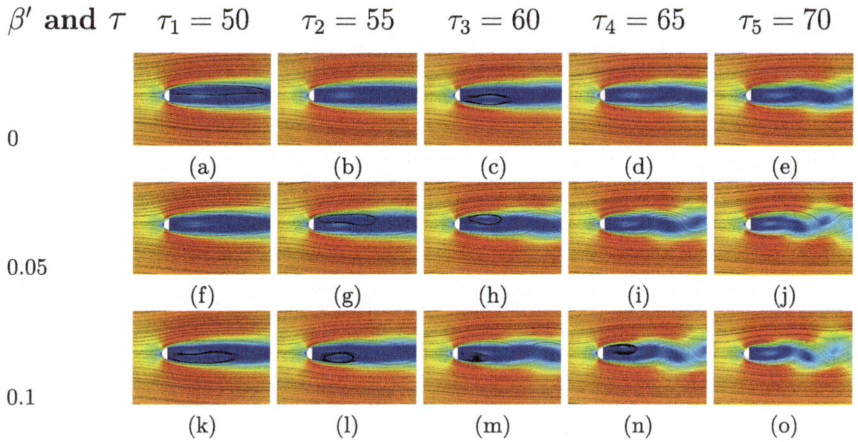

Fig. 3 Velocity Contours for flow at $Re = 100$ across elliptical cylinder with $\beta' = 0, 0.05$ & 0.1 at various time intervals

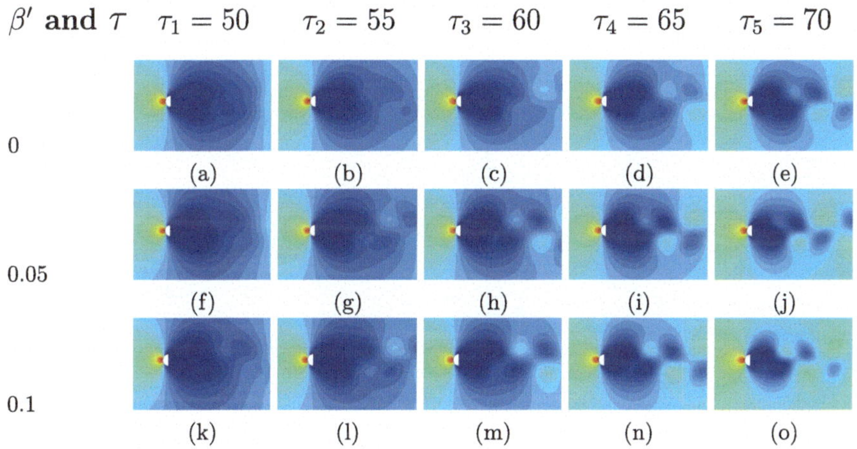

Fig. 4 Pressure Contours for flow at $Re = 100$ across elliptical cylinder with $\beta' = 0, 0.05$ & 0.1 at various time intervals

lift as even the flow separation doesn't even occur. Therefore, the slip has an effect on the lift amplitude but increased slip lengths have produced minimal changes in lift.

Figure 6 depicts the effect of slip lengths on the coefficient of drag at Re = 100 and it is plotted against non-dimensionalized time (τ). A measurement of the resistance an item encounters when traveling through a fluid is the drag coefficient. It displays the dynamic pressure and drag force ratio relative to the reference region. Less resistance to the passage of the item through the fluid is indicated by a lower drag coefficient. The fluid flow near the surface becomes less limited and the fluid velocity rises when

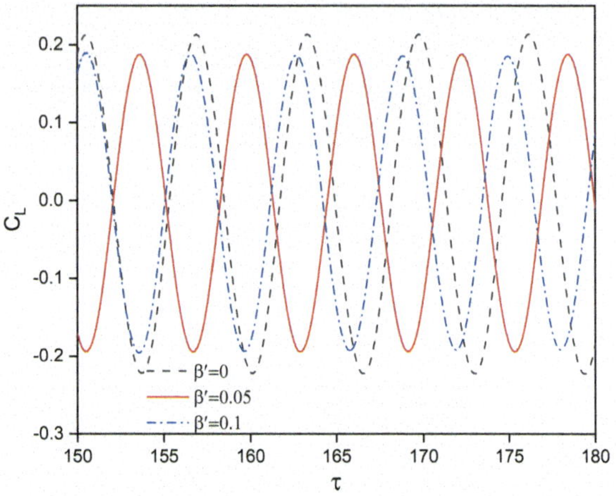

Fig. 5 Time history of coefficient of lift for a semi-circular cylinder at $Re = 100$ and $\beta' = 0, 0.05, 0.1$ portraying the effect of slip

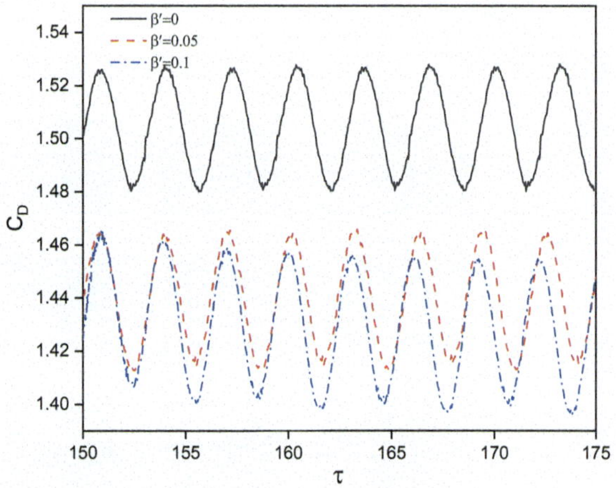

Fig. 6 Time history of coefficient of drag for a semi-circular cylinder at $Re = 100$ and $\beta' = 0, 0.05, 0.1$ portraying the effect of slip

there is a considerable slip length between the fluid and the surface of an object. The drag coefficient may drop as a result of this higher fluid velocity. Therefore the drag was found to decrease with an increase in slip length due to the larger surface area of impact unlike that of the lift coefficient and an increase of further slip does have an impact on the drag coefficient.

3.2 *Effect of Reynolds Number*

Figure 7 shows the time history of the lift coefficient for various Reynolds numbers at constant slip and this graph helps in demonstrating the effect of Re on the coefficients. It has been observed that as the Reynolds number increases which in turn means the

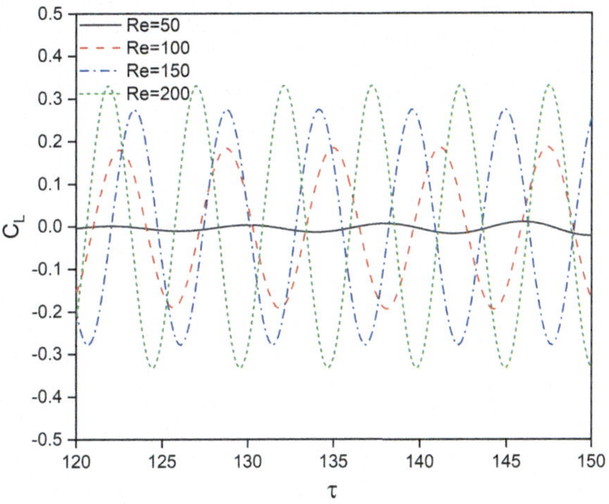

Fig. 7 Time history of coefficient of lift for a semi-circular cylinder at different $Re = 50, 100, 150, 200$ and $\beta' = 0.05$ portraying the effect of Reynolds number

velocity increases, increases the lift as well. At a lower Reynolds number of 50 due to the absence of vortex shedding phenomenon flow is more laminar and it therefore produces less lift compared to the others. The boundary layer which is the thin layer of fluid next to the object's surface, tends to be thinner with higher Reynolds numbers. The semi-circular cylinder can maintain attached flow and produce greater lift when the adverse pressure gradient on its top surface is reduced by a smaller boundary layer.

Figure 8 portrays the variation of the drag coefficient plotted against non-dimensionalized time showing the effect of the Reynolds number. At lower Reynolds numbers when the phenomenon of vortex shedding is absent, the drag shows a really less amplitude but increased amplitudes are observed at higher Reynolds numbers

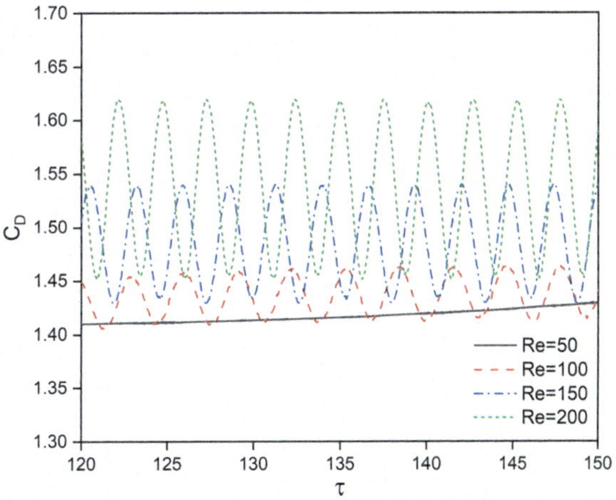

Fig. 8 Time history of coefficient of drag for a semi-circular cylinder at different $Re = 50, 100, 150, 200$ and $\beta' = 0.05$ portraying the effect of Reynolds number

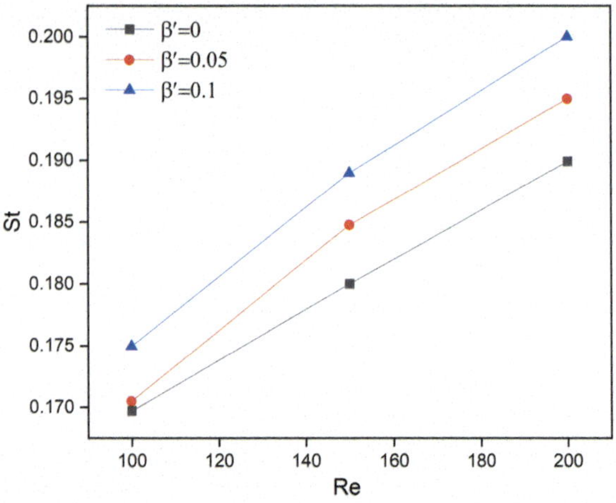

Fig. 9 Strouhal number variation for a semi-circular cylinder at different $Re = 50, 100, 150, 200$ and $\beta' = 0.05$

where shedding occurs. Higher Reynolds numbers cause a shift in the pressure distribution surrounding the semi-circular cylinder, increasing the pressure drag. The pressure differential between the object's front and back surfaces is what generates pressure drag and as the velocity of the fluid is increased the strong vortices produce a chaotic flow behind the object causing an increase in pressure drag, causing an increase in drag coefficient.

Figure 9 shows the effect of the Reynolds number as well as the slip on the Strouhal number. It has been observed that the Strouhal number increases in value with an increase in the Reynolds number portraying that the shedding frequency increases. The increase in velocity causes the flow to be more chaotic. The effect of slip is also evident in the graph and it is observed that as the slip increases the Strouhal number also increases. An increase in slip causes a decrease in drag forces acting on the body and thereby enhancing the shedding of vortices which thereby increases the frequency of the vortex shedding.

4 Conclusion

The numerical investigation on flow past a hydrophobic semi-circular cylinder was performed under finite element methods solving for momentum and mass conservation equations; considering that the flow is laminar and incompressible in nature. Three slip lengths were introduced and were tested under four flow velocity conditions. The effect of slip was studied for the parameters coefficient of lift and drag. The lift coefficient showed a decrease in amplitude initially with slip introduction but showed no significant further changes in amplitude with further increase in slip length possibly due to not enough surface area to generate enough lift. The drag coefficient showed a decrease even with the increase in slip due to the different surface areas in the front as well as the back portions and the reduction in skin friction caused by the increase in slip. As the slip increased, the vortex shedding was found to have more frequency and the phenomenon of boundary layer separation does not occur due to insufficient surface area on the top and bottom areas of the cylinder. As the Reynolds number was increased, the drag coefficient as well as the lift coefficient showed a rise in amplitude due to the increase in flow velocity. At Re = 50, both the lift and drag coefficient showed very low values due to the absence of vortex shedding, and shedding was seen in the higher Reynolds number we also see the increase in amplitude caused by it. The strouhal number showed an increase with slip length which indicates that the frequency of vortex shedding had increased with slip length and the same trend was found for the increase in Reynolds number as well, an increase was noted with an increase in Re. This study also portrays the effect of the shedding on the hydrodynamic parameters. Therefore, the performed numerical investigation helps to shed light on the flow physics in the simplest manner portraying the effect of slip as well as Reynolds number.

References

Amal, Joseph, K. K. Krishnaram, and S. Kumar Ranjith: Numerical study on the thermo-hydrodynamics of an inclined hydrophobic ellipsoid in steady uniform flow: Thermal Science and Engineering Progress 39: 101721 (2023).

Bhinder, Amrit Pal Singh, Sandip Sarkar, and Amaresh Dalal: "Flow over and forced convection heat transfer around a semi-circular cylinder at incidence." :Int. J of Heat Mass Transfer 55.19–20: 5171–5184 (2012).

Chandra, Avinash, R.P. Chhabra: Flow over and forced convection heat transfer in Newtonian fluids from a semi-circular cylinder: Int. J. Heat Mass Transfer 54 225–241 (2011)

Chandra, Avinash, R.P. Chhabra: Momentum and heat transfer characteristics of a semi-circular cylinder immersed in power-law fluids in the steady flow regime: Int. J. Heat Mass Transfer 54 2734–2750 (2011).

M. Coutanceau, J.R. Defaye: Circular cylinder wake configurations: a flow visualization survey, Appl. Mech. Rev. 44 255–305 (1991).

A. Dalal, V. Eswaran, G. Biswas: A Finite-Volume Method for Navier-Stokes Equations on Unstructured Meshes: Numer. Heat Transfer B 54 238–259 (2008).

Gu, X. J., Barber, R. W., John, B., Emerson, D. R: Non-equilibrium effects on flow past a circular cylinder in the slip and early transition regime: Journal of Fluid Mechanics, 860, 654–681 (2019).

Kim, N., Kim, H., Park, H: An experimental study on the effects of rough hydrophobic surfaces on the flow around a circular cylinder : Physics of Fluids, 27(8), 085113 (2015).

C.F. Lange, F. Durst, M. Breuer, Momentum and heat transfer from cylinders in laminar crossflow at 10<Re<200: Int. J. Heat Mass Transfer 41 3409–3430 (1998)

Muralidhar, P., Ferrer, N., Daniello, R., Rothstein, J. P. :Influence of slip on the flow past superhydrophobic circular cylinders: Journal of Fluid Mechanics, 680, 459–476 (2011).

Rajani, B. N., A. Kandasamy, and Sekhar Majumdar: "Numerical simulation of laminar flow past a circular cylinder.": Applied Mathematical Modelling 33.3 : 1228–1247 (2009).

Seo, Il Won, and Chang Geun Song. "Numerical simulation of laminar flow past a circular cylinder with slip conditions." International Journal for Numerical Methods in Fluids 68.12 (2012): 1538–1560.

B.M. Sumer, J. Fredsoe,: Hydrodynamics Around Cylindrical Structures: World Scientific, Singapore, (2006)

C.H.K. Williamson: Vortex dynamics in the cylinder wake :Annu. Rev. Fluid Mech. 28 477–539 (1996)

M.M. Zdravkovich: Flow Around Circular Cylinders, Applications, vol. 2: Oxford University Press, New York (2003)

M.M. Zdravkovich: Flow Around Circular Cylinders, Fundamentals, vol. 1: Oxford University Press, New York (1997).

Particulate Laden Exhaust Gas Spread from Jet Engines

Akshay Prakash, Alok Garg, Prashant Vasistha, and Chetan Mistry

Abstract More stringent emission norms are being put in place to curb environmental impact of pollution. Financial penalty in form of carbon currency is being proposed. While there have been major improvements in automobile and aviation industry in reducing pollution, quantification of its local dispersion and impact on environment locally has not been reported adequately. This study simulates a jet exhaust under partial load conditions and quantifies the spread over a relatively large domain (100 m). Particulate matter is injected with the jet as per known levels generated in a jet engine and its impact on the jet spread is shown. PM10 and PM2.5 sizes are simulated. Simulations are performed using OpenFOAM. In OpenFOAM two solvers are used – 'simpleFoam' and 'twoPhaseEulerFoam'. PM2.5 is found to disperse more than PM10. Both particulates induce unsteadiness in the jet and suppress the jet closer to ground.

Keywords CFD · Pollution · Jet exhaust

1 Introduction

With increasing emphasis on reducing the negative human footprint on the environment, more stringent emission norms are being put in place. Currently BSVI (Bharat Stage Emission 6) norms have been enforced for vehicles plying on the road with BSVI-phase 2 coming in force since April 1 2023. The phase-2 norms will require Real Driving Emissions being monitored and reported. The idea of countries following "carbon currency" is being floated and will likely be imposed in future (Liu et al. 2022). Enforcing the 'currency' will require better monitoring techniques (and possibly curbing methodology) for pollutants. Currently most of the pollution data are measured from source and at ideal conditions. Classification of localized

A. Prakash (✉) · C. Mistry
Indian Institute of Technology Kharagpur, Kharagpur, WB, India
e-mail: akshayprakash@gmail.com

A. Garg · P. Vasistha
Defence Labs Jodhpur, Jodhpur, Rajasthan, India

G. Rajesh et al. (eds.), *Proceedings of the 15th International Symposium on Experimental and Computational Aerothermodynamics of Internal Flows*,
https://doi.org/10.1007/978-981-96-4082-9_6

spatial impact of a pollution source is not mainstream and is done on a case-by-case basis as per severity. Study of dust winds impacting forests in Amazon basin, or burning of stubble in Punjab and Haryana, as one of the major causes of severe pollution spike in Delhi are some examples of this case-by-case analysis (Kanawade et al. 2020; Swap 1992).

Fossil fuel technology has matured over the years and is currently the cheapest option for developing economies. More than 70% of electrical power in India comes from coal-based plants. Almost all the transportation is based on either fossil fuel or electricity produced by power plants. Transition to EV (electric vehicles) has been fastest in China (incentivized) followed by UK and US (Palmer et al. 2018; Singh 2021). Other possible contenders for providing clean power – nuclear, solar, hydro and others, are either at a nascent stage of development or have not been scaling fast enough (Grover 2013). Developing technologies to reduce the impact of pollution might provide some relief till some alternative comes up.

Quantification of pollution in real case scenarios is the first step in developing strategy for repressing pollution. Flow through chimneys, exhaust pipe at lower speeds and higher speeds would represent most of the conventional contributors at a very broad level – thermal power generation, automobiles and aviation. Attempts have been made in understanding chimney flow (Hanna and Paine 1989; Said et al. 2005) and jet exhaust (Cummings and Chanson 1997). Efforts to quantify exhaust includes muffler simulations (Synylo et al. 2020; Capros et al. 2020), wind tunnel experiments (Hoglund-Isaksson et al. 2050) and numerical simulations of dispersion over large domains (Rumsey 2002; Xia 2002). Model based software such as PRIMES (Nagarajan et al. 2022), GAINS (Fernandez-Yanez 2017), etc., use economic spending data / emission norms data and measurement data with statistical tools to estimate pollution levels and have been used across the globe in absence of more accurate methods.

Use of CFD (Computational Fluid Dynamics) has been gaining traction among researchers and has become mainstream in various fields (Robins 1978; Toja-Silva et al. 2018). CFD simulations can give much details about the spread and mixing of pollutants, including particulate matter (Taghipour et al. 2005; Sungkorn et al. 2011; Venier et al. 2020). Any innovative setup to neutralize pollution could also be simulated and fine-tuned. However, CFD must be carefully performed. Mesh quality may drastically effect the solution quality. Obtaining controlled experimental data for validation is extremely difficult at large scales. Setting up a CFD simulation itself involves numerous assumptions on initial and boundary conditions and must be carefully examined.

Synylo et al. (Xia 2002) simulated jet exhaust of an aircraft at partial loads with and without temperature effects. The simulations quantified maximum velocity and jet core height over a length of 220 m. OpenFOAM's simpleFoam (SF) solver and SolidWorks flow simulation program were used for the simulations. Current research extends the study to include more details and simulation of particulate matter – PM10 and PM2.5. SF and twoPhaseEulerFoam (TPEF) is used for the simulations. Injection of particles is based on pollution data from Aircrafts (General 2020). Landing Take-Off (LTO) emissions have been monitored for contribution to pollution in and around airports. Typical emission data from subsonic engines for the duration of LTO (~33 min) are 462 g/Kg HC, 7834 g/Kg CO and 2809 g/Kg of NOx. For large commercial aircrafts the total particulate matter mass (tPMmass) is typically from 0.24 to 1.58 kg/LTO. In particulate matter, the non-volatile particulate corresponds mainly to black carbon mass. The volatile components consist of volatile organic compounds including non-methane hydrocarbons. The volatile and non-volatile together make up the PM2.5 and PM10 particulate.

2 Methodology

CFD simulations are performed using OpenFOAM's (Weller et al. 1998) (v2106) SF and TPEF. SF is based on the SIMPLE scheme or its variants. The momentum equations (Eq. 2) are discretized and solved with assumed pressure field to get *predicted* velocity. Using continuity equation (Eq. 1) the momentum equations are reduced to an algebraic equation in pressure which gives the corrected velocity and pressure (Weller et al. 1998). These steps are iterated till convergence is reached. PISO (Pressure-Implicit with Splitting of Operators), with an addition loop for pressure correction, is used in the current simulations. SIMPLE and its variants have been widely used in the CFD community. Documentation is readily available on tssheory and implementation of the scheme.

$$\vec{\nabla} V = 0 \tag{1}$$

$$\vec{V} \cdot \vec{\nabla} \vec{V} = -\vec{\nabla} P + \nu \Delta \vec{V} \tag{2}$$

TPEF uses compressible flow equations to solve for a mixture of two phases. They may be two fluids or particulate dispersed in a continuous media. The conservation of momentum and mass are given by Eqs. 3, 4.

$$\frac{\partial \alpha_\Psi \vec{U}_\Psi}{\partial t} + \vec{\nabla}\left(\alpha_\Psi \vec{U}_\Psi \vec{U}_\Psi\right) + \vec{\nabla}\left(\alpha_\Psi \vec{R}_\Psi^{eff}\right) = -\frac{\alpha_\Psi}{p_\Psi} \vec{\nabla} P + \alpha_\Psi \vec{g} + \frac{\vec{M}_\Psi}{p_\Psi} \tag{3}$$

$$\frac{\partial \alpha_\Psi}{\partial t} + \vec{\nabla}\left(\alpha_\Psi \vec{U}_\Psi\right) = 0 \tag{4}$$

Here subscript ψ denotes the phase, α is the volume fraction of the phase, $\overrightarrow{U}_\psi$ is the velocity of the phase, $\overrightarrow{R}_\psi^{eff}$ is combined Reynolds and viscous stress, ρ_ψ is phase density, P is system pressure, $\overrightarrow{g}$ is gravity and $\overrightarrow{M}_\psi$ is average inter-phase momentum transfer having contributions from drag, lift and virtual mass for particulate flow. Turbulence is modelled based on the Kinetic Theory of granular molecules (Patankar 1972). Simulation parameters are listed later in Table 1. More details on the solver can be found in OpenFOAM documentation.

Table 1 Particulate phase parameters

Section	Particular entity	Value
Phase properties	Diameter Density Alpha max Drag Virtual mass	$10 \times 10^{-6} m$ $2 \times 10^{-6} m$ 2000 kg/m^3 0.6 GidaspowErgunWenYu Constant, 0.5
Turbulence properties	Model E (restitution) Viscosity model Granular pressure model Frictional stress model	RAS, kinetic theory 0.8 Gidaspow Lun JohnsonJackson
Gravity		–9.8 (y direction)

Other options for simulation of granular flow include Lagrangian method where particulate is treated as such and is advected in accordance with the forces acting. Typically, Lift, drag, gravity, added mass force and Basset history forces are taken into account (Sungkorn et al. 2011). This class of methods is suitable for low densities where total number of particles is low. For the current simulations, due to the large domain, the number of particles could have reached hundreds of millions and hence the Eulerian method was chosen.

3 Numerical Setup

The domain was setup following Synylo et al. (Xia 2002). Stream wise direction is 220 m, 30 × 30 m in cross section. The jet is set as velocity inlet at 3.5 m above the ground at 100 m/s. The mesh is clustered around the jet and rapidly coarsens towards the domain boundaries. The grid is about 2.5 million mesh points. Figure 1 shows the computational domain in perspective view and the zoomed inlet area where the circular jet inlet may be seen. Multiple grids were used to test the reliability of the computations and are not discussed.

The domain and mesh were generated using the blockMesh utility of OpenFOAM. Synylo et al. used about 5.2 million mesh points for the OpenFOAM simulations and 2.1 million mesh for Solidworks simulation. They used uniform finer mesh for a much larger domain around the jet inlet. For the current simulations, mesh below the jet in vertical direction is kept uniform at about 10 cm per mesh point. Above the jet the mesh is stretched so that close to the jet it has 10 cm per node spacing but gradually increases to 1 m per division near the boundary.

Two sets of boundary conditions were used. For comparison with Synylo et al. and validation the following were imposed: Inlet is set to be "fixedValue" at 100 m/s in z direction (stream wise direction). The side curved exit, back wall except the

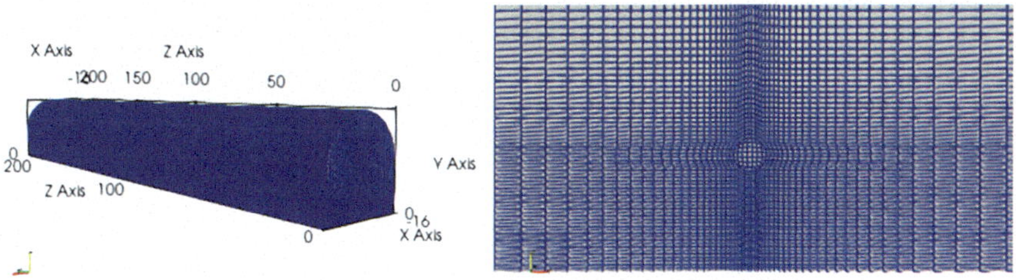

Fig. 1 Computational domain. Left: 220 m long domain with circular top. Right: zoomed in circular inlet

inlet region are set at "fixedValue" at 2 m/s in stream wise direction (as per Synylo et. al). The outlet is set to be "pressureInletOutletVelocity" which corresponds to the "outflow" boundary conditions (interpolating from interior for outgoing flow). The bottom boundary is set to be wall with "noSlip".

The second set of boundary conditions was used for particulate flow simulations where the curved side boundary is set at "pressureInletOutletVelocity". This was done to alleviate numerical instability with particulate flow simulation. Flow without particulate with both these boundary conditions was compared and it was found that this change in boundary condition did not alter the solution in the region of interest (near and around the jet core). *Please note that length of the domain in displayed in most of the figures in meters (Fig. 1 left,* Figs. 5, 6, *and others).*

4 Results

The computational setup was validated against results from Synylo et al. (Xia 2002). Simulations using SF and TPEF are compared. Velocity decay and location of jet core is compared with published data for isothermal flows. TPEF is validated against particulate flow in fluidized bed reactor (Venier et al. 2020). Some results are shown here; detailed results are shown in an accompanying article. Results for particulate flow and jet spread due the particulate is shown thereafter. While the computational domain is 220 m long, the results are shown for relevant domain containing the extent of particulate flow (~ 100 m).

4.1 Validation Using SimpleFoam: Synylo et al. (Xia 2002)

Figure 2 shows the variation of maximum velocity (left) and the height of the jet core along the stream wise direction. While the maximum velocity falls sharply, the jet core dips and then again rises up. The dip is due to lower pressure below the jet due to the present of wall and recirculation. Data from Synylo et al. is shown along

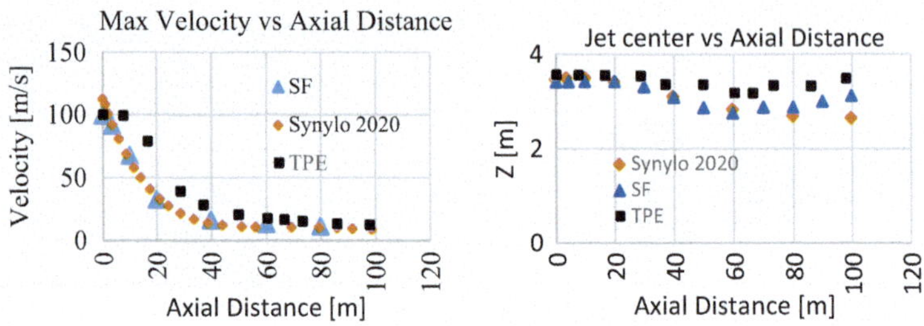

Fig. 2 Max. Velocity and jet core height. Comparison with Synylo 2019 (Synylo et al. 2020)

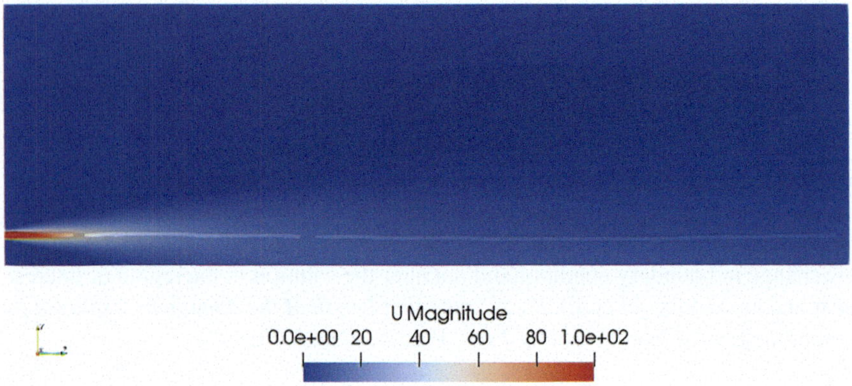

Fig. 3 Velocity contour in the mid-plane (X = 0). Also shown is the stream trace starting from middle of the jet

with current simulations with SF solver and TPEF solver. While SF uses the same numerical method as Synylo et al., TPEF is a compressible flow solver setup for multiphase simulations. More details about the TPEF solution will be shown in the following section.

Figure 3 shows velocity contour in the mid-plane (X = 0) for SF solution. Also shown is a stream trace starting from center of the jet (at 3.5 m above ground). This is corresponding to the data presented in Fig. 2 (right). Figure 4 shows cross sections view at Z = 10, 40 and 80 m from the nozzle exit (Z = 0). Stream traces were initiated from 5 locations: center of the jet and 4 points on the periphery of the jet. These stream traces are indicative of the periphery of the jet as it proceeds downstream. The black ring is indicative of the jet at Z = 0 (at 3.5 m from ground and 1 m diameter). It can be seen that the middle stream trace moves down the ring and the peripheral traces move outwards indicating spreading of the jet.

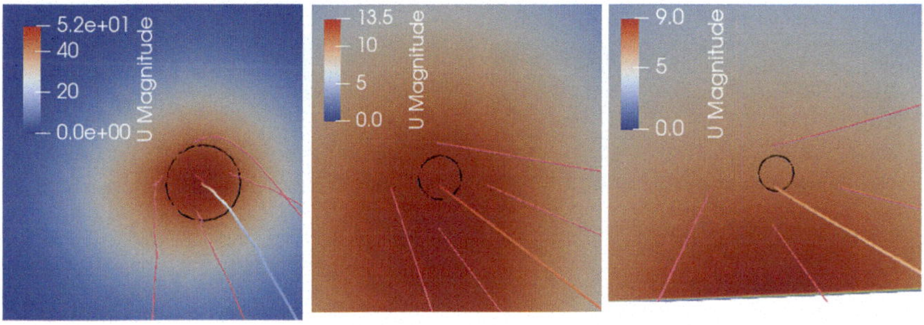

Fig. 4 Velocity contours and stream traces crossing XY planes at Z = 10 m (**left**), Z = 40 m (middle) and Z = 80 m (**right**). Note that black ring indicative of original location of jet inlet (1 m dia.)

4.2 Particulate Flow Simulations Using TPEF

Solution without particulate (only air flow) was performed for validation. Figure 2 shows the comparison with Synylo et al. results. The match is not perfect. The jet seems to hold off to the core velocity for a few meters before sharply dropping off to about 10 m/s. Similarly, the core of the jet seems to be slightly (0.5 m) above the SF or Synylo et al. results. It is believed that this small difference is due to the transient compressible solver.

Two sets of particulates were simulated – diameter $10 \times 10^{-6} m$, and $2 \times 10^{-6} m$, representing PM10 and PM2.5 particulate respectively. Due to the large domain being simulated, 33 min simulation (total time for LTO (General 2020)) was computationally too expensive. For the current study the mass injection rate was kept at 0.2 kg/33 min (total emission over total time) and simulations were carried out till solution appeared statistically stationary. This means an injection of rate of $55.6 \times 10^{-9} m^3/s$ for particulate. This was set as an appropriate volume fraction at the jet. The particulate velocity was set equal to air (100 m/s) at the jet exit. Other parameters for simulation are mentioned in Table 1.

Figure 5 shows the average velocity contours for particulate flow (10-micron left and 2-micron right) along with the stream trace from the center of the jet. The image on right is inverted to make the stream trace visible (for this case the center tends to move a little towards left causing the stream trace to be behind the mid-plane). The jet is weighed down by the particulate and is seen touching the ground a little further than 80m for 2-micron particulate and just ahead of 80m mark for the 10-micron case. It is expected that the heavier particle will cause the jet to dip more, however the difference here is marginal.

Figure 6 shows the instantaneous velocity and pressure contours for the two cases. The scaling for the contours is kept same for both the cases. While there are marginal differences, the trend remains the same. Most of the particulate falls on the ground and

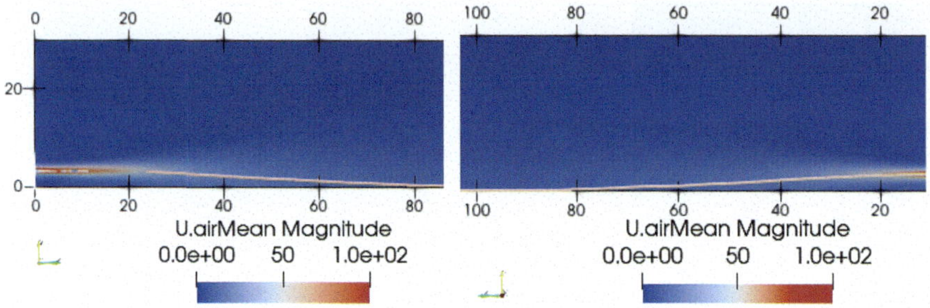

Fig. 5 Average velocity contours with jet centre stream trace for 2-micron diameter (**left**) and 10-micron diameter (**right**) particulate

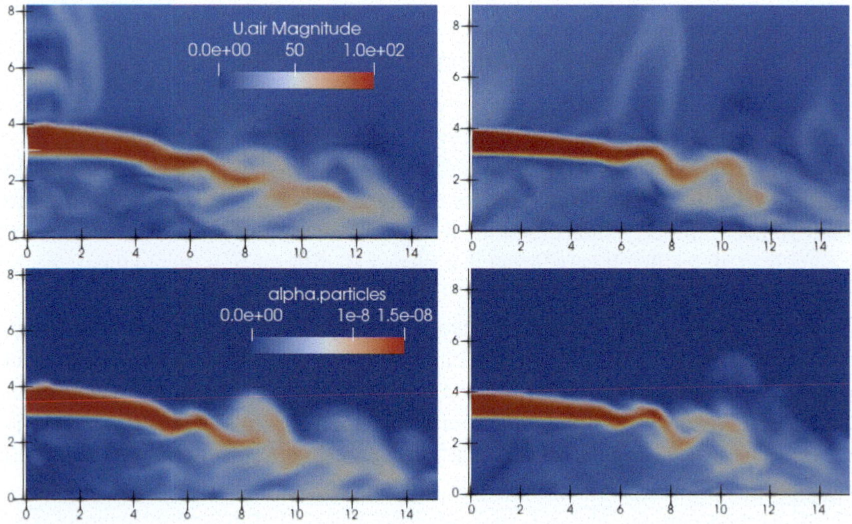

Fig. 6 Instantaneous velocity contours and particle volume fraction contours in the mid-plane. **Top:** Velocity contours, **Bot:** particle volume fractions. **Left:** 2-micron case, **Right:** 10-micron case

the particle concentration drops rapidly thereafter. For both cases noticeable volume fractions are visible till 30 m, reducing to unnoticeable levels by 40 m. Marginal differences in the cases include slight overshoot of the 2-micron particles near 8 m, a straighter jet for the 10-micron particle before the jet starts breaking (this is due to the higher inertia of the 10-miron particles), and a slightly straighter downward path for the 2-micron particles (see between 10 and 14 m in the volume fraction contours) due to their lighter weight.

Figure 7 shows the lateral spread of particulate at 6m (top), 10m (middle) and 14m (bottom) from the jet. The left images are for 2-micron case and the right images are

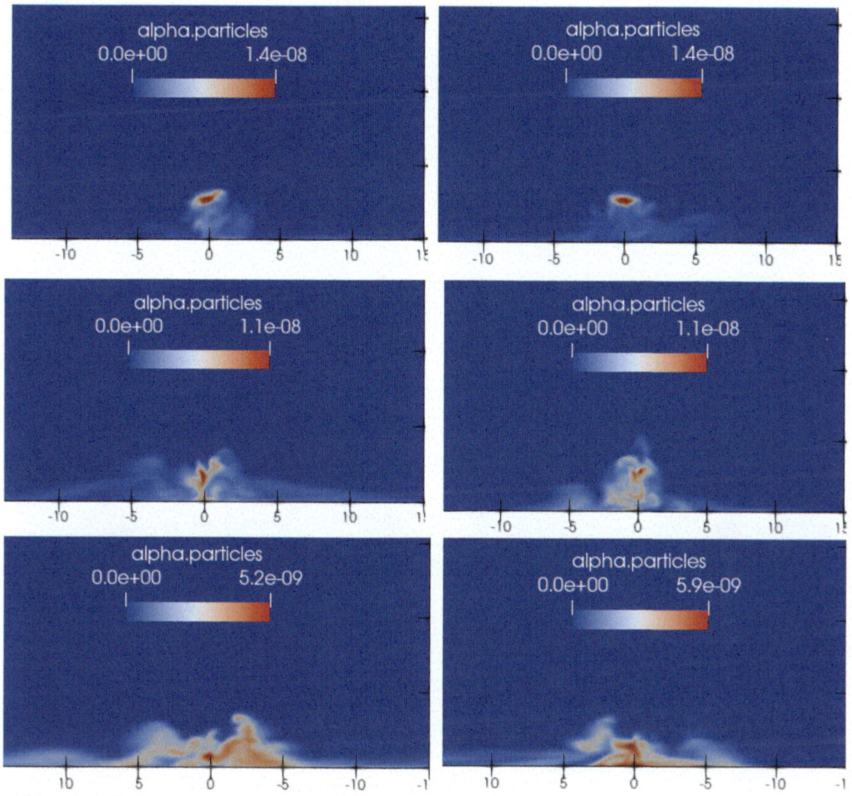

Fig. 7 Lateral spread of particulate. **Left:** 2-micron case, **Right:** 10-micron case; **Top:** Z = 6 m, **Mid:** Z = 10, **Bot:** Z = 14

for 10-micron case. While the cases are qualitatively similar, 10-micron case shows contours more confined to the space below the jet. The 2-micron particles spread over the bottom surface. This trend is expected due to the smaller mass of the 2-micron particle.

Figure 8 shows the stream traces (one from center of the jet and four from jet periphery) passing through a Z=10m slice with instantaneous velocity contours. The top left image is for case with no particulate. The stream traces follow the streamlines and spread as they move downstream as per the results in Figs. 3 and 4. The bottom left image is for the case with 10-micron particle injection. The right images are from 2-micron cases at 1sec apart. The flow field is unsteady. While the stream traces pass through same region in the section at Z=10 m, the path becomes chaotic downstream. Particulate gets scattered through this chaotic flow.

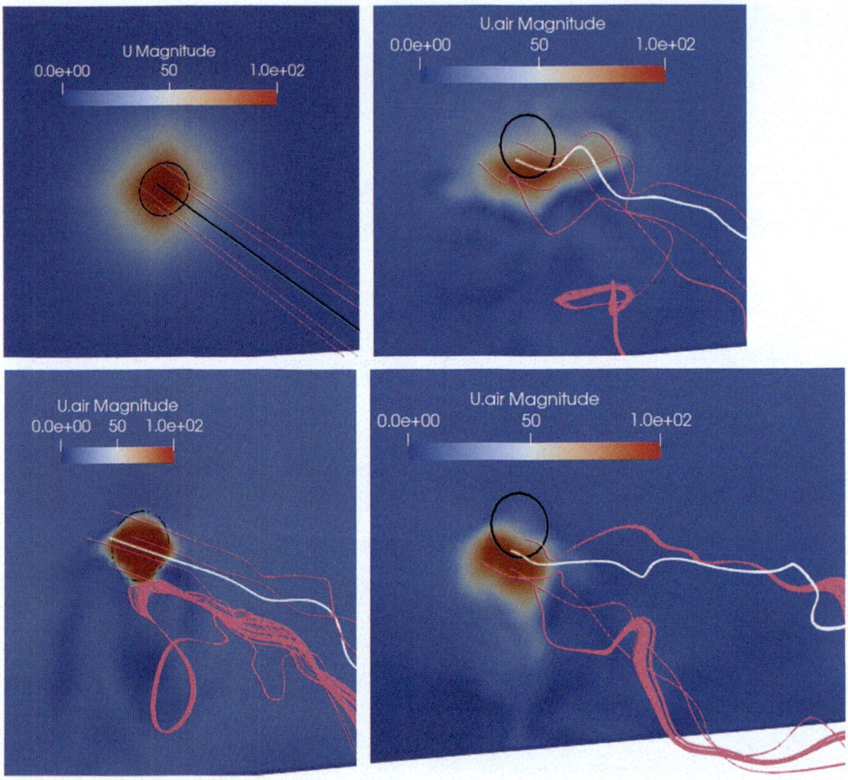

Fig. 8 Stream traces for TPEF cases. **Top left:** Without particulate, **Bot left:** 10-micron case, **Top right:** 2-micron case at 28 s, **Bot right:** 2-micron case at 29 s. The contours for velocity magnitude are at Z = 10 m slice

5 Conclusion

Jet exhaust consisting of PM2.5 and PM10 particulate was simulated using OpenFOAM solver. Validation was done through comparison with published data from Synylo et al. (Xia 2002) The spread of the jet was compared for flow without the particulate and with the particulate matter. The ambient air was forced at 2 m/s (7 km/hr.). In absence of particulate matter, the jet continued downstream, the jet core maintaining a consistent height above the ground. In presence of the particulate the jet was suppressed. The jet core broke into unsteady flow and dissipated away before the particulate diffused towards the ground at about 10 m from the jet exhaust. PM2.5 spread across the computational domain, while PM10 particulate was confined within 10 to 15 m of the jet location in the transverse direction.

The study did not take into account heated exhaust. This would have allowed the jet core to remain airborne longer and hence would have spread the particulate farther. Typically, ambient air fairly strong and will cause the pollutants to disperse farther.

While particulate matter was simulated, spread of gaseous pollutants like NOx and CO2 was not considered. Structures surrounding an airfield would also affect the extent of the spread. Another major factor is the induced flow due to the aircraft itself. The vortex shed from the wings tends to stay longer keeping the particulate airborne. These factors will be taken up in subsequent studies.

References

P. Capros, L. Mantzos, E. L. Vouyoukas and D. Petrellis, "European Energy and CO2 Emissions trends to 2020, with Primes V.2," *Bulletin of Science, Technology and Society,* vol. 19, no. 6, 1999.

P. D. Cummings and H. Chanson, "Air Entrainment in the Developing Flow Region of Plunging Jets - Part 1: Theoretical Development," *Journal of Fluids Engineering* , vol. 119, pp. 597-602, 1997.

P. A. O. G. A. G. A. Fernandez-Yanez, "Developing Computational Fluid Dynamics (CFD) Models to Evaluate Available Energy in Exhaust Systems of Diesel Light-Duty Vehicles," *Applied Sciences,* vol. 7, 2017.

R. Grover, "Green growth and role of nuclear power: A perspective from India," *Energy Strategy Reviews,* vol. 1, no. 4, pp. 255-260, 2013.

S. R. Hanna and R. J. Paine, "Hybrid Plume Dispersion Model (HPDM) Development and Evaluation," *Journal of Applied Meterology,* vol. 28, pp. 206-224, 1989.

L. Hoglund-Isaksson, A. Gomez-Sanabria, Z. Klimont, P. Rafaj and W. Schopp, "Technical potentials and costs for reducing global anthropogenic methane emissions in the 2050 timeframe - results from the GAINS model," *Environmental Research Communications,* vol. 2, 2020.

V. P. Kanawade, A. K. Srivastava, K. Ram, E. Asmi, V. Vakkari, V. K. Soni, V. Varaprasad and C. Sarangi, "What caused severe air pollution episode of November 2016 in New Delhi?," *Atmospheric Environment,* vol. 222, p. 117125, 2020.

Q. Liu, Z. Chen and S. X. Xiao, "A theory of carbon currency," *Fundamental Research,* vol. 2, pp. 375 - 383, 2022.

C. K. K. Lun, S. B. Savage and D. J. Jeffrey, "Kinetic Theories for granular flow: inelastic particles in Couette flow and slightly inelastic particles in a general flowfield," *Journal of Fluid Mechanics,* vol. 140, pp. 223-256, 1984.

P. C. Nagarajan, H. Strom and J. Sjoblom, "Transient Flow Uniformity Evolution in Realistic Exhaust Gas Aftertreatment System Using 3D-CFD," *Emission Control Science and Technology,* vol. 8, pp. 154-170, 2022.

K. Palmer, J. E. Tate, Z. Wadud and J. Nellthorp, "Total cost of ownership and market share for hybrid and electric vehicles in the UK, US and Japan," *Applied Energy,* vol. 209, pp. 108-119, 2018.

S. V. S. D. Patankar, "A calculation procedure for heat mass and momentum transfer in three dimensional parabolic flows," *International Journal of Heat and Mass Transfer,* vol. 15, p. 1787, 1972.

Robins, "Plume Dispersion From Ground Level Sources in Simulated Atmospheric Boundary Layers," *Atmospheric Environment* , vol. 12, pp. 1033–1044, 1978.

C. Y. S. X. Rumsey, "Prediction of high lift: review of present CFD capability," *Progress in Aerospace Sciences,* vol. 38, no. 2, pp. 145-180, 2002.

N. M. Said, H. Mhiri, G. L. Palec and P. Bournot, "Experimental and numerical analysis of pollutant dispersion from a chimney," *Atmospheric Environment,* vol. 39, pp. 1727-1738, 2005.

Secretary General, Airport Air Quality Manual, Montreal, Quebec: International Civil Aviation Organization, 2020.

V. S. V. V. S. Singh, "Analysis of electric vehicle trends, development and policies in India," *Case Studies on Transport Policy,* vol. 9, pp. 1180-1197, 2021.

R. Sungkorn, J. J. Derksen and J. G. Khinast, "Modeling of turbulent gas-liquid bubbly flows using stochastic Lagrangian model and lattice-Boltzman scheme," *Chamical Engineering Science,* vol. 66, pp. 2745 - 2757, 2011.

R. G. M. G. S. T. R. K. P. Swap, "Saharan dust in the Amazon Basin," *Tellus B,* vol. 44, no. 2, pp. 133-149, 1992.

K. Synylo, A. Krupko, O. Zaporozhets and R. Makarenko, "CFD simulation of exhaust gasses jet from aircraft engine," *Energy,* p. 118610 , 2020.

F. Taghipour, N. Ellis and C. Wong, "Experimental and computational study of gas-sold fluidized bed hydrodynamics," Chemical Engineering Science, vol. 60, pp. 6857 - 6867, 2005.

F. Toja-Silva, C. Pregel-Hoderlein and J. Chen, "On the urban geometry generalization for CFD simulation of gas dispersion from chimneys: Comparision with Gaussian plume model," *Journal of Wind Engineering and Industrial Aerodynamics,* vol. 177, pp. 1-18, 2018.

C. M. Venier, A. R. Urrutia, J. P. Capossio, J. Baeyens and G. Mazza, "Comparing ANSYS Fluent and OpenFOAM simulations of Geldart A, B and D bubbling fluidized bed hydrodynamics," *International Journal of Numerical Methods for Heat and Fluid Flow,* vol. 30, no. 1, pp. 93-118, 2020.

H. G. Weller, G. Tabor, H. Jasak and C. Fureby, "A tensorial approach to computational continuum mechanics using object-oriented techniques," *Computers in Physics,* vol. 12, no. 6, 1998.

B. S. D.-W. Xia, "Applications of computational fluid dynamics (CFD) in the food industry: a review," *Computers and Electronics in Agriculture,* vol. 34, no. 1-3, pp. 5-24, 2002.

Fuel-rich H_2–O_2 Flame Dynamics in a Crossflow, with and without Direct Igniter/Injector Jets Interaction, Applied to Air-Heater Configurations

Sekhar Susheel Kumar, Deepak Kumar Agarwal, T. John Tharakan, and S. Sunil Kumar

Abstract Numerical simulations of the firing of H_2–O_2 igniters in the presence of an air-O_2 crossflow are conducted with a specific focus on air-heater configurations with faceplates. Simulations are conducted for three igniter chamber pressures (22, 41 and 55 bar) to investigate penetration lengths of igniter flames into the crossflow. Two alignments of the air-heater faceplate are assessed: one where the igniter-injector jets exhibit a direct interaction (Case 1), and one where the igniter jet is fired in-between two adjacent injector jets (Case 2). For all three igniter chamber pressures, air-heater faceplate heating is lesser for Case 1 compared to Case 2 because the igniter flame is naturally deflected away from the faceplate here. Moreover for Case 1, the 41 bar igniter chamber pressure shows the least peak heating of 68 W/cm^2. For Case 2, with the 22 bar igniter chamber pressure, the igniter flame is deflected markedly towards the faceplate because of entrainment-instigated suction between two injector jets. This shows the highest peak heating on the faceplate of 467 W/cm^2. When the igniter chamber pressure is increased, the extent of deflection reduces, as does the peak heating on the faceplate. Notwithstanding, hot-spots near the centre of the faceplate are still evident even at higher igniter chamber pressures.

Keywords Fuel-rich H_2–O_2 flame · Jets interaction · Vitiated air-heater assembly

1 Introduction

An air-heater is used to generate vitiated air mixure of desired high temperature and pressure for various space applications, including preheating fuel and for combustors. It is important to design appropriate air-heater assemblies that operate within acceptable temperature limits. Numerous experimental studies have been undertaken during

S. Susheel Kumar (✉) · D. K. Agarwal · T. J. Tharakan · S. Sunil Kumar
Liquid Propulsion Systems Centre, ISRO, Thiruvananthapuram, India
e-mail: susheelkumar_s@lpsc.gov.in

G. Rajesh et al. (eds.), *Proceedings of the 15th International Symposium on Experimental and Computational Aerothermodynamics of Internal Flows*,
https://doi.org/10.1007/978-981-96-4082-9_7

the past several decades to estimate the effects of the vitiated air mixtures on different aspects of air breating propulsion system test facilities, including ignition characteristics, combustion instablity and performance (Edelman and Spadaccini 1969; Stallings and Burt 1984; Hashimoto and Yoshida 1991; Guimareãs, et al. 1994; Pellett, et al. 2002; Saari, et al. 2009; Kang, et al. 2010; Fetterhoff and Burfitt 2011; Feiteng et al. 2012; Bao 2013; Wang et al. 2015; Lee et al. 2023). Recent computational studies of such systems usually include detailed LES, which are used to further investigate and understand complex ignition and combustion aspects (Wolf et al. 2012; Urbano et al. 2016; Karaca et al. 2019; Hwang et al. 2021). The current numerical study aims to assess convective heating on sub-systems of a typical air-heater assembly, which can be used to design appropriate cooling systems, and qualify such assemblies for testing operations.

2 Computational Model and Numerical Set-Up

For a typical air-heater assembly, numerical simulations of the firing of H_2–O_2 igniters in the presence of an air-O_2 crossflow are conducted with a specific focus on air-heater configurations with faceplates. Fig. 1 shows an air heater configuration with a faceplate having 28 circular air injector ports and two identical fuel rich igniter tubes placed diametrically opposite to each other. Simulations are conducted with three different igniter chamber pressures to investigate penetration lengths of the corresponding flames into the crossflow. For each igniter setting, two alignments of the faceplate that result in contrasting igniter-injector jets interactions are assessed: one where the jets exhibit a direct interaction (Case 1), and one where the igniter jet is fired in-between two adjacent injector jets (Case 2). An additional set of purging ports are also included in the assembly.

7.2 kg/s of air-O_2 mixture at 300 K is set to flow through 28 injector ports in an air-heater of diameter 260 mm, with each injector port of 18 mm diameter, for the crossflow. 10 g/s of N_2 gas is assumed to flow through the purging ports of 6 mm

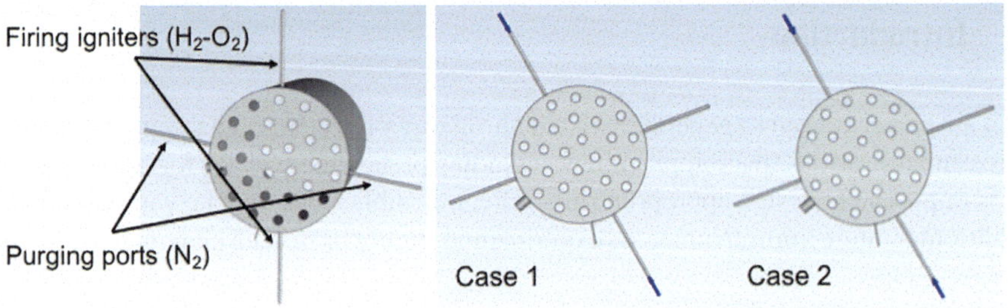

Fig. 1 Schematic of a typical air-heater assembly seen from faceplate (left), and two faceplate alignments in air-heater assembly showing relative positions of igniters marked with blue arrows and injector ports (right)

Table 1 Fuel-rich H_2–O_2 igniter conditions

$P_{chamber}$ (bar)	Mixture ratio (MR)	$\dot{m}$ H_2 (g/s)	$\dot{m}$ O_2 (g/s)
22	1.15	2.7	3.1
41	1.46	4.6	6.7
55	1.27	7.0	8.9

diameter. Table 1 gives the gas flow parameters through the igniter ports of 6 mm diameter for three different chamber pressures used in the analysis.

For these conditions, NASA's CEA (McBride and Sanford 1996) is used to determine igniter combustion chamber temperatures and combusted gas species mass fractions at the igniter ports. These are presented in Table 2.

CFD simulations of this assembly are run using ANSYS Fluent's (2022 R1) P-based solver using the P-v coupled scheme. The pseudo-time method is used to run these simulations to steady state, with 2nd order spatial accuracy and least-square cell-based gradients. Higher order terms are set for a relaxation factor of 0.25. Turbulence in the domain is modelled using Menter's k-ω SST model with default coefficients. 3D polyhedral meshes with 3.2 million grid cells are prepared for the simulations with 0.5 mm sized cells in the igniter/purge ports and 2 mm cells for the 200 mm region, where combustion of the excess H_2 is expected. The domains are extended by an additional 300 mm for a buffer region with coarser cells (4 mm), to ensure numerical stability. 2D slices of the polyhedral meshes cut through the H_2–O_2 igniter axes are shown in Fig. 2.

For modelling the combustion of excess H_2 from the igniters, the eddy-dissipation concept (EDC) model is used with the stiff-chemistry solver in Fluent. EDC is an improved form of the eddy dissipation model (EDM) which assumes that the fuel and oxidizer carried by separate turbulent eddies in a diffusion flame react as soon as they mix on a molecular scale (infinitely fast chemistry). EDC, alternatively, calculates detailed chemical kinetics by using empirical expressions for mean reaction rates based on the assumption that chemical reactions occur in the regions where the turbulence energy dissipation occurs. This approach uses a probability density function to model reacting turbulent flows, and has been shown to provide an effective solution to closure problems arising from filtering highly nonlinear chemical energy source terms, and modelling the effects of convection and body forces (Vankova 2021). This is used to model finite-rate combustion of H_2 using a simplified 1-reaction model (ANSYS Fluent Theory Guide 2022). Temperature-dependent specific heat capacity

Table 2 Fuel-rich H_2–O_2 igniter combustion chamber conditions from NASA CEA (McBride and Sanford 1996)

$P_{chamber}$ (bar)	Mixture ratio (MR)	$T_{chamber}$ (K)	%m H_2	%m H_2O	γ
22	1.15	1,413.14	39.772	60.228	1.32
41	1.46	1,668.98	33.172	66.828	1.3
55	1.27	1,514.42	37.004	62.996	1.31

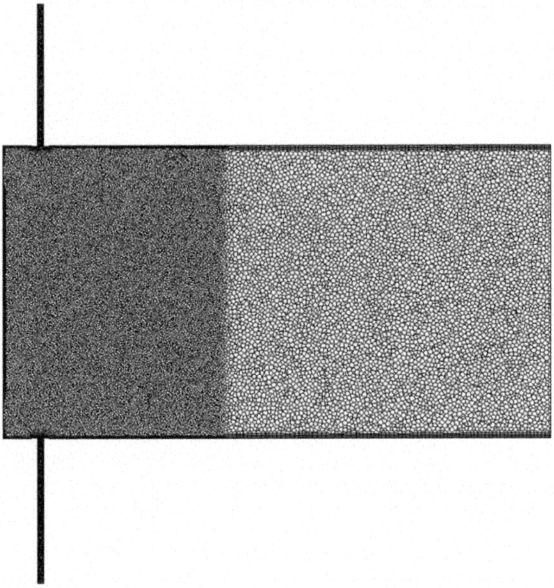

Fig. 2 2D profile view of 3D polyhedral mesh on a plane through H_2–O_2 igniters' axis

(Cp) for each species is modelled as a curve fit, and for the mixture, a mixing law is used. For the mixture transport properties (k and μ), mass-weighted mixing laws are used with constant transport properties for each species. A unity Lewis number is assumed to estimate mass diffusivities.

3 Results and Discussion

Results of numerical simulations of the firing of H_2–O_2 igniters in the presence of an air-O_2 crossflow for three igniter chamber pressures (22, 41 and 55 bar) and two air-heater faceplate alignments (Case 1 and Case 2) are presented here.

3.1 Chamber Pressure = 22 bar

In Fig. 3, velocity vectors on a vertical plane through the igniter axes are shown for Cases 1 and 2 with a chamber pressure of 22 bar. Even though mass flow rates of the gases are the same for both cases, stark differences in the flow dynamics due to contrasting interactions with the injector jets are clearly evident. Whereas in Case 1, the injected air-O_2 mixture clearly deflects the igniter gas away from the faceplate, in Case 2, the igniter gas is deflected towards the faceplate due the suction created by entrainment in the gap between adjacent injector jets. This is also evident in

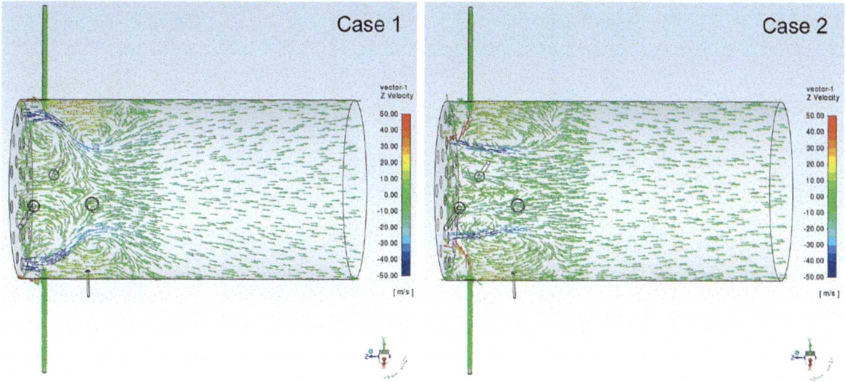

Fig. 3 Velocity vectors on a plane through igniters' axis for $P_{chamber} = 22$ bar (Case 1: left, Case 2: right)

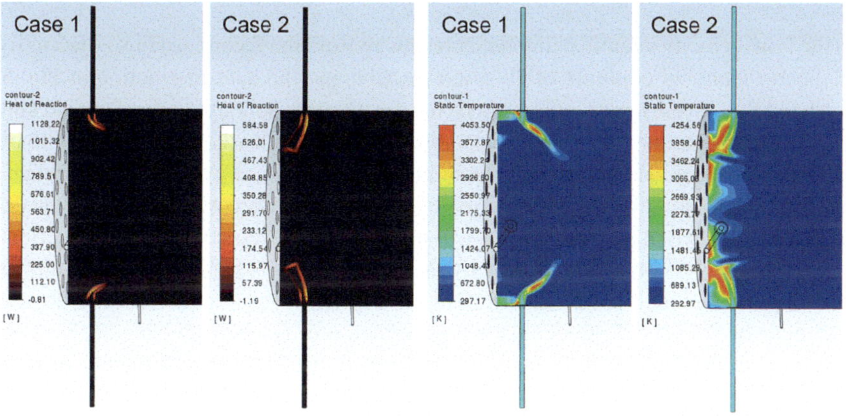

Fig. 4 Heat of reaction (left) and temperature contours (right) on a plane through igniters' axis for $P_{chamber} = 22$ bar

Fig. 4, where heat of reaction and temperature contours are presented for both cases. These contours represent the flame front and highlight high temperature spots, and are indicative of the regions of excessive H_2. Differences between the configurations are strikingly evident in the temperature plots, with excessive high temperature gas present in the vicinity of the faceplate for Case 2, and localized combustion away from the face and chamber walls for Case 1.

Figure 5 shows the species mass fractions of O_2 (left) and H_2 (right) for Cases 1 and 2. A direct correlation between the O_2 contours shown here and the high temperature regions shown in Fig. 4 is clearly evident. Similarly, the H_2 contours shown in Fig. 5 conform to the heat of reaction zones highlighted in Fig. 4, indicating the regions where the H_2 combustion reactions occur. For Case 1, the collective observation of

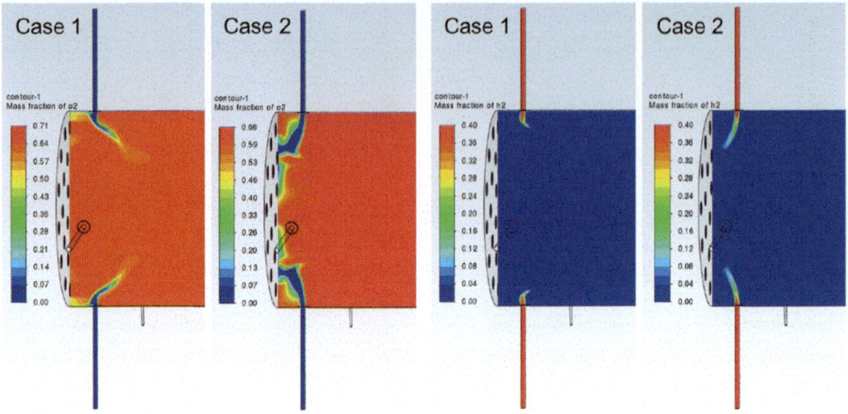

Fig. 5 Mass fractions of O_2 (left) and H_2 (right) on a plane through igniters' axis for $P_{chamber} = 22$ bar

(i) the flow velocity vectors being deflected away from the faceplate (Fig. 3, left), (ii) the nearly identical contours of H_2 mass fraction and the heat of reaction in Fig. 5 (left) and Fig. 4 (left), respectively, and (iii) the corresponding O_2 mass fraction and temperature contours in Fig. 5 (right) and Fig. 4 (right), clearly suggest that bulk of the combustion of excess H_2 from the igniter ports occurs away from the faceplate. Prominently, a small recirculation bubble near the perimeter wall of the air-heater, just behind the exit of the igniter port, is evident, and is attributed to the suction effect created due to the entrainment of air by the injector jet just below it. This results in a small pocket of high temperature gas in this region.

Similarly, for Case 2, the collective observation of (i) the flow velocity vectors being deflected towards the faceplate (Fig. 3, right), (ii) the nearly identical contours of H_2 mass fraction and the heat of reaction in Fig. 5 (left) and Fig. 4 (left), respectively, and (iii) the corresponding O_2 mass fraction and temperature contours in Fig. 5 (right) and Fig. 4 (right), clearly suggest that bulk of the combustion of excess H_2 from the igniter ports occurs in the vicinity of the faceplate.

Figure 6 shows the heat flux estimates on the faceplate, which is 35 mm away from the axis of the igniter ports, for both Cases 1 and 2. The heat flux magnitudes for Case 1 are significantly lower than that for Case 2. Peak heating for Case 1 (~169 W/cm^2) is seen in the region of entrainment between the perimeter wall of the air-heater and the first injector jet. On the other hand, for Case 2, the peak heating happens closer to the centre of the air-heater assembly, and the magnitude is nearly three times that of Case 1 (~467 W/cm^2), clearly showing the effect of the location of the H_2 combustion zone.

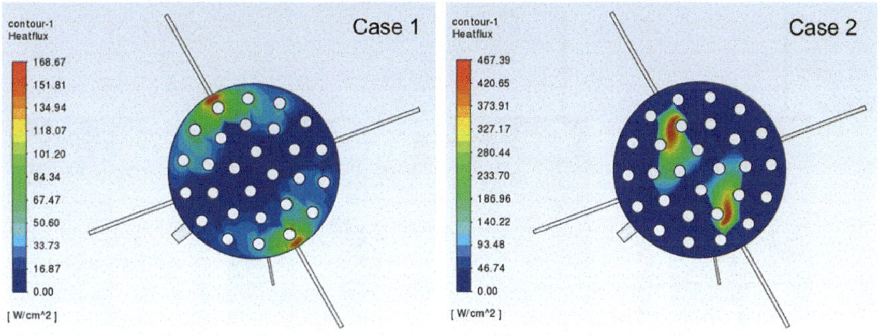

Fig. 6 Heat flux on faceplate for $P_{chamber} = 22$ bar (Case 1: left, Case 2: right)

3.2 Chamber Pressure = 41 bar

Figure 7 shows the velocity vectors on the same vertical plane through the igniter axes for Cases 1 and 2 with a chamber pressure of 41 bar. Compared to the case with a chamber pressure of 22 bar, the penetration extent of the igniter jet is significantly more pronounced for both Cases 1 and 2, highlighted by the shape of the blue coloured arrows. Even for Case 2, the extent of deflection towards the faceplate is much less prominent compared to the lower chamber pressure case.

The deeper penetration of the H_2 flame for a chamber pressure of 41 bar compared 22 bar is clearly seen in Fig. 8 and Fig. 9, which present the contours of heat of reaction (Fig. 8, left), temperature (Fig. 8, right), mass fractions of O_2 (Fig. 9, left) and H_2 (Fig. 9, right). Here, Case 1 does not exhibit the recirculation bubble seen before, near the air-heater perimeter behind the igniter ports because the momentum of the igniter jets are significantly higher here. For Case 2, the reduced deflection towards

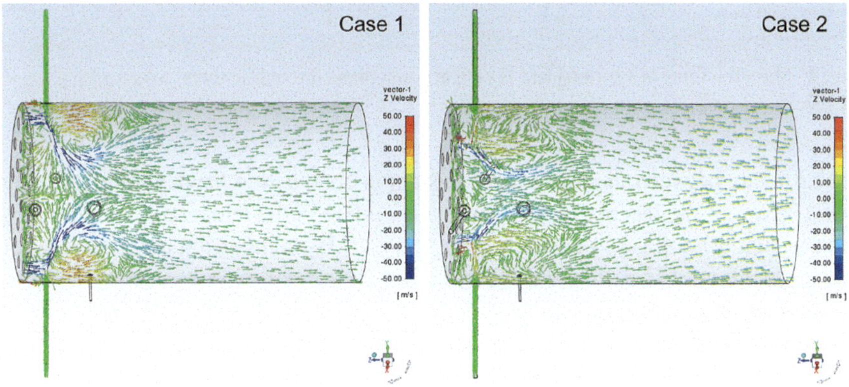

Fig. 7 Velocity vectors on a plane through igniters' axis for $P_{chamber} = 41$ bar (Case 1: left, Case 2: right)

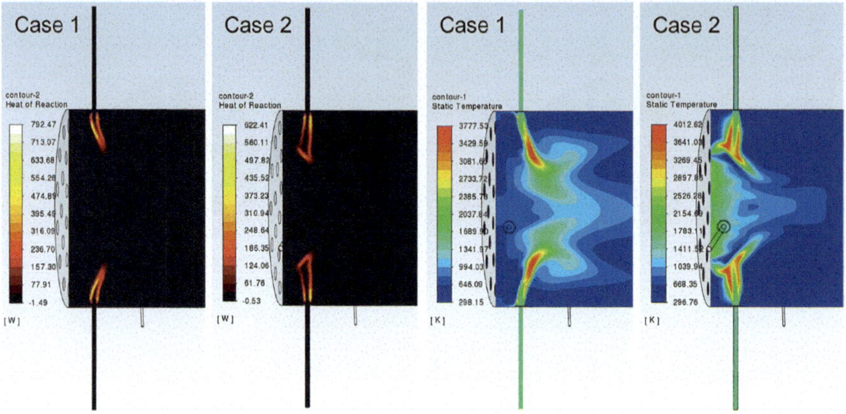

Fig. 8 Heat of reaction (left) and temperature contours (right) on a plane through igniters' axis for $P_{chamber} = 41$ bar

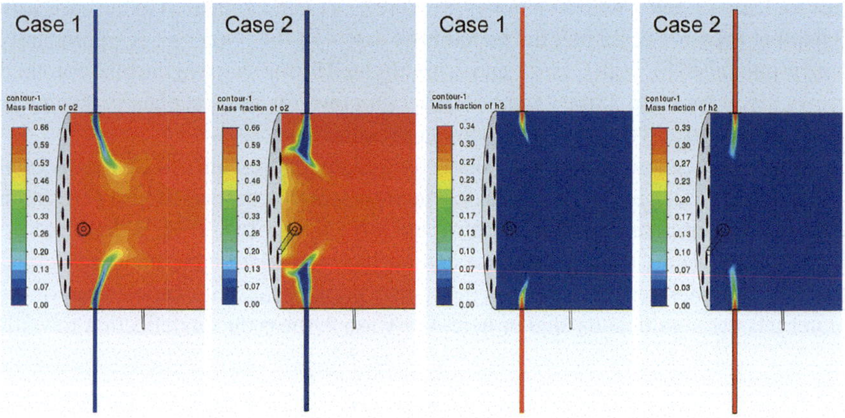

Fig. 9 Mass fractions of O_2 (left) and H_2 (right) on a plane through igniters' axis for $P_{chamber} =$ 41 bar

the faceplate between the injector jets results in combustion happening marginally downstream of the faceplate. Both these are evident in the temperature magnitudes seen in Fig. 8 for Cases 1 and 2. Larger zones of H_2 combustion are indicative of higher quantum of H_2 being present in this set-up (refer Table 1).

Furthermore, heat flux estimates on the faceplate in this case are significantly lower than that for a chamber pressure of 22 bar, shown in Fig. 10. Whereas Case 1 shows a peak heating of only about 68 W/cm^2, the peak heating for Case 2 also drops to about 347 W/cm^2, seen locally near the centre of the faceplate. Here too, there is a significant difference in the peak magnitudes between Case 1 and Case 2, where Case 2 has a peak value that is more than five times that of Case 1.

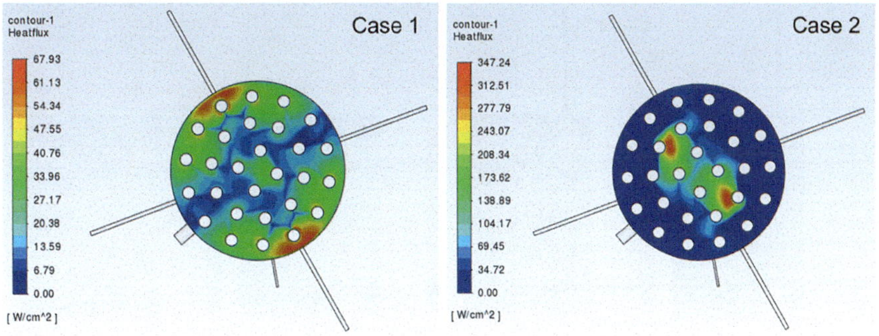

Fig. 10 Heat flux on faceplate for $P_{chamber} = 41$ bar (Case 1: left, Case 2: right)

3.3 *Chamber Pressure = 55 bar*

Figure 11 shows the velocity vectors on the same vertical plane through the igniter axes for Cases 1 and 2 with a chamber pressure of 55 bar. Compared to the case with a chamber pressure of 41 bar, the penetration extent of the igniter jet is significantly more pronounced for Case 1, but is somewhat similar for Case 2. For Case 1, the injector jets from the top and the bottom ports interact near the centre of the air-heater, and this also seen in Case 2, but to a lesser extent.

Figures 12, 13 present the contours of heat of reaction (Fig. 12, left), temperature (Fig. 12, right), mass fractions of O_2 (Fig. 13, left) and H_2 (Fig. 13, right) for a chamber pressure of 55 bar. A deeper penetration of the flame is clearly seen for Case 1 compared to that for a chamber pressure of 41 bar. This results in combustion of the excess H_2 occurring nearer to the faceplate, compared to the previous case, which is evident in the size of the high temperature zone seen in Fig. 12 (right)

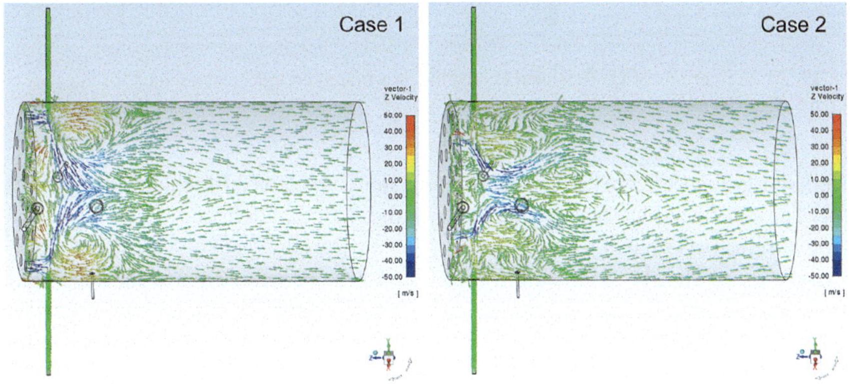

Fig. 11 Velocity vectors on a plane through igniters' axis for $P_{chamber} = 55$ bar (Case 1: left, Case 2: right)

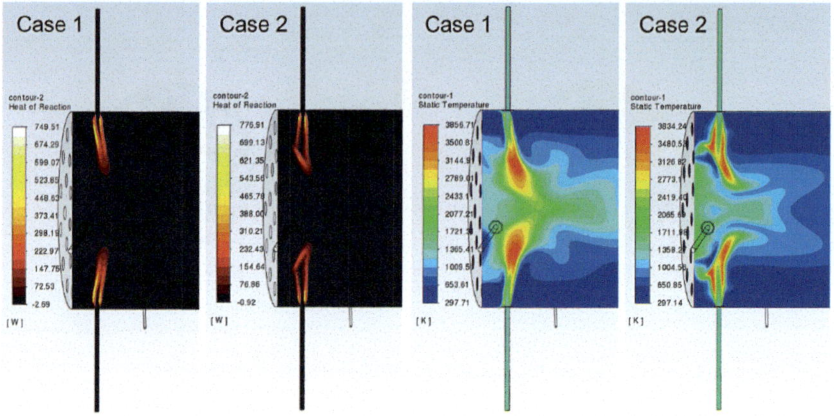

Fig. 12 Heat of reaction (left) and temperature contours (right) on a plane through igniters' axis for $P_{chamber} = 55$ bar

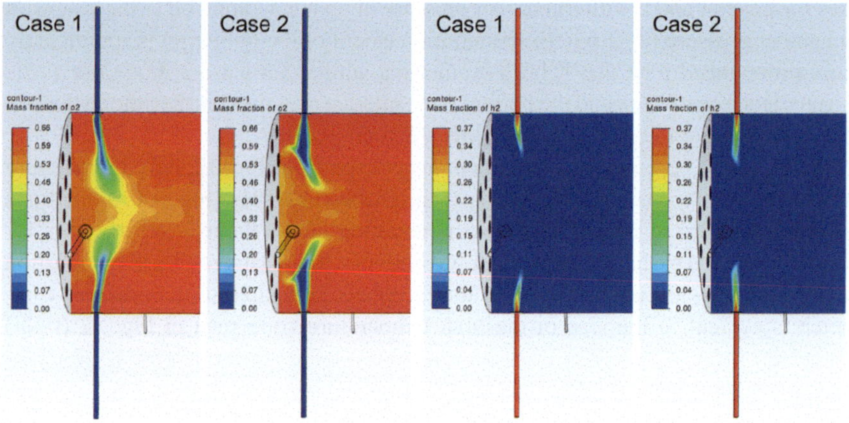

Fig. 13 Mass fractions of O_2 (left) and H_2 (right) on a plane through igniters' axis for $P_{chamber} =$ 55 bar

compared to Fig. 8 (right). The penetration extents in Case 2 for chamber pressures of 41 bar and 55 bar are nearly similar, with the marginal deflection seen for 41 bar being absent for 55 bar (Fig. 14).

Considering that H_2 combustion for Case 1 here occurs closer to the face plate centre compared to the previous set-up, a higher peak heating is seen here (~136 W/cm^2). For Case 2, a further drop in the peak heating is also evident (~298 W/cm^2).

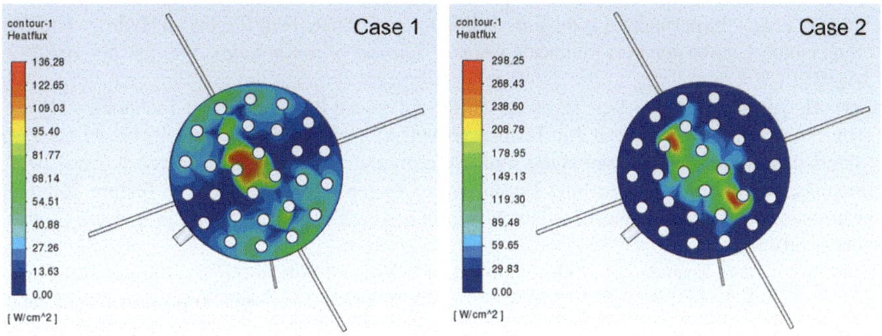

Fig. 14 Heat flux on faceplate for $P_{chamber} = 55$ bar (Case 1: left, Case 2: right)

4 Conclusions

Numerical simulations of the firing of H_2–O_2 igniters in the presence of an air-O_2 crossflow were conducted for three igniter chamber pressures (22, 41 and 55 bar). Two alignments of the air-heater faceplate were assessed: one where the igniter-injector jets exhibit a direct interaction, and one where the igniter jet was fired in-between two adjacent injector jets. For all three igniter chamber pressures, faceplate heating was lesser for Case 1 compared to Case 2 because the igniter flame was naturally deflected away from the faceplate here. Moreover for Case 1, the 41 bar igniter chamber pressure showed the least peak heating of 68 W/cm^2. For Case 2, with the 22 bar igniter chamber pressure, the igniter flame was remarkably deflected towards the faceplate because of the entrainment-instigated suction between two injector jets. This showed the highest peak heating on the faceplate of 467 W/cm^2. When the igniter chamber pressure was increased, the extent of deflection was reduced, as did the peak heating.

Obviously, it is important to consider the alignment of the injector ports with respect to the igniter ports, and a quantitative assessment aids in gauging the magnitude of difference between cases where the jets interact directly and when they do not. Such estimates guide the design of appropriate cooling systems for similar air-heater assemblies with vitiated airflow.

References

ANSYS Fluent Theory Guide, Release 2022 R1, January 2022.

Bao, W. *et al.*, "Ignition Characteristics of a Liquid-Kerosene-Fueled Scramjet during Air Throttling Combined with a Gas Generator," *Journal of Aerospace Engineering*, Vol. 27, No. 5, January 2013.

Edelman, R.B. and Spadaccini, L.J., "The Effect of Vitiated Air Contamination on Ground Testing," AIAA Paper 1969-0456, AIAA 5th Propulsion Joint Specialist Conference, June 1969.

Feiteng, L. *et al.*, "Experimental and Numerical Studies of Vitiated Air Effects on Hydrogen-fueled Supersonic Combustor Performance," *Chinese Journal of Aeronautics*, Vol. 25, pp. 164-172, April 2012.

Fetterhoff, T.P. and Burfitt, J.W., "Overview of the Advanced Propulsion Test Technology Hypersonic Aero Propulsion Clean Air Testbed (HAPCAT)," AIAA Paper 2011-2279, AIAA 17th International Space Planes and Hypersonic Systems and Technologies Conference, April 2011.

Guimareãs, A.L. *et al.*, "Liquid Fuel Burner with Oxygen Replenishment for Testing Scramjet Combustors," AIAA Paper 1997-3020, AIAA/SAE/ASME/ASEE 33rd Joint Propulsion Conference, July 1994.

Hashimoto, T. and Yoshida, M., "Development of a Vitiated Air Heater for Engine Test Facilities," AIAA Paper 1991-2501, AIAA/SAE/ASME/ASEE 27th Joint Propulsion Conference, June 1991.

Hwang, W.-S. *et al.*, "Read-Gas-Flamelet-Model-Based Numerical Simulation and Combustion Instability Analysis of a GH_2/LO_X Rocket Combustor with Multiple Injector," *Energies*, Vol. 14, No. 2:419, January 2021.

Kang, S.H. *et al.*, "Scramjet Engine Combustor Tests in a Supersonic Wind Tunnel with a Vitiated Air Heater," AIAA Paper 2010-7123, AIAA/SAE/ASME/ASEE 46th Joint Propulsion Conference, July 2010.

Karaca, M. *et al.*, "Implicit large eddy simulation of vitiation effects in supersonic air/H_2 combustion," *Aerospace Science and Technology*, Vol. 89, pp. 89-99, June 2019.

Lee, J.-H. *et al.*, "A Study on a Vitiated Air Heater for a Direct-Connect Scramjet Combustor and Preliminary Test on the Scramjet Combustor Ignition," *Aerospace*, Vol. 10, No. 5:415, April 2023.

McBride, B.J. and Sanford, G., "Computer Program for Calculation of Complex Chemical Equilibrium Compositions and Applications II. Users Manual and Program Description," NASA RP 1311, 1996.

Pellett, G.L. *et al.*, "Review of Air Vitiation Effects on Scramjet Ignition and Flameholding Combustion Processes," AIAA Paper 2002-3880, AIAA/SAE/ASME/ASEE 38th Joint Propulsion Conference, July 2002.

Saari, D.P. *et al.*, "Clean Air Regenerative Storage Heater Technology for Propulsion Test Facilities," AIAA Paper 2009-7377, AIAA/DLR/DGLR 16th International Space Planes and Hypersonic Systems and Technologies Conference, October 2009.

Stallings, D.W. and Burt, R.H., "Demonstrated Performance Capabilities of the Aerodynamic and Propulsion Test Unit (APTU) with a Vitiated Air Heater," AIAA Paper 1984-0605, AIAA 13th Aerodynamic Testing Conference, March 1984.

Urbano, A. *et al.*, "Exploration of combustion instability triggering using Large Eddy Simulation of a multiple injector liquid rocket engine," *Combustion and Flame*, Vol. 169, pp. 129-140, 2016.

Vankova, O.S., "Comparison of turbulence/chemistry interaction models in the problem of ignition a parallel hydrogen jet in a supersonic air flow," AIP Conference Proceedings 2351, 040052, 2021.

Wang, C. *et al.*, "Experimental investigation on detonation combustion patterns of hydrogen/vitiated air within annular combustor," *Experimental Thermal and Fluid Science*, Vol. 66, pp. 269-278, September 2015.

Wolf, P. *et al.*, "Using LES to Study Reacting Flows and Instabilities in Annular Combustion Chambers," *Flow, Turbulence and Combustion*, Vol. 88, pp. 191-206, 2012.

Heat Transfer Characteristics of a Semi-circular Cylinder with Hydrophobic Coating A Numerical Study

Collin Joseph, Megha Paul, K. Nandakumar Chandran, K. K. Krishnaram, and Ranjith S. Kumar

Abstract In this paper, the effect of hydrophobicity on heat transfer for a flow past an isothermal semicircular cylinder is studied. Hydrophobicity is implemented using the Navier slip model where partial slip is employed on the semicircular cylinder. Here the heat transfer is characterised in terms of local Nusselt number and average nusselt number of the surface. The variation of local Nusselt number at points on the upper surface of the semi-circular cylinder and the average Nusselt number of the entire surface of the body is studied for various slip values and explained. The local Nusselt number follows a trend where Nu decreases from the front stagnation point upto the edges and a local maxima is generated at the edges of the semicircular cylinder (singularity points) and thereafter the Nu minimally increases along the flat surface of the body. When slip is introduced in the body, improvement in Nu is observed along the surface of the semicircular cylinder except at the singularity point which is located at the edge of the body. The average Nusselt number $\overline{Nu}$ of the isothermal body increases with introduction of slip and becomes higher for higher values of slip length and Reynolds number of the fluid.

Keywords Navier-slip model · Singularity points · Slip length

1 Introduction

Flow past bluff bodies provide knowledge for various real life applications such as heat exchangers, aerodynamics for flights and vehicles, design for bridges, design of wind turbines etc. Also the quest for the geometry of the bluff body that provides the desired hydrodynamic characteristics such as lift and drag and heat transfer through convective capabilities is more prevalant than ever due to the requirement of higher efficiencies in modern day equipments. Several studies on hydrodynamics (Lu et al.

C. Joseph (✉) · M. Paul · K. N. Chandran · K. K. Krishnaram · R. S. Kumar
Micro/Nano-fluidics Research Laboratory, Department of Mechanical Engineering, College of Engineering Trivandrum (Government of Kerala), Thiruvananthapuram, Kerala, India
e-mail: collinjoseph398@gmail.com

G. Rajesh et al. (eds.), *Proceedings of the 15th International Symposium on Experimental and Computational Aerothermodynamics of Internal Flows*,
https://doi.org/10.1007/978-981-96-4082-9_8

2014; Ong et al. 2010; Mikheev et al. 2017) as well as heat transfer (Bharti et al. 2007; Chandra and Chhabra 2011; Singha et al. 2007) is done on circular cylinders. Similarly studies on hydrodynamics (Fonseca et al. 2013; Mittal and Balachandar 1996) and heat transfer (Paul et al. 2013; Berbish 2011) for elliptical cylinders are also performed widely. Elliptical bluff bodies are mostly used as ideal geometries for tubular heat exchangers due to their low pressure drop compared to circular cylinders and lower fouling on the outer surface. Effect of hydrophobic coating on these bluff bodies is a fascinating study which provides improved hydrodynamic as well as heat transfer characteristics caused by the friction-less surface of the wall of the body over which fluid tends to flow with higher velocity compared to a non hydrophobic body. This increased wall velocity inadvertently causes reduced drag and lift and improved convective heat transfer from the body to the fluid. Effectiveness of hydrophobicity on heat exchangers is discussed in Kafarakis et al. (2016) where the bluff body used is a hydrophobic circular cylinder.

In this paper we attempt to identify the effectiveness of hydrophobic coating on heat transfer characteristics when a semi-circular cylinder undergoes cross flow of a fluid under unsteady conditions. Here the cylinder wall is isothermal and hydrophobicity is provided in the form of partial slip by slip length which is based on Navier-slip model. The significance of hydrophobicity is discussed under the influence of Reynolds number of the fluid to identify the relationship between Reynolds number and slip length.

2 Methodology

This section covers the specifics of numerical modelling of the flow past an isothermal-surfaced, hydrophobic semicircular cylinder. The associated flow physics are assumed to be two-dimensional, incompressible, unsteady, newtonian and laminar in order to simplify the problem. The nature of the flow is represented by the Reynolds number and is given by the relation $Re = U * D_h/\nu$ where U represents the inlet velocity of the fluid, D_h represents the equivalent diameter of the semi-circular cylinder and ν represents the kinematic viscosity of the fluid. Navier slip model is devised for modelling the hydrodynamic resistance of the hydrophobic semi-circular cylinder and is given by the relation $\beta = u_{wall}/(\delta u/\delta y)_{wall}$. By using the above assumptions and modelling techniques a two dimensional domain is generated as shown in Fig. 1 which demonstrates the geometrical elements and boundary conditions used. The fluid is provided with an inlet on the left of the channel having a uniform velocity U and fluid temperature T_∞. The outlet of the channel is present on the right of the channel with gauge pressure (P=0) condition which indicates that the outlet is open to the atmosphere. The surface of the semi-circular cylinder made isothermal by providing surface temperature T_s. Here $T_s > T_\infty$.

Governing Equation and Boundary Conditions For simulating the flow past a semicircular cylinder, COMSOL software based on Finite element method(FEM)

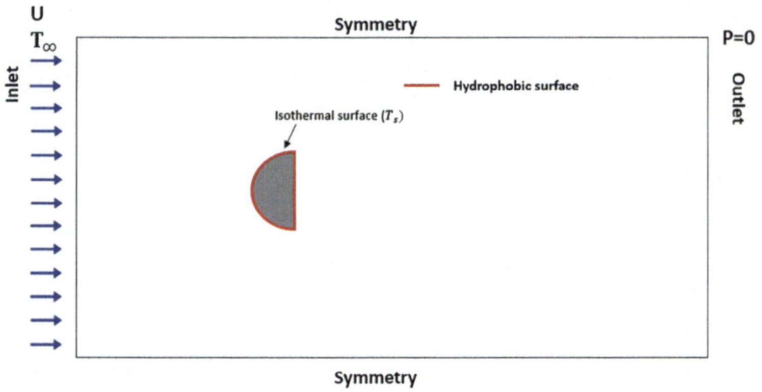

Fig. 1 Schematic of computational domain with a hydrophobic semi-circular cylinder in uniform flow at isothermal conditions

is used for creating the 2 dimensional domain. The governing equations used for this two phase problem involves the conservation of mass, momentum and energy equations.

Here a non-uniform structured mesh is used to discretize the computational domain. The continuum based mass, momentum and energy balance equations provide the governing equations that explain such physical phenomena, which are given by:

$$\nabla.\mathbf{v} = 0, \tag{1}$$

$$\frac{\partial \mathbf{v}}{\partial t} + \mathbf{v}.\nabla\mathbf{v} = -\frac{1}{\rho}\nabla P + \nu\nabla^2\mathbf{v}, \tag{2}$$

$$\frac{\partial T}{\partial t} + \mathbf{v}.\nabla T = \alpha\nabla.\nabla T. \tag{3}$$

Inorder to evaluate the local Nusselt number for CWT (Constant Wall Temperature) on the surface of the semi-circular cylinder, the following expression is used:

$$Nu = \left(\frac{\partial T}{\partial x}\right)_{local} * \frac{D_{eq}}{T_s - T_\infty} \tag{4}$$

where $(\frac{\partial T}{\partial x})_{local}$ represents the local temperature gradient at a particular point on the surface of the semi-circular cylinder and D_{eq} represents the equivalent diameter of the semicircular cylinder.

For evaluating the average Nusselt number for CWT of the entire surface of the semi-circular cylinder, the following expression is used:

$$\overline{Nu} = \left(\frac{\partial T}{\partial x}\right)_{avg} * \frac{D_{eq}}{T_s - T_\infty} \tag{5}$$

where $(\frac{\partial T}{\partial x})_{avg}$ represents the averaged temperature gradient for the entire surface of the semi-circular cylinder.

2.1 Validation

To validate the hydrodynamic characteristics of the flow past a bluff body studies were done to compare the coefficient of lift for various time constants of the simulated present study with the numerical study done by Seo et al. (2012) for slip lengths $\beta = 0$ and $\beta = 0.0135$ which is illustrated in Fig. 2. It is observed that the present study is highly in agreement with the theoretical study and hence the hydrodynamic characteristics are thereby validated.

For validation of heat transfer characteristics of the flow past a bluff body, validations for local Nusselt number and average Nusselt number were performed. Here the local Nusselt number distribution on the upper surface of an isothermal circular cylinder was compared with the numerical study by Hsu (2021) and experimental study by Eckert and Seohngen (1952) which is illustrated in Fig. 3 and found to be in good agreement.

For validation of Average Nusselt number, $\overline{Nu}$ of the entire surface of an isothermal elliptical cylinder when placed in cross flow of air is compared with the numerical study of Khan et al. (2005) which is illustrated in Fig. 4 and found to be in agreement and hence the heat transfer characteristics are thereby validated.

Inorder to ensure grid independency, a grid independency study was performed for the above computational domain for various number of mesh elements and the corresponding average nusselt number is calculated and compared.

Table 1 shows the values of average nusselt Number for the above problem at various number of mesh elements i.e. 100000, 79522, 40000 and 28000 where the

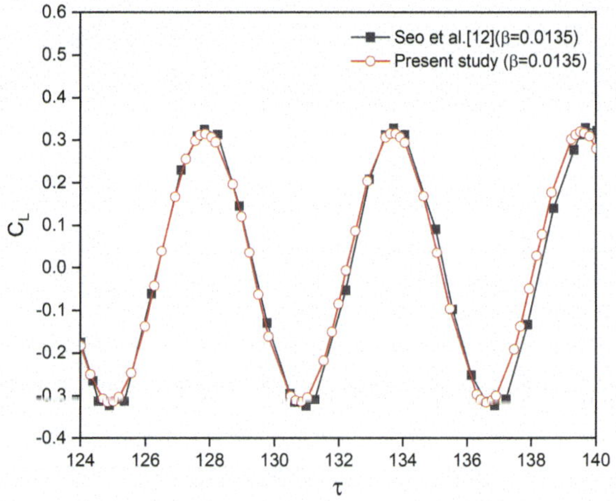

Fig. 2 Variation of coefficient of lift with respect to slip length. Present results are plotted along with numerical (Seo and Song 2012) results

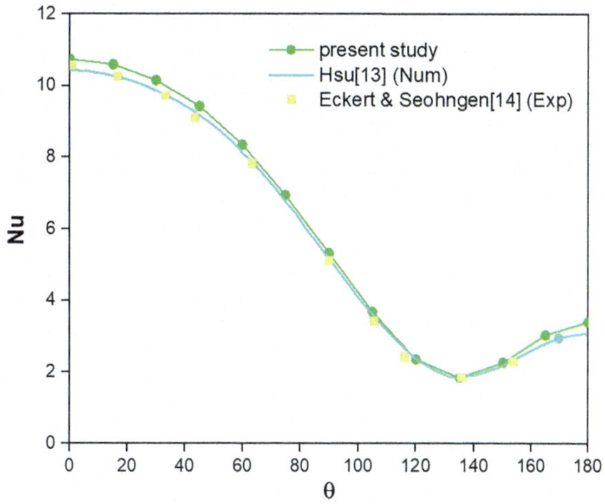

Fig. 3 Variation of local Nusselt number on the upper surface of a cylinder. Present results are plotted along with numerical (Hsu 2021) and experimental (Eckert and Soehngen 1952) results

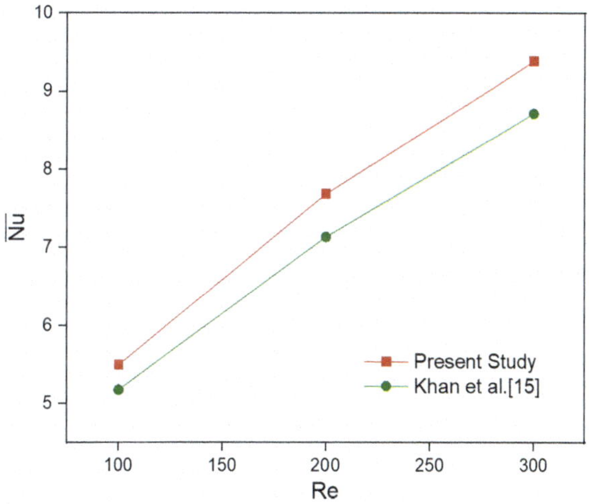

Fig. 4 Variation of average Nusselt number with respect to Re. Present results are plotted along with numerical (Khan et al. 2005) results

Table 1 Grid independent study conducted for the numerical investigation at Re = 100

Mesh elements	$\overline{Nu}$
100000	14.052
79522	14.049
40000	14.036
28000	14.0178

values of $\overline{Nu}$ undergoes very minimal variation and thereby the grid independence is found to be satisfactory and a mesh size of 79522 is selected for the investigation of results.

3 Results and Discussion

Here simulations are performed and are discussed in the upcoming sub sections. These include the effect of slip length and the effect of Re on the heat transfer characteristics. Both local and average Nusselt numbers are monitored to understand the heat transfer phenomena.

3.1 *Effect of Slip Length*

In this section the effect of β on Nu and $\overline{Nu}$ is discussed. For visualising these effects velocity streamlines and temperature contours for various slip lengths are plotted. For numerical understanding, Nu and $\overline{Nu}$ is plotted with respect to various slip lengths namely $\beta = 0$, $\beta = 0.05$ and $\beta = 0.1$.

Figure 5 shows the velocity streamlines for Re = 100 at 5 time instances for no slip, $\beta = 0.05$ and $\beta = 0.1$. Here we can observe that when slip is implemented the velocity of the fluid in the rear of the semi-circular cylinder becomes lower by the introduction of more bluish-green region which indicates the fluid undergoes higher amount of mixing thereby enhancing the heat transfer from the body to the fluid.

Figure 6 shows the temperature contours for Re = 100 at 5 time instances for no slip, $\beta = 0.05$ and $\beta = 0.1$. Here we can observe that when slip is introduced in the isothermal semi-circular cylinder, the fluid surrounding the body lies at a higher temperature (lightbluish-green region) compared to the no slip condition. This indicates that the fluid absorbed more amount of heat from the body when slip

Fig. 5 Velocity streamlines for Re = 100 at various time instances for **a** No slip, **b** $\beta = 0.05$ and **c** $\beta = 0.1$

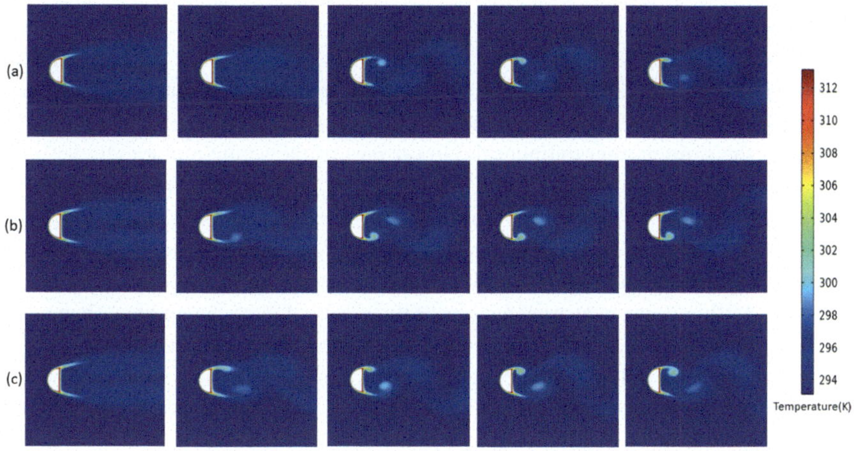

Fig. 6 Temperature contours for Re = 100 at various time instances for **a** No slip, **b** $\beta = 0.05$ and **c** $\beta = 0.1$

is introduced and thereby we can conclude qualitatively that the heat transfer from the heated body to the fluid is improved with the application of slip.

Figure 7 represents the local Nusselt Number distribution at various pts on the upper surface of the isothermal semi-circular cylinder at Re = 80 for various slip lengths. Here Nu for 10 points on the surface of the semi-circular cylinder is plotted for $\beta = 0$, $\beta = 0.05$ and $\beta = 0.1$. In this figure for no slip condition, we can observe that the local Nusselt number is highest at point 1 which represents the front stagnation point and it gradually decreases along points 2, 3, 4, 5, 6 and finally at point 7 which is the edge of the semicircular cylinder as it is a singularity point it will have

Fig. 7 Variation of local Nusselt number on the upper surface of the semicircular cylinder for various slip lengths

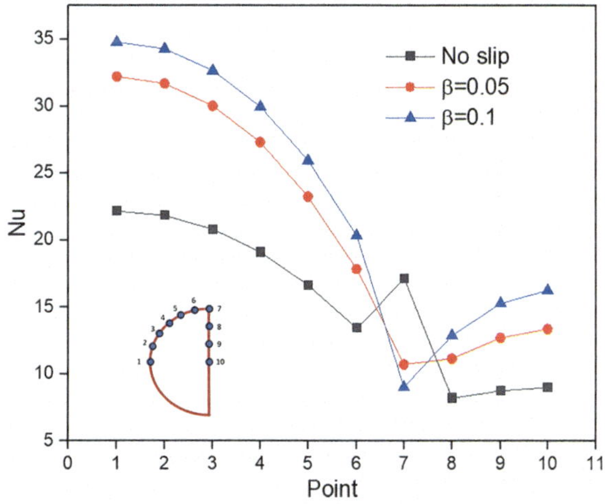

instantaneous increase in Nu. The points along the flat rear surface of the cylinder i.e. 8, 9 and 10 will have nearly same values of Nu as the presence of fluid in these regions are nearly the same and have minimal heat transfer from these points on the cylinder to the surrounding fluid. As the flow is symmetric along the upper and lower regions of the semicircular cylinder the lower part of the semicircular cylinder also conforms to the same trend as that of the upper part. This is similar to the observations by Chandra and Chhabra (2011) for semicircular cylinders. As the slip length is increased the values of Nu at all these points increases due to the increase in the velocity of the fluid near the fluid-cylinder interface except at the singularity point (point 7). At this point heat transfer primarily occurs due to the sudden change in the surface orientation and thereby when slip is introduced flow becomes partially frictionless thereby reducing the roughness of the surface at this point and hence has lower Nu values compared to no slip condition. As a result, improved convection occurs from the isothermal cylinder to the surrounding fluid causing increased temperature gradients at all points except at the singularity point where the implementation of hydrophobicity reduces the impact of change in surface orientation on heat transfer along the surface of the isothermal hydrophobic semi-circular cylinder.

Figure 8 represents the variation of $\overline{Nu}$ with respect to β namely $\beta = 0$, $\beta = 0.05$ and $\beta = 0.1$. As mentioned earlier, when the slip length increases the velocity of the fluid near the surface from zero with enhances the local Nusselt number of all the points on the surface of the semicircular cylinder. As a result, the average nusselt number value increases and hence improvement in heat transfer can be concluded with increase in slip length.

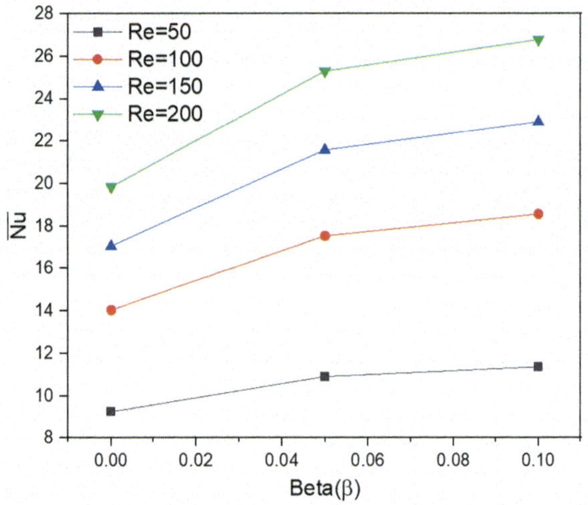

Fig. 8 Variation of average Nusselt number of the semicircular cylinder with respect to β, varying Re

3.2 *Effect of Re*

In this section the effect of *Re* on $\overline{Nu}$ is discussed. Inorder to visualise and understand its effects, Nu and $\overline{Nu}$ is plotted with respect to Re namely Re = 50, 100, 150 and 200 for various slip lengths namely $\beta = 0$, $\beta = 0.05$ and $\beta = 0.1$.

Figure 9 represents the variation of $\overline{Nu}$ with respect to Re at Re = 50, Re = 100, Re = 150 and Re = 200 for slip lengths $\beta = 0$, $\beta = 0.05$ and $\beta = 0.1$. Here as the Reynolds number increases, increase in $\overline{Nu}$ is observed for a given β. This occurs because increase in Re causes the increase in inlet velocity of the fluid which creates more unsteady motion while interacting with the bluff body and thereby causes higher mixing of heat within the fluid layers. It is also observed that implementation of slip is more effective at higher Re. Hence Re also contributes to the improved effectiveness of hydrophobicity for heat transfer purposes.

4 Conclusion

In this study a flow past an isothermal semi-circular cylinder having a hydrophobic coating was simulated where the cylinder is at a higher temperature than the surrounding fluid. Studies on heat transfer characteristics were conducted by monitoring both local as well as average Nusselt numbers inorder to identify the effectiveness of the coating under various slip lengths and Re values of the fluid.

We can conclude that increasing the slip as well as Re of the fluid enhances the heat transfer capabilities from the heated semicircular cylinder to the surrounding fluid with improvement in local and as well as average Nusselt numbers.

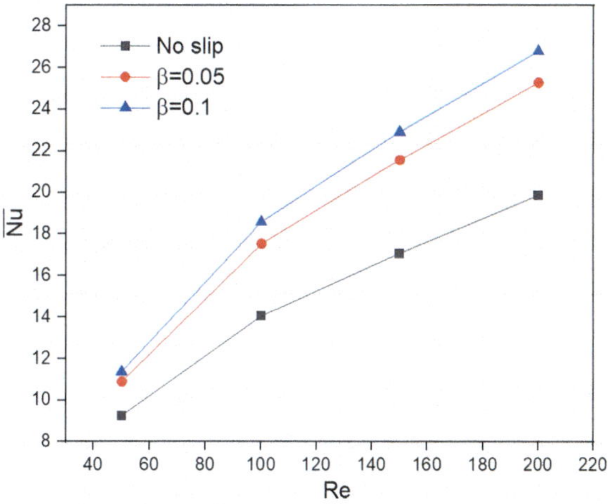

Fig. 9 Variation of average Nusselt number of the semicircular cylinder with respect to Re, varying β

References

Berbish, N. S. (2011). Heat transfer and flow behavior around four staggered elliptic cylinders in cross flow. Heat and mass transfer, 47, 287–300.

Bharti, R., Chhabra R.P., and Eswaran, V., (2007),Steady forced convection heat transfer from a heated circular cylinder to power-law fluids *International Journal of Heat and Mass Transfer*

Chandra, A., & Chhabra, R. P. (2011). Flow over and forced convection heat transfer in Newtonian fluids from a semi-circular cylinder. International Journal of Heat and Mass Transfer, 54(1-3), 225–241.

ERG Eckert and E Soehngen, Distribution of heat-transfer coefficients around circular cylinders in crossflow at reynolds numbers from 20 to 500, Transactions of the American Society of Mechanical Engineers 74 (1952), no. 3, 343–346.

Fonseca, F. B., Mansur, S. S., & Vieira, E. D. R. (2013). Flow around elliptical cylinders in Moderate Reynolds Numbers. In Proc. Of the 22 nd International Congress of Mechanical Engineering (COBEM 2013), Nov (pp. 3–7).

Hsu, Li-Chieh, Heat transfer of flow past a cylinder with a slit, International Journal of Thermal Sciences 159 (2021), 106582

S.T Kafarakis, M.E Mastrokalos, C.I Papadopoulos, L. Kaiktsis (2016),Heat Transfer Enhancement for Heat Exchanger Applications: Use of Surface Hydrophobicity *ASME Pressure Vessels and Piping Conference*

Lu, L., Liu, M. M., Teng, B., Cui, Z. D., Tang, G. Q., Zhao, M., & Cheng, L. (2014). Numerical investigation of fluid flow past circular cylinder with multiple control rods at low Reynolds number. Journal of Fluids and Structures, 48, 235–259.

Mikheev, N. I., Molochnikov, V. M., Mikheev, A. N., & Dushina, O. A. (2017). Hydrodynamics and heat transfer of pulsating flow around a cylinder. International Journal of Heat and Mass Transfer, 109, 254–265.

Mittal, R., & Balachandar, S. (1996). Direct numerical simulation of flow past elliptic cylinders. Journal of Computational Physics, 124(2), 351–367.

Ong, M. C., Utnes, T., Holmedal, L. E., Myrhaug, D., & Pettersen, B. (2010). Numerical simulation of flow around a circular cylinder close to a flat seabed at high Reynolds numbers using ak-ϵ model. Coastal Engineering, 57(10), 931–947.

Paul, I., Arul Prakash, K., & Vengadesan, S. (2013). Forced convective heat transfer from unconfined isothermal and isoflux elliptic cylinders. Numerical Heat Transfer, Part A: Applications, 64(8), 648–675.

Seo, Il Won and Chang Geun Song, Numerical simulation of laminar flow past a circular cylinder with slip conditions, International Journal for Numerical Methods in Fluids 68 (2012), no. 12, 1538–1560.

Singha, A. K., Sarkar, A., & De, P. K. (2007). Numerical study on heat transfer and fluid flow past a circular cylinder in the vicinity of a plane wall. Numerical Heat Transfer, Part A: Applications, 53(6), 641–666.

Waqar A Khan, JR Culham, and Michael M Yovanovich, Fluid flow around and heat transfer from elliptical cylinders: analytical approach, Journal of thermophysics and heat transfer 19 (2005), no. 2, 178–185.

Aeroacoustics Characteristics of the Supersonic Free Jet at Mach Number 1.8

K. Manikandan, Mahadevan Shankar, D. Devanathan, and Das Rajarshi

Abstract Computational analysis is used to investigate the aeroacoustics characteristics of supersonic jets with design Mach numbers of 1.8. To account for nozzle expansion from over to optimal, the nozzle pressure ratio ranged from 2.75 to 5.75. For nozzle pressure ratios of 2.75 and 3.75, the shadowgraph results show complex shock cell structures as well as the Mach stem. However, once the NPR is significantly greater than highly over-expanded, the Mach stem disappears from the jet plume. Furthermore, even at the design operating condition, a convergent-divergent supersonic nozzle exhibits a throat-induced shock, but the shock from the nozzle exit dissipates when it works at the design NPR. Unsteady numerical shadowgraph is used to identify propagating sound waves emanating from the jet. Acquired time-series acoustic pressure data from the six different microphone locations are utilized to calculate the sound pressure variations across the frequency scale. Processed SPL values show that when the jet expands for 2.75 and 3.75, its magnitude increases at 90 and 115° where shock-associated noise tends to dominate. But for the angles from 30 to 75° magnitude of the variation is smaller than the other microphone locations. Once the operating NPR is close to the design value, the sound pressure level values increase for the angles of 30 to 75° where mixing noise dominates, but for the upstream angle, SPL magnitude is notably lesser than the other angles. Hence it can be inferred that highly over-expanded flow tends to emit more intense shock noise than the mixing noise, but the opposite is true for the operating nozzle pressure ratios 4.75 and 5.75.

Keywords Aeroacoustics · Free jet · Nozzle pressure ratio · Core jet · Shock structure

K. Manikandan (✉) · M. Shankar · D. Devanathan · D. Rajarshi
Department of Aerospace Engineering, SASTRA University, Thanjavur, India
e-mail: manikandan@mech.sastra.edu

G. Rajesh et al. (eds.), *Proceedings of the 15th International Symposium on Experimental and Computational Aerothermodynamics of Internal Flows*,
https://doi.org/10.1007/978-981-96-4082-9_9

1 Introduction

Quite a lot of research work is happening on the acoustics analysis of supersonic jets. Its complex flow structure and relatively higher exit velocity prompt intense noise radiation. During take-off and landing, aircraft with C-D Nozzles tend to produce extreme acoustic radiation when it is operating at over-expanded conditions. A better understanding of the acoustic behavior of supersonic jets is necessary for that. The noise characteristics of supersonic jets are quite a bit different from those of subsonic jets. The primary source of noise from the supersonic-free jets is broadband shock-associated noise and turbulent mixing noise. Broadband shock-associated noise, which dominates the acoustic characteristic in the direction upstream of the jet flow, is produced by the interaction of an unsteady shock wave in the shock cell with a turbulence structure in the shear layer. The interaction of the large-scale turbulence structure downstream produces mixing noise, which becomes very effective noise sources when their phase speeds (relative to the ambient speed of sound) become supersonic. Mixing noise produces a broad peak that dominates the spectrum in the downstream quadrant and radiates most strongly from the end of the jet potential core (Henderson and Bridges 2010; Munday 2012).

Theoretical research on supersonic jet noise was also carried out by many researchers for a deeper understanding of the causes of noise generation in supersonic aircraft exhaust jets. Turbulent mixing acoustic waves are categorized into large-scale turbulent acoustic waves that are generated from a large-scale turbulent structure advecting with supersonic Mach number and fine-scale turbulent acoustic waves that are generated from the fine structure of the turbulent flow. Large-scale turbulent acoustic waves have a low-frequency spectrum with strong directivity on the downstream side, and fine-scale turbulent acoustic waves have a high-frequency spectrum with strong directivity on the upstream (Tam 1995). The upcoming theory will address the relevant work carried out by many researchers, which provides valuable inputs for effective computational work and its relevant challenges.

Munday et al. (2009) investigated the experimental and computational study of the flow field and acoustics of convergent-divergent nozzles for design Mach numbers of 1.30, 1.50, and 1.65, and the shadowgraph technique is used for flow visualization. For the numerical analysis, a monotonically integrated large eddy simulation (MILES) approach was used. The unsteady three-dimensional inviscid, compressible flow equations are solved with a finite element code FEFLO on unstructured tetrahedral grids. The results indicate that shock-free condition is never attained at the design Mach number and the formation of shock at the throat section. The intensity of the screech noise has been noticed to be slightly reduced.

Barbarino et al. (2017) computationally analyzed the prediction and reduction of noise from subsonic jets by reconstructing the turbulent fields from the Reynolds Averaged Navier Stokes (RANS) calculations. This computational approach is called Stochastic Noise Generation and Radiation. A 3-D axisymmetric model using a second-order upwind scheme coupled with RANS equations and the turbulence is accounted for through K-ε and K-ω Shear Stress Transport (SST) models. The results indicate that the RANS approach validates both experimental and Large-Eddy Simulations (LES) results. Chen et al. (2018) studied the high-temperature effects on aerodynamic and aeroacoustic characteristics in a rectangular supersonic jet using the compressible unsteady Navier-Stokes equations with the LES approach in Cartesian coordinates. A multi-block structured mesh was used. The grid points were densely packed in the jet region where large flow gradients prevail and stretched slowly in both the axial and radial directions. The first layer thickness of the mesh has a y+ ~1 in the wall-normal direction, and the growth ratio of the grid is controlled below 5% to avoid dispersion errors.

Martelli et al. (2010) investigated the turbulent flow separation in over-expanded rocket nozzles numerically in a subscale parabolic nozzle. A 2-D axisymmetric time-accurate perfect gas Reynolds averaged Navier-Stokes solver is used to compute the nozzle flow field. Subsonic inflow at the nozzle feeding chamber, symmetry axis, adiabatic wall, subsonic inflow boundary condition on the left side of the external region (main flow direction is from left to right), assigned pressure, and non-reflecting boundary conditions on the upper and right sides, respectively. Thus, complicated phenomena can be approximated with reasonable accuracy using the 2D axisymmetric model. In addition, wherever the viscosity and gradient effects are strong, a minimal initial cell distance normal to the nozzle wall is desirable. Bogey et al. (2003) used a subsonic circular jet with a Mach number of 0.9 and a Reynolds number of 6.5×104 to study the flow field and acoustic radiation, computing with the LES approach. The computational parameters are in good agreement with the experiments. Sound pressure levels and directivities obtained in the acoustic field also agree with experimental results.

Massey et al. (2003) simulated the fluid flow by solving the asymptotically steady, compressible, Reynolds Averaged Navier Stokes (RANS) equations using an implicit, up-wind, flux difference splitting finite volume scheme and k-ε turbulence model. The CFD results were seen to be in good to excellent agreement with the NASA Langley Jet Noise Laboratory data. Andersson et al. (Andersson and Eriksson 2008) studied the Prediction of the flow field and acoustic signature of a coaxial jet using Reynolds Averaged Navier Stokes (RANS). Acoustic signature obtained using a method based on RANS and Lighthill's acoustic analogy is compared with the flow field and radiated sound predictions obtained using LES combined with Kirchhoff surface integration, and the k-ε turbulence model is adapted for this work. Results show that the difference between the RANS and LES predictions was that the turbulence kinetic energy is higher close to the nozzle in the LES simulations, leading to more intense mixing and thus shorter potential core regions.

Tide et al. (2009) have combined unsteady RANS with the Ffowcs Williams-Hawkings (FWH) method. A three-dimensional, round, compressible, turbulent jet using the Shear Stress Transport model was carried out. On comparison of mean axial velocity along the jet centerline and along the radial direction at several locations shows that the Unsteady Reynolds averaged Navier-Stokes (URANS) predictions are, in general, comparable with and surprisingly better than the LES results. The simulation result under-predicted the sound pressure level (SPL) compared with experiments, but the calculation can predict the trends fairly accurately. In addition, from calculations, the SST k–ω model can predict mean turbulence quantities.

2 Computational Methodology

The research technique is divided into two parts. The first part involves running steady and unsteady simulations to record mean-flow characteristics and time-series acoustic pressure data, respectively. The SPL of input from all microphones is calculated in the second section using time-series pressure data using fast Fourier transform (FFT). To reach the design Mach number of 1.8, the nozzle contour was created using the method of characteristics (MOC) for an area ratio of 1.439. Using the Area-Mach number relation, the area ratio for the relevant design Mach number has been determined. Four different Nozzle Pressure Ratios (i.e., NPR) of 2.75, 3.75, 4.75, and 5.75 are used to simulate the three over-expanded and one ideal-expanded jet. For this investigation, a 2D axisymmetric model is preferred. In addition, to ensure the far field, the computational domain is expanded up to 30De on the XY plane. Figs. 1, 2 show the entire domain as well as near-wall mesh details. Similarly, Fig. 3 depicts the boundary condition in further detail. Near the wall, a fine mesh is desirable because it allows the solver to realistically capture the boundary layer growth. Similarly, fine mesh is used to anticipate free-jet growth downstream in places such as the core jet and shear layer. Correspondingly, a smaller first cell distance along the nozzle wall is desirable (Chen et al. 2018; Martelli et al. 2010) and in this study, it was kept below or equal to 5 (i.e., $y+ \sim 5$). The nozzle inlet was subjected to a pressure inlet condition, and the total pressure was computed using the isentropic relation equation (1). The nozzle wall has a no-slip adiabatic wall condition, whereas the remaining boundaries get an atmospheric condition. With an ideal gas assumption of a total temperature of 300K (i.e., Cold flow), the density-based solver is used; Sutherland's law with a three-coefficient model for viscosity estimation is enabled throughout the simulation. The RANS equation, in combination with the SST K-ω model, can reasonably predict the aerodynamic and acoustic characteristics of a supersonic free-jet, as per literature (Andersson and Eriksson 2008; Barbarino 2017; Mankbadi et al. 1994; Tide and Babu 2009). Also, 2-D axisymmetric simulations were found to be reliable for analyzing the compressible jets' aero-acoustic characteristics. As a result, over the entire simulation, the RANS equations with turbulence accounted for using the K-ω SST model is used. Up to a residual of 10−6, all equations are solved. TECHPLOT 360 was used to post-process the solution after it had converged. For all the nozzle's NPR values, the numerical shadowgraph is used to illustrate the shock pattern, shock-cell length, and shear layer development. For all situations, an Unsteady RANS simulation is used with a time step of $\Delta t = 1 \times 10{-}6$ sec and a total duration of 0.025 sec. According to the literature (Andersson and Eriksson 2008; Barbarino 2017; Chen et al. 2018; Tide and Babu 2009), the microphones should be placed at a specific distance from the Free-jet to avoid the near-field effect. To ensure the far-field impact, R/De should be adequate (i.e., $30 \geq R/De \leq 50$). The acoustic

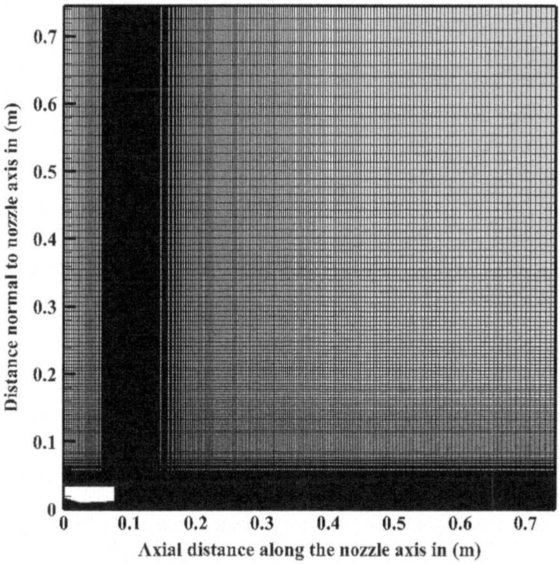

Fig. 1 Domain mesh details

pressure (i.e., dynamic pressure) data is captured on the microphones at R/De = 30 for each time step (i.e., Far-field). There are six microphones in total, each having a 15∘ angle of separation. The corresponding microphone arrangements can be seen in Fig. 3. Time-series acoustic pressure values were processed in MATLAB using the FFT approach to calculate an Amplitude-frequency spectrum of the acoustic signal, and the corresponding SPL was calculated using the equation (2). For NPR values of 2.75, 3.75, 4.75 (i.e., Over-expansion), and 5.75 (i.e., Ideal-expansion), the free-jet velocity is plotted along the nozzle axis.

$$NPR = \frac{P_0}{P_\infty} = \left(1 + \frac{\gamma - 1}{2} M^2\right)^{\frac{\gamma}{\gamma - 1}} \tag{1}$$

$$SPL(dB) = 20\log_{10}\left[\frac{P'_{rms}}{P_{ref}}\right] \tag{2}$$

3 Results and Discussion

The TECHPLOT 360 is used to get numerical shadowgraph images of the steady-state results for all NPR values. The obtained results are utilized for exploring the flow features of the supersonic-free jet embedded with the induced shock cells. The NPR is varied from 2.75 to 5.75 to explore the effect of highly over-expanded flow and moderately over-expanded jet flow . From Fig. 5a, it can be observed that shock with the strong Mach stem appears inside a divergent section of the nozzle exit with a

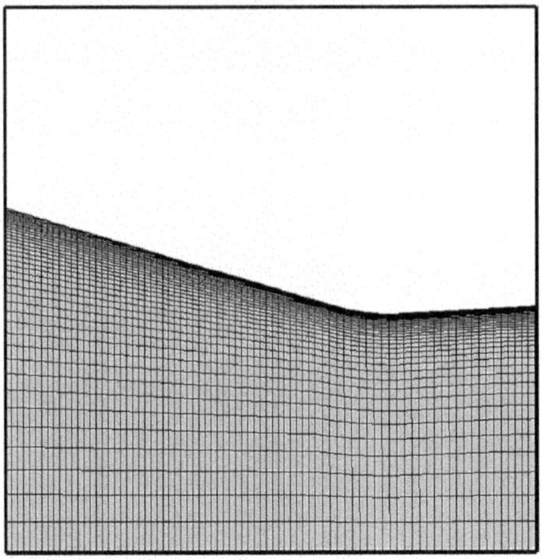

Fig. 2 Near wall mesh details

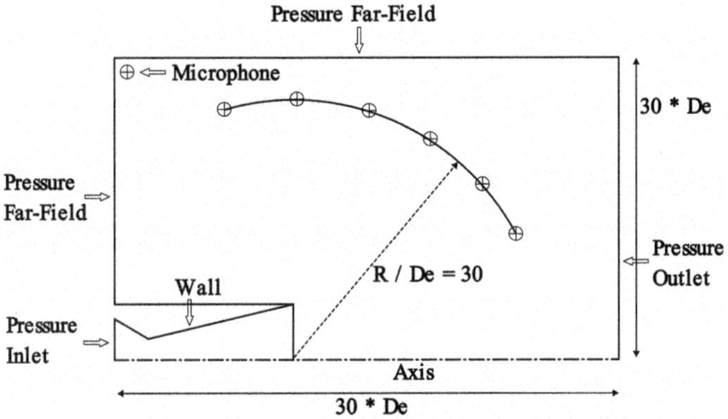

Fig. 3 Boundary condition with domain size and microphone arrangement

shock angle of 45°. Fig. 5b exhibits the shock structure of the NPR of 3.75, the shock with an angle of 42° and a relatively smaller Mach stem can be observed at the nozzle exit. Hence, it can be inferred that highly over-expanded jets consist of a combined strong oblique shock with a Mach stem expected to emit strong acoustic radiation. Once the nozzle operates at NPR 4.75 which is close to the design NPR of 5.75, from Fig 5c it can be witnessed that the shock cell appears without the Mach stem, also calculated shock angle gets reduced to 40° which is notably lower than the former two NPR values. Likewise, it can be witnessed that shock-cell length increases with the decrease in variation between operating NPR and design NPR of the nozzle.

Once the jet expands at the design NPR 5.75, the shock at the nozzle exit tends to disappear due to negligible/minimal pressure mismatch between the jet and atmospheric pressure. Hence, from Fig. 5d it can be inferred that the jet plume is free from induced shock at the nozzle exit. But for all the NPR values mild shock appears immediately downstream of the nozzle throat and gets multiple identical reflections from the solid wall before it reaches the nozzle exit, a similar phenomenon can be seen even at the ideal. The analogous findings were reported by Munday et al. (2009), and it was concluded that military-style convergent-divergent supersonic conical nozzles are never shock-free due to the geometrical nature of the nozzle contour. Hence, shock is always embedded with the turbulent mixing noise. They have also mentioned that C-D nozzle contours are not as smooth as rocket engine nozzle; also, military engines need to house the variable area mechanism to enable the engine to perform optimally at different operating conditions. The measured values of sound pressure levels at the six different microphone locations are given in Fig. 6a, b, c and d, for the respective NPR values from 2.75 to 5.75. The microphones are located between 30 to 115° with an interval of 15°.

Figure 4 shows the velocity decay characteristics for the NPR values from 2.75 to 5.75. At NPR 2.75, a steep fall of velocity from 485 to 213 m/s can be seen at 74 mm, which is happening due to the formation of strong shock with Mach-stem, subsequent velocity recovery led to the formation of multiple shock cells within the core jet. The velocity fluctuation nearly ends at 348 mm. If the jet operates at the NPR 3.75, a similar velocity drop can be detected at 87 mm from 486 m/s to 226 m/s. The core jet for the NPR of 3.75 is notably larger than that of the NPR of 2.75. The jet expansion at NPR 3.75, exhibits the shock formation with the less Mach stem portion exactly at the nozzle exit which can be witnessed from the figure. The core jet length at the NPR of 3.75 is larger than the core jet length of the NPR of 2.75. The Jet expansion at NPR of 4.75 shows a lesser velocity variation along the jet axis than the three former NPR values. The initial drop in velocity can be seen at 90 mm from 484 m/s to 405 m. Finally, when the jet develops at optimal expansion (i.e., NPR = 5.754) minor velocity deviations can be witnessed along the nozzle axis. Even this fluctuation is only because of induced throat shock.

Figure 6a clarifies the SPL variations for the NPR of 2.75, which is considered a highly over-expanded flow. At the 30° location, SPL starts with a value of 104 dB and with an end value of 45 dB in the frequency scale. Similarly, at the 45° position, the SPL begins with 94 dB and ends with 31 dB. For the 60° location Microphone records the SPL value of 94 dB at the beginning and reduces to 39 dB at 50,000Hz. The magnitude of SPL variations along the frequency scale for 60° is relatively lesser than the former two locations (i.e., 30 and 45°). The 75° location also exhibits nearly identical characteristics to 60°. Finally, for the 90 and 115° locations, the maximum SPL values are slightly higher than the values of 45, 60, and 75° locations. For 90° SPL reaches the values of 103 dB and for 115° it records 107 dB, respectively. In addition, the end values are also considerably higher than the previous two microphone angles. 90° shows the final value as 36 dB, and for the 115° location, it is 47 dB. Hence, it can be concluded that the upstream quadrant 90 to 115° experiences a significant increase in the SPL values of maximum to minimum than the other microphone locations. As mentioned earlier high strength shock mechanism at the nozzle can be a possible reason for these elevated SPL values at the highly over-expanded condition.

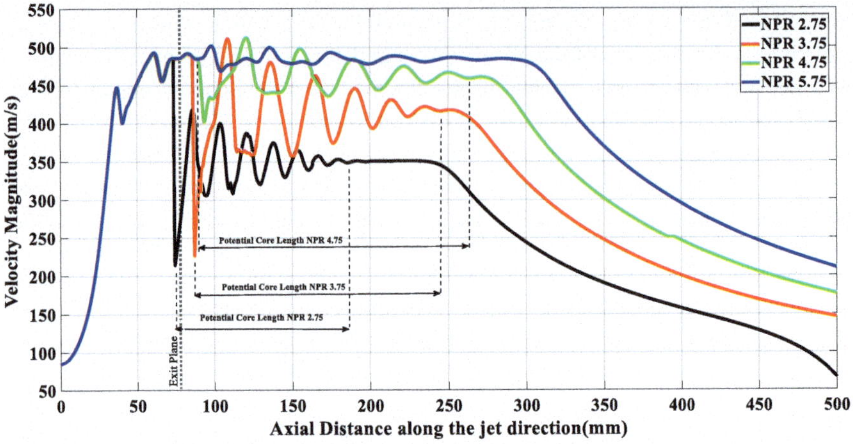

Fig. 4 Velocity decay of the jet at different nozzle pressure ratios

Figure 6b explains the SPL characteristics of NPR 3.75, at 30 ° microphone records 104 dB and 54 dB as maximum and minimum SPL in the frequency scale. Also, it can be seen that the SPL fluctuates all along the frequency scale. At the 45° location, maximum and minimum SPL values are around 92 and 40 dB, which is notably lesser than the 30° location. In addition, a notable peak can be observed at 3418Hz for the SPL value of 95 dB, after this SPL decays to its minimum value of 40 dB without any fluctuation. Likewise, at 60° fluctuation in SPL can be seen up to 8300Hz after that it exhibits smooth decay to its minimum of 55 dB. It can be inferred that the 60° location shows only a marginally smaller dip in SPL of 95 to 55 dB, than that of 92 to 40 dB at 45°. At 75° also microphone exhibits nearly identical SPL maximum (i.e., 97 dB) and minimum (i.e., 55 dB) values as 60°. Finally, the 90 and 115° also experience a relatively higher SPL maximum of 92 dB and a minimum of around 56 dB, and for both locations' fluctuations can only be observed up to 12210 Hz. Hence, it can be inferred that at microphone locations 75, 90, and 115° maximum and minimum SPL values are notably higher than the NPR of 5.75. It clearly shows that these angles are dominated by the shock-associated noise rather than the mixing noise; high strength-induced shocks in the jet can be a possible reason for this increase in shock-associated noise components.

The SPL changes for the NPR = 4.75 can be clarified from Fig 6c; at 30° it reads 106 dB as the maximum SPL and around 60 dB as the end value. The SPL variations are notable all along the frequency spectrum with a decrease in amplitude. But for 45° the SPL fluctuations are notably more significant than the previous location. Similarly, at 60° maximum and end SPL values are 93 and 42 dB, which is lesser than the previous microphone locations. In addition, almost five dips can be seen from the same Fig. 6c. The 75 and 90° also show slightly lesser SPL maximum and slightly higher SPL end values observed, and both the SPL values show minimal fluctuation in the frequency spectrum. In the end, 115° shows an SPL maximum

Fig. 5 Shadowgraph images at different Nozzle pressure ratio and emanating sound waves from the jet flow **e** emanating sound waves captured at an instance

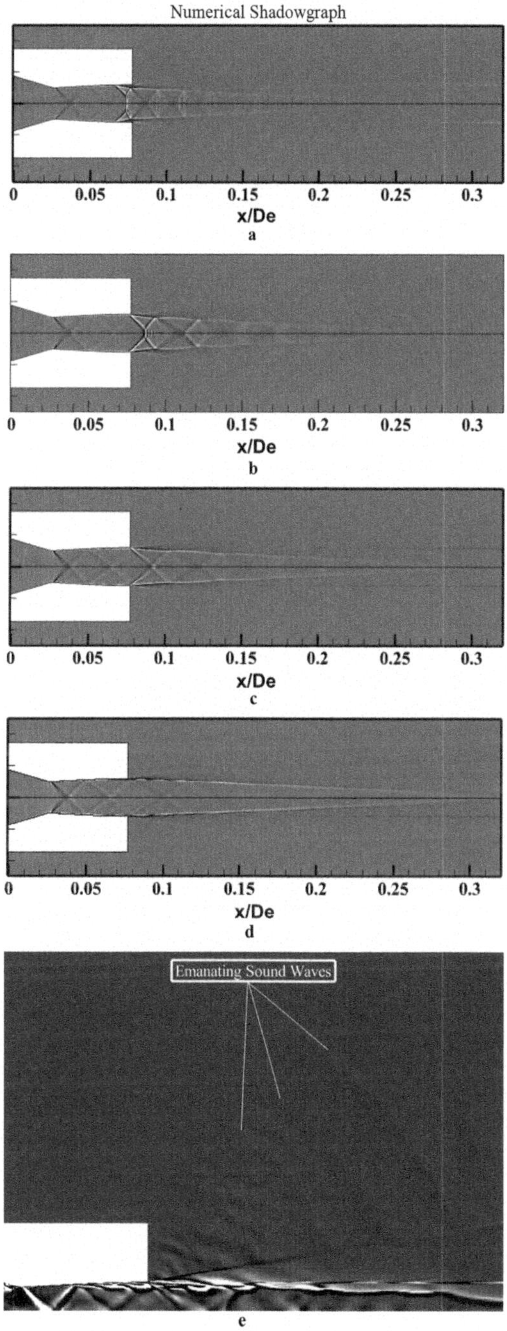

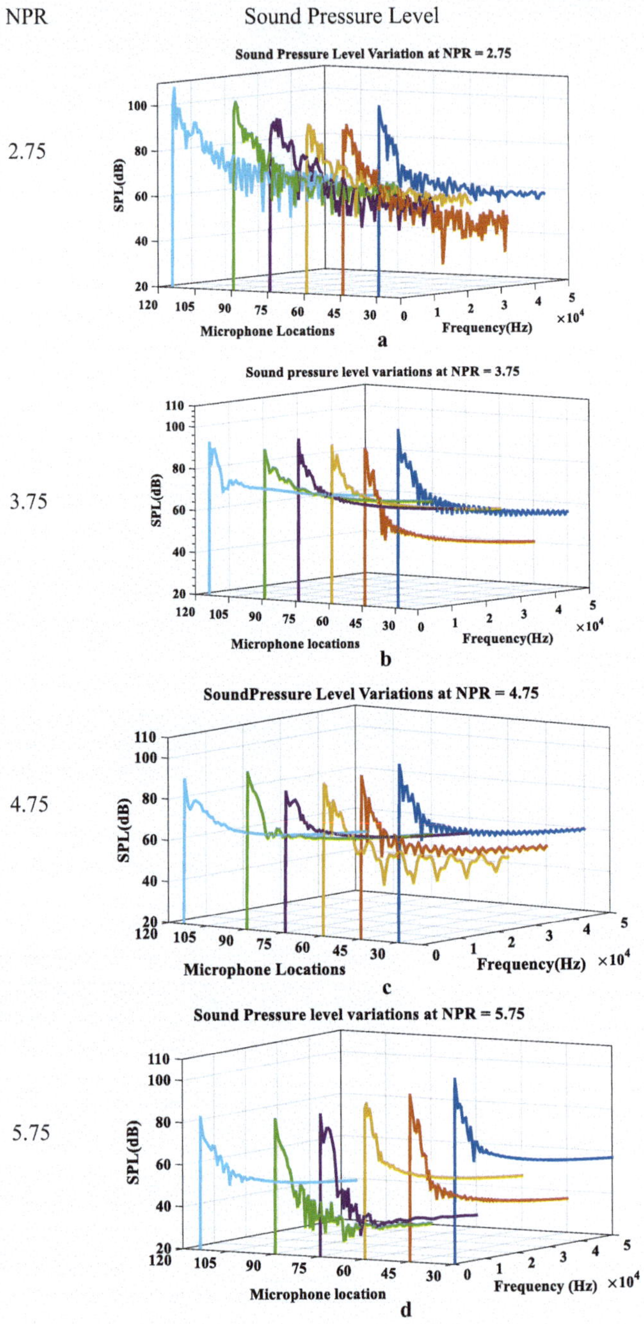

Fig. 6 Sound pressure level variations for NPR values from 2.75 to 5.75

of 90 dB and a minimum of 50 dB, a notable pump can be seen between 2400 to 5800Hz. Hence, it can be concluded that an increase in NPR shows a decrease in shock-associated noise in the upstream quadrant, but in the aft quadrant mixing noise tends to increase due to a higher Mach number floe at the nozzle exit.

Figure 6d shows the SPL characteristics of NPR 5.75, microphone located at 30° exhibits an SPL maximum of 107 dB as well as a minimum of 57 dB at the end of the frequency spectrum (i.e., 50000 Hz). Also, notable fluctuation can be seen all along the frequency scale. Similarly, at 45° microphone records the SPL maximum of 99 dB and a minimum of 36 dB. Furthermore, a notable second peak can also be observed at 3420 Hz for the respective SPL value of 86 dB. Likewise, for the 60° position maximum and minimum SPL values are 93 and 46 dB, and the second observable peak is happening at 2440 Hz for the SPL value of 90 dB. Also, SPL fluctuates up to 9700Hz then falls monotonically along the frequency spectrum. At 75° witnessed SPL maximum and minimum are 87 and 25 dB, respectively. The major peak is observed between 1950 to 6340Hz, and secondary fluctuation can be witnessed between 6348 to 18500Hz. At 90°, recorded maximum and minimum SPL values are 83 dB and 20 dB respectively, which is the lowest among the other microphone locations. Finally, the 115° position microphone exhibits 83 and 39 dB as maximum and minimum SPL values and a notable peak can be witnessed at 5370Hz for the SPL value of 67 dB. Hence, it can be inferred that at 75 and 90° locations average maximum and minimal SPL values are SPL is 85 and 20 dB, these two locations show the highest possible decay of SPL values from their respective maximum values.

4 Conclusion

The 2-D axisymmetric model is selected to investigate the aerodynamic and acoustic features of a supersonic jet exiting with the design Mach number 1.8. Mainly NPR values are varied between 2.75 to 5.75, such that the jet undergoes from highly over-expanded expansion to optimal expansion. Acquired shadowgraph images show the complex flow features associated with the highly over-expanded jet at NPR of 2.75 and 3.75. These shock structures can incubate intense interaction between shock cells and turbulent structures within the jet plume, which can enable the jet to emit intense shock-associated noise in the upstream direction of the jet. The sound pressure level variation for the NPR of 2.75 experiences a higher magnitude at the microphone locations of 90° and 115°. In addition, from all six microphone locations sound pressure level fluctuation is intense all along the frequency scale only for the NPR of 2.75, but for the other NPR values the same cannot be observed from the available graph. Similarly, for the other NPR values like 4.75 and 5.75 sound pressure level magnitude increases in the downstream direction where turbulent mixing noise tends to dominate. Because at higher NPR values jet tent experiences a relatively higher Mach number, which can allow the downstream propagating turbulent structure to propagate relatively at a higher speed. The sound pressure level variation for the higher NPR values also shows the increase in SPL magnitude for the corresponding microphone locations of 60 to 30°.

References

Andersson, N. and Eriksson, L.-E. 'Prediction of Flowfield and acoustic signature of a coaxial jet using RANS-based methods and large-eddy simulation,' International Journal of Aeroacoustics, 7(1), pp. 23–40 (2008).

Barbarino, M. et al. 'Combined CFD-stochastic analysis of an active fluidic injection system for jet noise reduction,' Applied Sciences, 7(6), p. 623 (2017).

Bogey, C., Bailly, C. and Juvé, D. (2003) 'Noise investigation of a high subsonic, moderate Reynolds number jet using a compressible large eddy simulation', Theoretical and Computational Fluid Dynamics, 16(4), pp. 273–297 (2003).

Chen, S., Gojon, R. and Mihaescu, M. 'High-temperature effects on aerodynamic and acoustic characteristics of a rectangular supersonic jet,' 2018 AIAA/CEAS Aeroacoustics Conference [Preprint] (2018).

Henderson, B. and Bridges, J. 'An MDOE investigation of chevrons for Supersonic Jet Noise Reduction,' 16th AIAA/CEAS Aeroacoustics Conference [Preprint] (2010).

Mankbadi, R.R., Hayer, M.E. and Povinelli, L.A. 'Structure of supersonic jet flow and its radiated sound,' AIAA Journal, 32(5), pp. 897–906 (1994).

Martelli, E., Nasuti, F. and Onofri, M. 'Numerical calculation of FSS/RSS transition in highly overexpanded rocket nozzle flows,' Shock Waves, 20(2), pp. 139–146 (2010).

Massey, S.J., Thomas, R.H., Abdol-Hamid, K.S., and Elmiligui, A.A., Computational and Experimental FlowField Analyses and Pylon Interaction, AIAA Paper 2003-3212 (2003).

Munday, D. et al. 'Flow structure of supersonic jets from conical C-D nozzles,' 39th AIAA Fluid Dynamics Conference [Preprint] (2009).

Munday, D. et al. 'Acoustic effect of chevrons on supersonic jets exiting conical convergent-divergent nozzles,' AIAA Journal, 50(11), pp. 2336–2350 (2012).

Nonomura, T. et al. 'Large eddy simulation of acoustic waves generated from a hot supersonic jet,' Shock Waves, 29(8), pp. 1133–1154 (2019).

Tam, C.K. 'Supersonic Jet Noise,' Annual Review of Fluid Mechanics, 27(1), pp. 17–43 (1995).

Tide, P.S. and Babu, V. 'Numerical predictions of noise due to subsonic jets from nozzles with and without chevrons,' Applied Acoustics, 70(2), pp. 321–332 (2009).

Improvement of Aerodynamic Characteristics at Leading-Edge of a High-Speed Train

Yosuke Shiromoto, M. M. A. Alam, and Shinya Mashimo

Abstract Tunnel micro-pressure waves are one of the most important environmental issues associated with the aerodynamics of current high-speed trains in Japan. When a train enters a tunnel, compression waves are formed. This pressure waves then travel through the tunnel at the speed of sound, and finally, they radiate as pulsed pressure waves when they reach the tunnel exit; the larger the peak value, the louder the sound impact becomes near the tunnel mouth. In this study, an experimental work was conducted with a new reduced-size model of a train car and a tunnel model, and in this experiment, running experiments were performed to reproduce the formation of micro-pressure waves in the tunnel. The pressure inside the tunnel was measured, and finally, the data was analysed using a data acquisition software. Moreover, a three-dimensional numerical experiment replicating the experimental conditions was carried out to simulate the flow around a high-speed train running through a tunnel. The rhoPimpleFoam solver, which uses the flexible PIMPLE algorithm for time-resolved, pseudo-transient simulations, in OpenFOAM was used for simulating the flow field.

Keywords Acoustic noise · Compressible flow · Micro-pressure wave · Environmental impact

1 Introduction

Currently, high-speed trains are being developed throughout the world as a means of transportation that can transport people and goods at a high speed and in large capacity, and they are functioning well as one of the main means of transportation for

Y. Shiromoto
Graduate School of Engineering, Osaka Sangyo University, Osaka, Japan

M. M. A. Alam (✉) · S. Mashimo
Department of Mechanical Engineering for Transportation, Osaka Sangyo University, Osaka, Japan
e-mail: alam@tm.osaka-sandai.ac.jp

G. Rajesh et al. (eds.), *Proceedings of the 15th International Symposium on Experimental and Computational Aerothermodynamics of Internal Flows*,
https://doi.org/10.1007/978-981-96-4082-9_10

long distances. In particular, the demand for railroads has been increasing in recent years, especially in Europe, due to the growing environmental awareness and the tendency to avoid the use of airplanes and automobiles as modes of transportation. In addition, the growth of emerging economies is creating demand for high-speed inter-city travel in these regions as well. High-speed train construction projects are now underway all over the world, and the marketing of standards for new railway lines is also intensifying, with countries developing higher-performance, less expensive high-speed rail systems. High-speed railroad construction projects are now underway all over the world, and the push for standardization of new railway lines is also gaining momentum, as countries are developing high-performance, low-cost, high-speed rail systems.

High-speed rail is defined internationally as a high-speed railroad that operates at speeds of 250 km/h on new high-speed lines and 200 km/h or more on improved conventional lines. In Japan, according to Chap. 1, Sect. 2 of the National Shinkansen Railway Development Act, a Shinkansen railway is defined as a railway that runs at least 200 km per hour in its main section. Although there is no excessive competition for high-speed trains in the world these days, trains are now operated at around 300 km/h consistently, and passenger trains are expected to be operated at 300 km/h on high-speed railroads to be built in the future, including those in emerging countries. In Japan, the trend toward higher speeds on major trunk lines has begun to stabilize, and the construction of sub-transit lines with a maximum speed of 260 km/h is expected in the near future.

However, one of the drawbacks of high-speed trains is that they cause vibration and noise pollution in the areas nearby, which increases with the higher speed of the operation. In particular, tunnel micro-pressure waves are the biggest problem in the Shinkansen system due to Japan's mountainous terrain and narrow tunnel cross-sections that are designed to reduce construction costs. As mention earlier, the tunnel micro-pressure waves are one of the most important environmental issues associated with the aerodynamics of current high-speed trains in Japan. When a train enters a tunnel, compression waves are formed. This pressure waves then travel through the tunnel at the speed of sound, and finally, they radiate as pulsed pressure waves when they reach the tunnel exit (Kurita 2012), as illustrated schematically in Fig. 1; the larger the peak value, the louder the sound impact becomes near the tunnel mouth. One of the important countermeasures for this adverse effect is to increase the length of the railcar tip and sharpen its shape. However, the magnitude of the micro-pressure wave is highly dependent on the train speed, and there is a limitation to the countermeasure using a sharp-edge geometry, therefore, a new method of countermeasure is desirable.

In this study, a new reduced-size model of a train car and a tunnel model were constructed, and the running experiments were conducted to reproduce the formation of compressional waves in the tunnel, which is the cause of micro-pressure waves. The pressure fluctuation in the tunnel model was measured using pressure sensors, and an attempt was made to verify the effect of the shape by a technique to improve the aerodynamic characteristics of the vehicle frontal nose, regardless of the effect of the sharp shape and any improvement of the shape, and finally, the validity of the

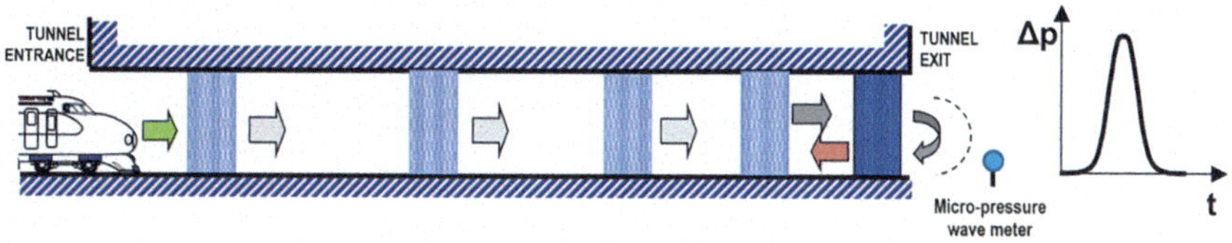

Fig. 1 A figure caption is always placed below the illustration. Short captions are centered, while long ones are justified. The macro button chooses the correct format automatically

experiment and the effect by the shape of the model were evaluated. Moreover, a Computational Fluid Dynamics (CFD) analysis replicating the experimental conditions was performed to simulate the flow around a high-speed train running through a tunnel.

2 Experimental Work

2.1 Experimental Apparatus

Vehicle Model. A 1/10 scale model of the vehicle as shown in Fig. 1 was fabricated for the purpose of this research and to reduce the weight of the vehicle. The entire framework of this vehicle model is made of aluminum alloy, and the outer panel is also made of aluminum composite plate, attached with cloth tape. In addition, a cylinder with a diameter of 150 mm was placed inside the vehicle from the tip to the end in order to provide ventilation holes for later use. The nominal height and length of the model car are 0.41 and 2.0 m, respectively.

Tunnel Model. Figure 2 shows the tunnel model that was fabricated using aluminum composite plates. The gaps were filled with fabric tape to make it airtight, and L-shaped metal fittings were fixed inside the tunnel model to stabilize the entire tunnel model. Here, the nominal height of the tunnel is 0.55 m, while the maximum length of the model tunnel is 14.56 m.

Fig. 2 Train car model

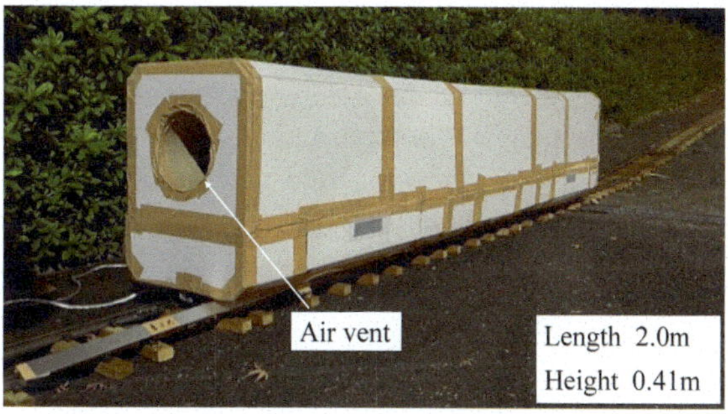

Fig. 3 Tunnel model

Models of the Modified Vehicle Frontal Nose. In this study, two modified models of the vehicle frontal nose shown in Fig. 3 (hereafter referred to as "nose models") were prepared and installed at the tip of a vertically truncated blunt-head vehicle model. The short model on the left is referred to as "nose model 1" and the long model on the right as "nose model 2" because these two nose models have different horizontal lengths when observed from the side. The horizontal length of the nose model 1 is 0.168 m and that of the nose model 2 is 0.29 m. Moreover, the ventilation holes were drilled in such a way that the position of the holes are aligned with that of the vehicle models mentioned above.

Measuring Instrument. In this experiment, the pressure inside the tunnel model was measured using a measurement system consisting of a transducer controller (EDX-10B), transducer voltage measuring instrument (EDX-12A), input adapter (UI-52A), and micro differential pressure sensor (PDV-10GA) all manufactured by Kyowa Dengyo, and then the data was analyzed using a data acquisition software (DCS-100 A). In addition, a track marker with a switch at 1 m from the rail edge was used to detect the vehicle position and to calculate the speed of the model train.

3 Experimental Procedure

Figure 4 shows the experimental configuration. In this experiment, the vehicle model was accelerated from a downhill slope and travelled through a tunnel. The actual acceleration distance was 7.5 m. In order to calculate the vehicle speed at the time of tunnel entry from the vehicle position, a rail with markers was installed in front of the tunnel entrance. The tunnel micro-pressure wave generated at the moment the accelerated vehicle model entered the tunnel was measured by placing a pressure introduction pipe 0.5 m from the tunnel entrance in the direction of the vehicle's travel. The model car was tested under six different conditions, with and without ventilation holes, and under three different nose configurations.

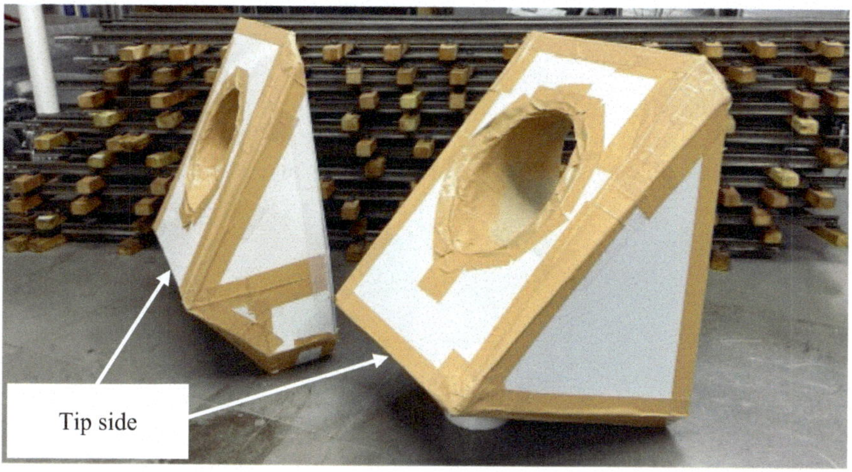

Fig. 4 Two models of the nose of the vehicle

4 Computational Analysis

4.1 Governing Equations

In this study, The Navier–Stokes equations for turbulent compressible flows are used, which are composed of a continuity equation (mass conservation law), a momentum conservation law, and an energy conservation law, and they are expressed by the following equations.

$$\frac{\partial \rho}{\partial t} + \nabla \cdot (\rho U) = 0 \tag{1}$$

$$\frac{\partial \rho U}{\partial t} + \nabla \cdot (\rho U U) = \nabla p + \nabla \underline{\tau} \tag{2}$$

$$\frac{\partial (\rho E)}{\partial t} + \nabla \cdot [U(\rho E)] + \nabla \cdot q + \nabla \cdot \left[(pI - \underline{\tau}) U \right] = 0 \tag{3}$$

where ρ is the density, t is an independent variable representing time, $\boldsymbol{U}$ is the velocity vector, and $\boldsymbol{I}$ is the unit tensor. The viscous tensor $\underline{\tau}$, the energy E, the heat flux $\boldsymbol{q}$ from Fourier's law are given by the following Eqs. (4) through (6).

$$\underline{\tau} = 2\mu dev(\underline{D}) \tag{4}$$

$$E = e_s + \frac{1}{2} U \cdot U \tag{5}$$

$$q = -\lambda T \quad (6)$$

Here, μ is the kinematic viscosity, λ is the thermal conductivity, and T is the temperature. The deformation gradient tensor $\underline{\mathbf{D}}$ is given by Eq. (8), where dev($\underline{\mathbf{D}}$) is its deviation, the sensible energy e_s is given by Eq. (9), and where h_s is the sensible enthalpy.

$$\underline{D} = \frac{1}{2}\left[\nabla U + (\nabla U)^T\right] \quad (7)$$

$$e_s = h_s - \frac{p}{\rho} \quad (8)$$

The pressure p is given by the equation of state.

$$p = \rho RT \quad (9)$$

Here, R is the gas constant.

The standard k-ε model (Launder and Sharma 1974; OpenFOAM: User guide, The OpenFOAM Foundation Ltd., UK (2022) for turbulence, is an eddy viscosity model which is a class of turbulence models used to calculate the Reynolds stresses, was used to simulate turbulent stresses. It consists of two transport equations representing the turbulent flow characteristics. The first transported variable is turbulent kinetic energy, k. The second transported variable in this case is the turbulent dissipation, ε. It is the variable that determines the scale of the turbulence, whereas the first variable, k, determines the energy in the turbulence.

4.2 Numerical Methods

The simulations were performed using OpenFOAM, a free and open source software for CFD with a large number of users in the commercial and academic worlds. The rhoPimpleFoam (OpenFOAM: User guide, The OpenFOAM Foundation Ltd., UK (2022) solver, a semi-implicit method for solving unsteady laminar or turbulent flow of compressible flows, which uses the flexible PIMPLE algorithm for time-resolved, pseudo-transient simulations in OpenFOAM was employed in this study for simulating the flow.

4.3 Numerical Methods

Figure 5 shows the schematics of the computational domain and the location of the model vehicle. The dimensions of computational space are 12.5L, 17W, and 12H when the vehicle length, width, and height are L, W, and H, respectively. The reason for considering the large computational domain and cylindrical shapes at the tunnel inlet and exit is to get a better computational stability. The non-slip condition is used for the car body and the ground, and the tunnel wall and the tunnel gate, while the atmospheric condition is used for the upstream and downstream of the domain. A 3D image of the computational domain is assumed to be as shown in Fig. 6, and the detailed configuration of the railcar is shown in Fig. 7.

The computational domain was meshed with SALOME (SALOME: Documentation, Open Cascade EDF CEA, France 2022), which is an open source software and integrated platform for numerical analysis. The mesh is created finer inside of the tunnel, around the railcar, at tunnel inlet and exit, as shown in Fig. 8, in order to observe the flow behavior in detail, in particular, inside the tunnel.

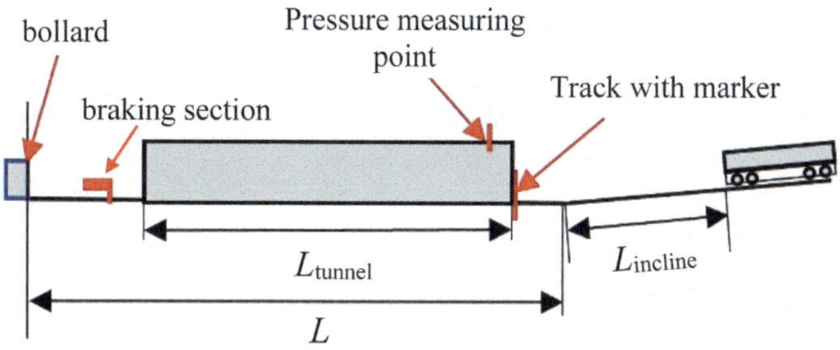

Fig. 5 Outline of train aerodynamic characteristics experiment

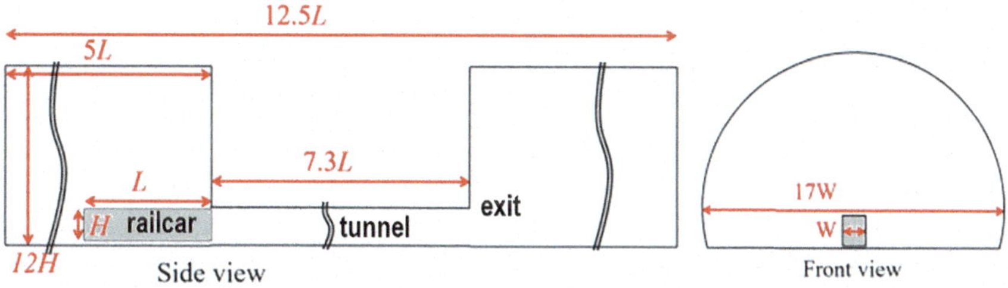

Fig. 6 Schematics of the computational domain

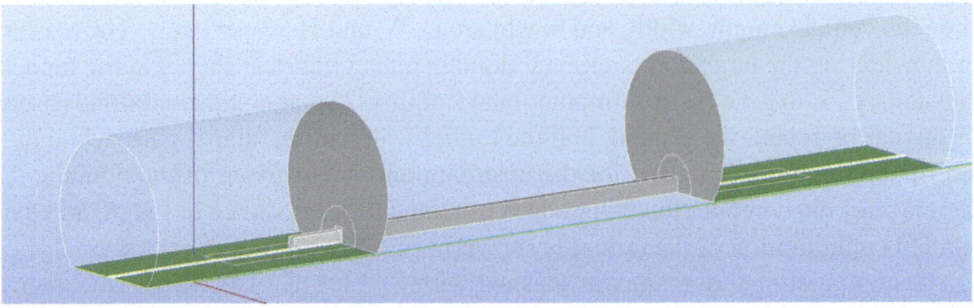

Fig. 7 A 3D view of the computational domain

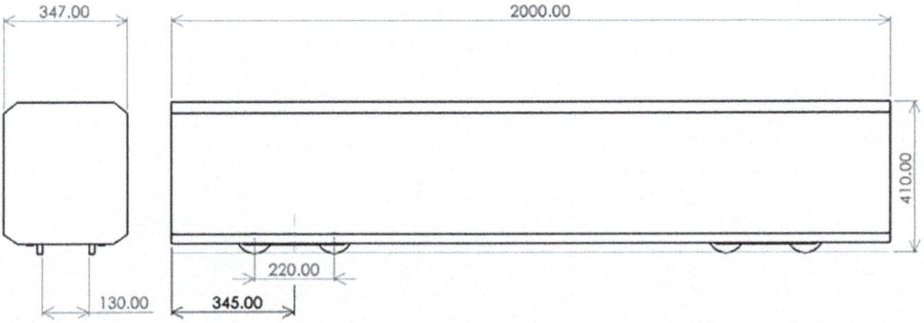

Fig. 8 Detailed configuration of the gabled roof car

5 Results and Discussion

5.1 Experimental Results

Compression Waveforms for Different Nose Models. Figure 9 compares the compression waveforms of the three shapes without the vent holes. The legend shows the velocity of the model train at the time it enters the tunnel, the horizontal axis is the elapsed time, which is set to 0 s at the moment the nose of the model train (excluding the lead model) enters the tunnel, and the vertical axis shows the measured pressure in the tunnel. In all cases, the compression waves caused the pressure to increase to a maximum value, followed by a slight increase or decrease, and then a large decrease at around 0.2 s. The compression waves propagate at the speed of sound, reflect at the tunnel exit, and return to the entrance as expansion waves, which reach the measurement point at 0.08 s. Thereby, the pressure is assumed to increase to a maximum value and then decrease slightly. The magnitude of the micro-pressure waves depends on the magnitude of the maximum pressure gradient of the compression waves at the entrance, and the sharp-nosed shape of the high-speed train is intended to reduce it. The magnitude of the wave front gradient is the smallest for the nose model 2 and the largest for the blunt profile, which confirms that the pressure gradient is reduced with a longer tip in this experiment as well.

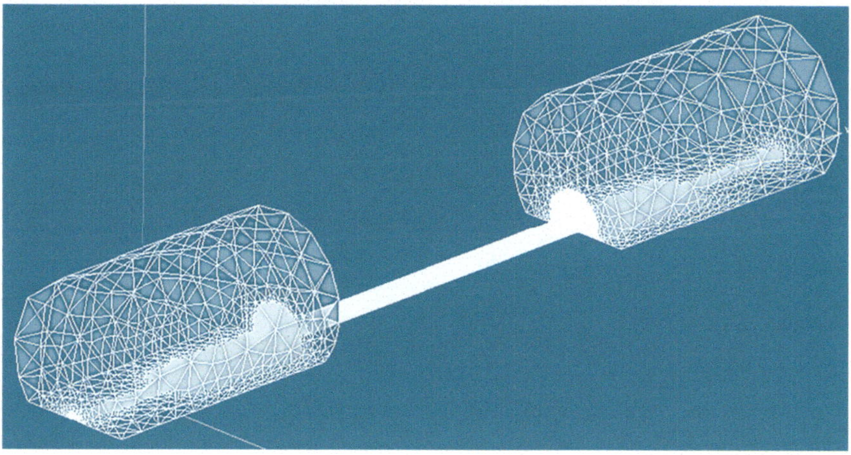

Fig. 9 Typical mesh

Effect of Nose Shape. In the compression waveform shown in Fig. 9, the pressure gradient was more gradual with the nose model than with the blunt shape. This is thought to be due to the fact that a longer time was required for the pressure to reach its maximum value from the start of the pressure increase due to the presence of the nose model. The main reason for the difference in pressure for the blunt shape and for the nose model 1 and nose model 2 is thought to be that the nose model 2, which has longer sides and a smaller slope angle, has less separation around the vehicle model, resulting in smoother airflow in the tunnel .

Compression Waveform with and without Air Vent. The following compression waveforms are compared for each of the three nose models with and without vents. Figure 10 shows the pressure history for a blunt nose shape, while Figs. 11, 12 show the pressure history for nose model 1 and nose model 2, respectively, for both the cases of with and without air vent at the tip of the train car. As seen from Fig. 11, the pressure gradient becomes slower due to the vents, while no significant change was observed in Figs. 10, 12.

Influence of the Installation of the Air Vents. Figure 11 shows that the pressure reduction effect of the vents in nose model 1 was greater than that of the other shapes, and the maximum value of the pressure gradient was also reduced. This result can be attributed to the fact that the air flows smoothly into the vents to some extent, reducing the amount of compression effect. In contrast, the compression waveforms in Figs. 10, 12 show almost no difference depending on the presence or absence of the vents. In Fig. 11, the effect was observed, which is considered to be due to differences in the shape of the inlet to the main flow and the occurrence of separation
.

Fig. 10 Compression waveforms by nose shape

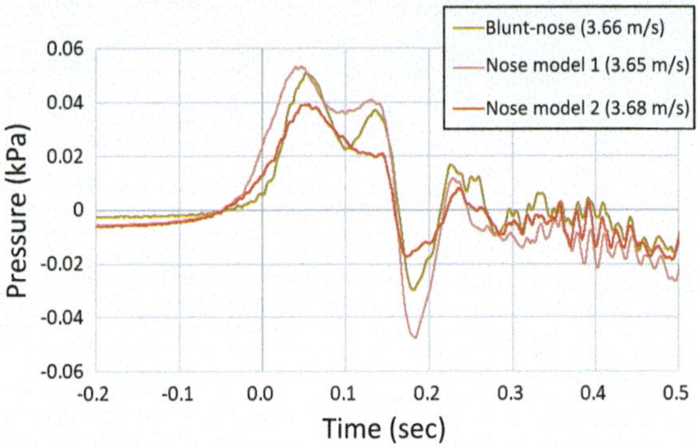

Fig. 11 Compression waveforms with air vents in the blunt head shape

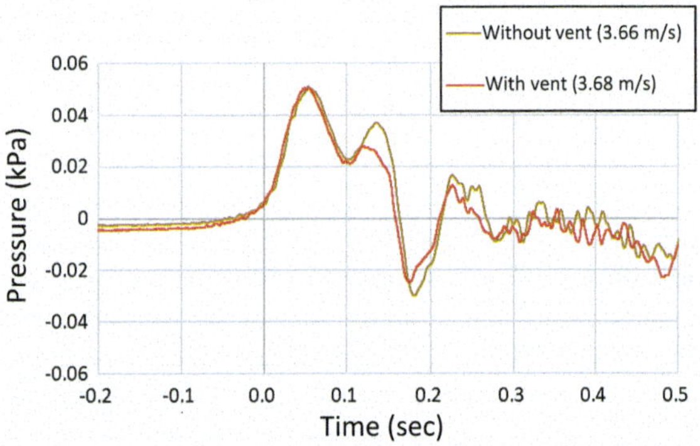

Fig. 12 Compression waveforms with air vent for nose model 1

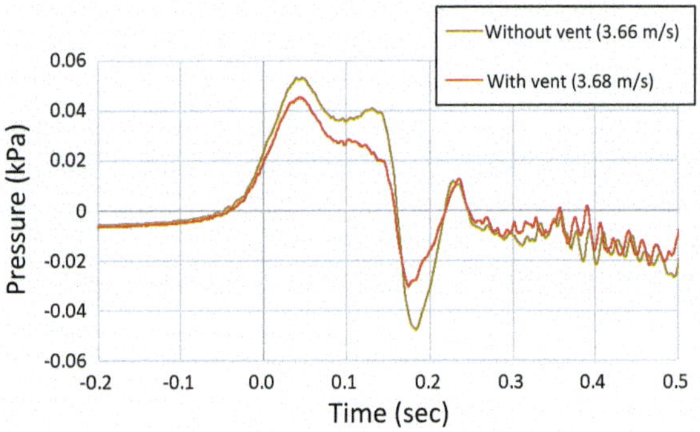

5.2 Computational Results

Figure 13 shows the pressure contours at different times, while Fig. 14 shows the velocity contours. In this study, the railcar was kept stationary in a position and the air was driven from the opposite direction. Moreover, the position of the railcar was changed along the line to simulate the flow field under different positions of

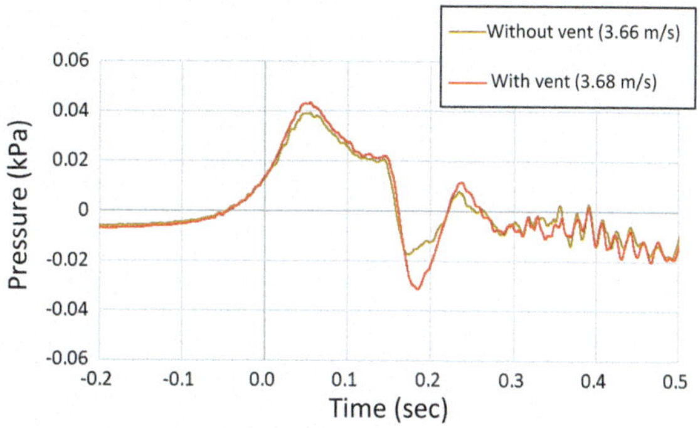

Fig. 13 Compression waveforms with air vent for nose model 2

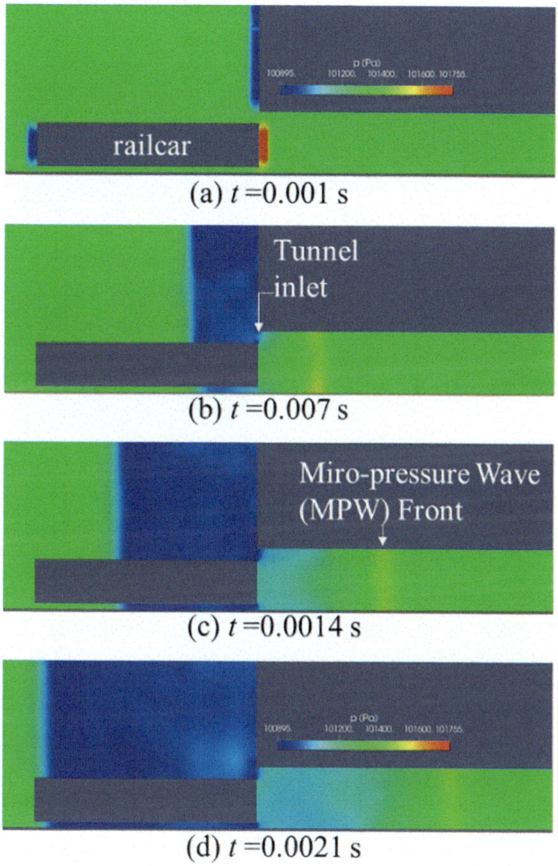

Fig. 14 Pressure contours

Fig. 15 Velocity contours

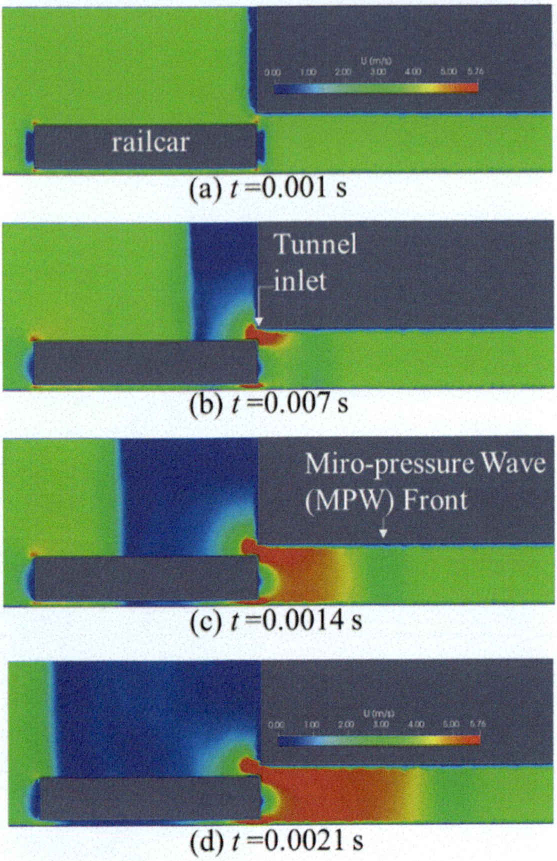

the railcar. As seen from Fig. 13, at the time when the railcar enters the tunnel a micro-pressure wave is generated in front of the car, and then, it propagates towards the tunnel exit. Here, the speed of the propagating is much higher than the speed of the railcar; the speed of the microwave is more than 300 km/h (Fig. 15).

6 Conclusions

The compression waves in a tunnel generated at the moment a train enters the tunnel were evaluated by running experiments and numerical simulation. The effect of railcar nose shape and ventilation holes were evaluated. The results confirmed that the maximum value of the pressure gradient can be reduced by introducing nose model to the railcar model. However, no significant change was observed in the case of the blunt nose shape or in the case of the nose model 2 with the air vents installed.

In the future, it is necessary to study the improvement of the efficiency of air flow into the vents.

References

Jones, W. P., Launder, B. E.: The Prediction of Linearization with a Two-Equation Model of Turbulence, Int. J. Heat Mass Trans., vol. 15, pp. 301-314 (1972).
Kurita, T.: Adjustment of Entrance Hood for Tunnel Micro-Pressure Waves, Report of the Railway Technical Research Institute (RRR), vol. 69(8), pp. 32 (2012).
Launder, B. E., Sharma, B. I.: Application of the Energy Dissipation Model of Turbulence to the Calculation of Flow Near a Spinning Disc, Let. Heat Mass Trans., vol. 1(2), pp. 131-138 (1974).
OpenFOAM: User guide, The OpenFOAM Foundation Ltd., UK (2022).
SALOME: Documentation, Open Cascade EDF CEA, France (2022).

Characterization of an Anechoic Chamber for Conducting Aeroacoustic Experiments at IIT Kanpur

Ambica Singh, Arun Kumar Perumal, and Arnab Samanta

Abstract The Jet Acoustics and Flow Control (JAFC) Laboratory is a mixed-use facility commissioned at IIT Kanpur. For jet acoustics study, the laboratory contains an open jet facility where nozzles of varying diameters can be mounted, which is located inside an anechoic chamber. The paper presents an overview of the calibration results of this jet acoustics facility: both the acoustics and aerodynamic characterizations are presented for a Mach 0.3 jet. The acoustics measurements were performed using free-field microphones, while the aerodynamic characterization used a hotwire anemometer. These benchmark experiments for the Mach 0.3 jet match with the results reported in the literature. The calibration results highlight the lower cut-off frequency of the open jet facility to be 500 Hz.

Keywords Anechoic chamber · Jet noise · Open jet facility · Strouhal number · Cut-off frequency

1 Introduction

Jet noise from civilian aircraft engines is predominant during take-off, adversely impacting the community near airports and thus calling for strict noise regulation measures. Correct physical understanding of jet noise requires twin approaches: first, turbulent dynamics in the jet shear layers, downstream of the jet potential core breakdown, need to be captured, and second, the spherically-symmetric small-amplitude acoustic perturbations from the corresponding vortical structures need to be measured. To accurately calculate these relatively small perturbations, an anechoic chamber is required that is designed to simulate a free-field environment, eliminating any background noise level and vibrations. Acoustic insulation lining in the form of wedges inside such an anechoic chamber absorbs sound reflections, thereby preserving the directivity patterns from the radiated aerodynamic sound sources over

A. Singh (✉) · A. K. Perumal · A. Samanta
Department of Aerospace Engineering, Indian Institute of Technology Kanpur, Kanpur, India
e-mail: ambica20@iitk.ac.in

G. Rajesh et al. (eds.), *Proceedings of the 15th International Symposium on Experimental and Computational Aerothermodynamics of Internal Flows*,
https://doi.org/10.1007/978-981-96-4082-9_11

a wide frequency range (Ahuja 2003). The cut-off frequency is defined as that frequency above which the test environment inside the chamber must be anechoic at the required range of frequencies. Additionally, it is required that the flow quality at the nozzle entrance, established through stagnation pressure and temperature distributions within the plenum chamber, be isotropic. Also, the core turbulence intensity level must be low (typically $< 0.1\%$) to minimize the rig noise and propagation of upstream disturbances while acquiring the static noise data (Viswanathan 2003). Generally, for successful aeroacoustic measurements, the anechoic chamber must be designed to permit measurements up to the geometric far-field from a suitably high-Mach number jet flow.

The paper describes the calibration of such an anechoic chamber, now commissioned at the Jet-Acoustics Flow Control (JAFC) Laboratory, Department of Aerospace Engineering, Indian Institute of Technology (IIT) Kanpur. It is an isolated, temperature-controlled room, preserving the characteristics of high-speed jets issuing from various nozzle configurations up to Mach 2. The objective here is to perform a critical investigation of jet mixing characteristics potentially responsible for the far-field aeroacoustics. For a high-speed axisymmetric jet, there are two well-known universal similarity spectra (Tam et al. 1996), one for the noise from large-scale turbulent structures or instability waves of jet flow and the other corresponding to the noise from jet flow fine-scale structures. The present facility is designed to obtain spectral data over a frequency range >10 kHz (the highest frequency for aircraft certification) to facilitate the comparison of scaled model data with actual engine noise data. The paper describes the aerodynamic and aeroacoustic qualification tests of the open jet facility at Mach $\geq$0.3 and Re $\approx 10^5 - 10^6$ that ensure the acquisition of high-quality jet noise spectra, capable of predicting the two distinct source mechanisms of jet noise.

2 Experimental Facility at the JAFC Laboratory

2.1 *Description of the Open Jet Facility*

The experimental facility at the JAFC lab is shown schematically in Fig. 1. The open jet facility provides a compact jet rig for low subsonic to supersonic jet experiments. This facility is designed for the experimental examination of the flow field and acoustic near and far-field generated by free and manipulated jets.

As described in Fig. 1, atmospheric air is first compressed by dual-reciprocating compressors to 8 bar and stored in storage tanks of 3000 ft^3 volume. The air is then fed to the air dryer, which uses a refrigerant circuit and heat exchangers to pre-cool the air to condense and remove the excess moisture. The dried air enters the filter assembly, where a pre filter removes the droplets of oil, any remaining water, and miscellaneous dust particles. Next, the dried air undergoes a filtration process to remove particles smaller than 0.01 microns. This clean and dry air is then supplied to

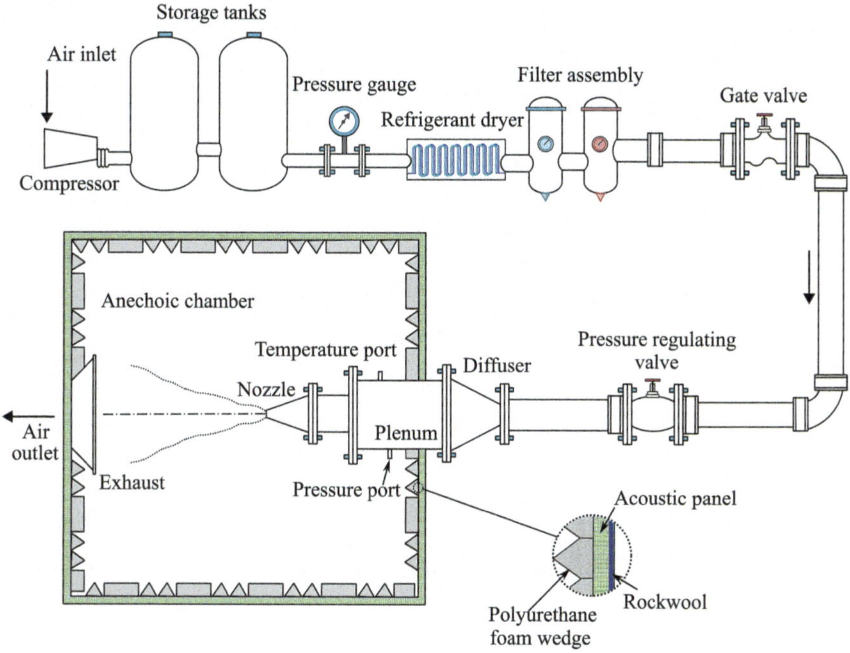

Fig. 1 Plan view of the open jet experimental facility and anechoic chamber at Jet-Acoustics and Flow Control (JAFC) Laboratory at Indian Institute of Technology, Kanpur

the open jet facility located inside the anechoic chamber, where a pressure regulating valve is used to maintain the desired stagnation pressure. The facility dimensions include a 1 m long mixing pipe, a wide-angled diffuser with an outer diameter of 220 mm, a cylindrical plenum chamber of 200 mm inner diameter that is 500 mm long between the two flanges, finally terminated by a nozzle at the exit. The nozzle contraction profile was designed based on a 5th order polynomial, which minimizes flow non-uniformity (Perumal and Zhou 2018). A wire mesh is placed inside the plenum chamber to smoothen out the unsteadiness in the upstream flow, thus making the flow at the nozzle entrance nearly isotropic. The tappings for stagnation pressure and temperature monitor the flow uniformity at the nozzle entrance using a pressure scanner and K-type thermocouple, respectively (Viswanathan 2003). At the other end, the jet flow exhausts into a square cross-section of 1.4 m × 1.4 m acoustically-treated diffuser that ensures smooth exit of the turbulent jet post-entrainment, thus preventing any flow recirculation.

2.2 Design of the Anechoic Chamber

Figure 1 also provides a plan view of the anechoic chamber with workspace dimensions of 4 m × 4 m × 2.3 m. Since, temperature, pressure, humidity, and background noise disturb the uniform sound radiation and propagation, an anechoic chamber is appropriately designed to eliminate such effects. Here, the chamber's 0.4 mm thick acoustic panel contains rockwool insulation, which along with 0.1 mm long wedge-shaped polyurethane foam structures absorb all the sound reflections and any external noise that can contaminate the small-amplitude sound measurements. Following recommendations based on ISO 3745 standard, the parametric height of the wedge needs to be larger than $1/4^{th}$ of wavelength at the lower cut-off frequency (Beranek and Sleeper 2005). These wedges forming the chamber lining have a sound absorption coefficient of up to 0.99 at normal incidence, ensuring 99 % absorption of the incident sound energy. Also, the chamber is equipped with a three-dimensional automated traverse system run by DC stepper motors for independent linear movement along x, y and z orthogonal directions with a resolution of 0.1 mm, controlled via the LabVIEW software. The resolution is adequate for the minimum required traverse to capture the jet's turbulent statistics (Cunefare et al. 2003).

3 Calibration Experiments

3.1 Calibration of the Anechoic Chamber

The chamber is calibrated as per ISO 3745 and ANSI S12.35 qualification test, defining the anechoic chamber's standardized calibration procedures. A spherical wave generated by a sound source should propagate without getting reflected by any obstruction in its path. Therefore, calibration determines the lower cut-off frequency and maximum allowable radius between a test source and a measurement location where the inverse square law holds. Cut-off frequency, theoretically estimated using the 1/4th wavelength rule and length of the wedges, is the threshold value above which the chamber shows its anechoic behaviour.

Here, to calibrate the chamber, a speaker with a maximum frequency of 18 kHz along with two 1/4 "microphones were mounted at its center. These 1/4" free-field, pre-polarized PCB microphones (378C01) with 165 dB re 20 μPa dynamic range have a flat frequency response up to 100 kHz. The microphones, calibrated by Larson Davis CAL300 for sensitivity with the output level of 94 dB at 1 kHz, were connected to a Data Acquisition System (cDAQ 9178 & 24-bit NI-9232 module) via PCB 482C-series signal conditioner, which yields a sensitivity of each microphone as 2.0 mV/Pa. The microphone signals are filtered at a cut-off frequency of 256 kHz based on Nyquist criteria and then digitized using a NI-9232 module at a sampling rate of 512 kHz for a sampling time of 10 s, as decided based on data repeatability tests. One microphone traversed radially, while the other was kept stationary, thus monitoring the speaker. The stationary reference microphone provided a quality check on the sound pressure levels of the traversing microphone. The traversing mechanism ensured that the speaker and microphones were positioned along the chamber's mid-height, coinciding with the nozzle centreline. Further, a function generator (Scientific) generated the excitation tones f_e required for 1/3rd octave-band frequencies in the frequency range of 350 Hz – 14 kHz. During experiments, the speaker faced the traversing microphone to ensure the sound source was always omnidirectional. The microphone was traversed in the horizontal plane of the noise source along radial lines extending from the noise source exit plane to the chamber walls.

Figure 2 shows the result of the characterization of the present anechoic chamber for inverse-square law based on ISO 3745. The law suggests the sound intensity to decay by a factor of 6 dB on doubling the distance from the source (i.e. speaker). Here, ΔSPL is the difference between the measured sound pressure levels at subsequent microphone locations, and r_1/r_2 is the distance ratio between them. The variation of ΔSPL for $f_e \geq 500$ Hz falls along the theoretical curve as seen in Fig. 2. However, for $f_e = 400$ Hz it shows a significant deviation from the prescribed ISO 3745 bounds. The results conclude the present anechoic chamber at the JAFC lab to be anechoic for any frequency above 500 Hz.

Table 1 highlights the allowable deviations based on 1/3rd octave-band frequencies from measured data at discrete radial locations (Cunefare et al. 2003). The ISO 3745 standard lower and upper bounds express the measured sound pressure levels deviating from the theoretical inverse square law. Figure 3 demonstrates these deviations of measured data from theoretical inverse square decay at the 90° sideline angle, where

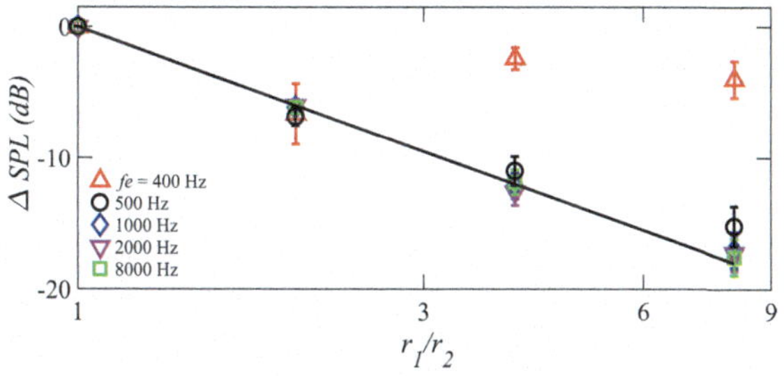

Fig. 2 Characterization of anechoic chamber based on ISO 3745 standard, shown as the solid line—while symbols indicate data at the indicated frequencies

Table 1 Maximum allowable differences in the anechoic chamber between measured and theoretical free-field levels as per ISO 3745 and ANSI S12.35 (Cunefare et al. 2003)

One-third octave band centre frequency (Hz)	Allowable difference (dB)
<630	±1.5
800–5000	±1.0
>6300	±1.5

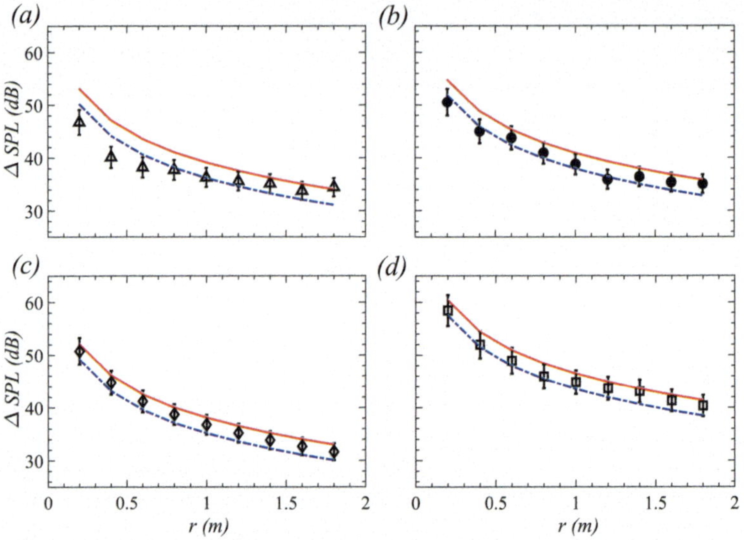

Fig. 3 Cut-off frequency analysis at **a** 400 Hz, **b** 500 Hz, **c** 1000 Hz, and **d** 8000 Hz at 90° sideline angle based on ISO 3745 deviations from the theoretical inverse square law decay. ISO 3745 lower and upper limits are evaluated based on formulation provided in Cunefare et al. (2003) and are indicated as dashed and solid lines, respectively, in each figure

the most significant amount of jet noise radiation directly impacts communities near airports.

The deviations are calculated for multiple excitation tones up to 14 kHz ensuring accurate acoustic measurements of jet noise at least up to 10 kHz (Viswanathan 2003), within the tolerance bounds of ISO 3745 standard. In Fig. 3, the solid lines represent the upper-bound ISO curve, and the dashed line is the lower-bound ISO curve. Uncertainty shown in each measured ΔSPL is based on several repeatability tests that were carried out. In Fig. 3a, ΔSPL at $f_e = 400$ Hz is clearly out of permissible deviations from free-field performance limits listed in Table 1, which reconfirms the cut-off frequency for the present anechoic chamber to be 500 Hz. In terms of non-dimensionalized frequency described by the Strouhal number $St = fD/\bar{U}_c$ (where D is the nozzle exit diameter and $\bar{U}_c$ is the mean jet centreline exit velocity), this cut-off frequency has $St \approx 0.036$ for the given nozzle for a Mach 0.8 jet, whose characteristics are evaluated later in this paper. Quite desirably, this value is 2.8 times lower than the lowest frequency ($St = 1$) used in the present experiments (Ahuja 2003). Also, the cut-off frequency analysis remained consistent even for other polar angles, not shown in the paper for brevity. Additionally, in Fig. 3, the maximum interference-free distance between source and microphone locations, also known as the "acoustic far-field", was found to be 100 D, extending up to near the wedge tips (Cunefare et al. 2003). The free-field environment thus simulated by the anechoic chamber was verified for upstream, sideline, and aft angles up to a maximum of 180° at an interval of 30° which span the entire range of observation angles about which jet noise predominantly radiates (Ahuja 2003). The calibration also established the symmetry of the axisymmetric jet along its centreline for spherical wave propagation in the acoustic far field. With the anechoic chamber calibrated, the low and high-frequency noise spectra from aerodynamic sources of jet turbulence can now be accurately measured and analyzed.

3.2 Calibration of the Nozzle

Instantaneous velocity distributions of low, medium, and high-speed subsonic jets can be acquired using a single platinum-plated-tungsten hotwire probe (55P11) of 5 μm diameter, 1.2 mm length, operated on a constant temperature anemometer circuit (Dantec Streamline Pro), as shown in Fig. 4a. It can measure the one-point and two-point statistics of the high turbulence intensity of cold, subsonic jets up to the self-similar regions. Here, the origin is defined at the centre of the jet nozzle. The hotwire in Fig. 4a was calibrated for the range of 0 – 30 m/s based on mean velocities measured using a standard Pitot static probe with an outer diameter of 3 mm, connected to the Testo-DP 440 digital manometer, placed adjacent to the hotwire near the centre of the nozzle exit. For higher velocities of 30 – 300 m/s, it was calibrated via the settling chamber pressure, measured using an Ethernet-based pressure scanner. The calibration coefficients were extracted from a 4th-order polynomial curve fit. Here, the hotwire signal was offset, filtered at a cutoff frequency of 5 kHz, amplified, and then digitized using a 12-bit NI-6143 A/D board at a sampling frequency of 10 kHz. It was also accompanied by a Dantec temperature probe to measure instantaneous local temperature fluctuations in its vicinity. In these experiments, the jet exit Mach number was defined using isentropic relations and nozzle exit pressure ratios with static pressure at the nozzle exit assumed to be equal to ambient pressure.

4 Benchmark Tests for Flowfield and Acoustic Measurements

Figure 4 shows the experimental arrangement of the jet rig mounted inside JAFC lab's anechoic chamber capable of performing flow-field and acoustic measurements from low subsonic to supersonic Mach numbers.

4.1 Hydrodynamic Nearfield Validation via Mach 0.3 Jet

Table 2 describes the operating conditions for the flow-field validation test. The test was carried out on a Mach 0.3 jet issued from a smoothly convergent nozzle of

Fig. 4 Experimental arrangement inside JAFC lab's anechoic chamber for **a** hotwire anemometry and **b** polar microphone array for acoustic measurement

Table 2 Description of exit parameters

Parameters (Symbols)	Magnitude
Nozzle exit diameter (D)	20 mm
Jet exit velocity (U_o)	103.5 m/s
Reynolds number (R_e)	1.4×10^5
Mach number (M)	0.3
Momentum flux/unit mass (M_o)	1.2566 m^4/s^2

exit diameter $D = 20\,\text{mm}$, (Cunefare et al. 2003) where the corresponding Reynolds number $Re = \bar{U}_c D/\nu$ was 1.4×10^5. Here, $\bar{U}_c$ is the jet centreline exit velocity (at $x^* = 0.05$), the overbar denotes time averaging, and ν is the kinematic viscosity of air. An asterisk denotes non-dimensionalization with D and U_c being the length and velocity scales, respectively. The near-field data of the Mach 0.3 jet was measured using a single hotwire probe. The signal was recorded for 60 s, ensuring u_{rms} to converge within 1% uncertainty.

Figures 5 and 6 show the streamwise mean velocity $\bar{u}^*$ and the fluctuations u^*_{rms} at two axial locations ($x^* = 1$ and 7), and characteristic decay along the jet centerline, respectively, which are compared against data reported by Iqbal and Thomas (2007) for a Mach 0.3 jet at $Re = 1.4 \times 10^5$. Few observations can be made. Firstly, the $\bar{u}^*$ at $x^* = 0.05$ shows a "top-hat" profile at the nozzle exit (not shown in the paper) with an initial turbulence level of 0.72%. The potential core of the Mach 0.3 natural jet is characterized by a low-turbulence level of about $< 0.1\%$. The mean radial profile at $x^* = 1$ has a top-hat velocity profile indicating a uniform, inviscid potential core flow wrapped by a thin axisymmetric shear layer. The increase in the fluctuating velocity component at the edge of the jet is due to the development of a thin shear layer just near the exit of the nozzle lip. Secondly, the variation of mean velocities $\bar{u}^*$ at $x^* = 1$ and $x^* = 7$ (Figs. 5a and 5b, respectively) agree well with the other study. The turbulent fluctuations u^*_{rms} at the jet core are larger near the exit (Fig. 5c), although by $x^* = 7$ when the jet would have been fully developed, this difference

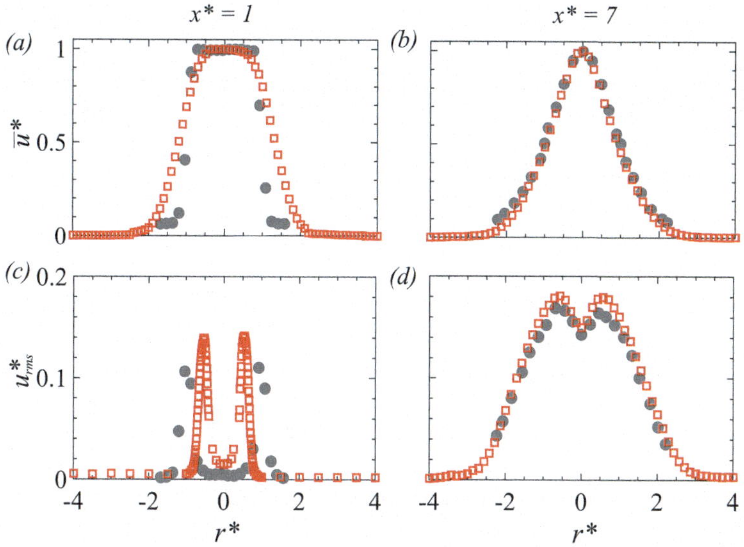

Fig. 5 Variation of $(a - b)$ mean velocity and $(c - d)$ turbulent intensity of the Mach 0.3 jet at x^* = 1 and 7. Present data (□); Iqbal and Thomas (2007) (●)

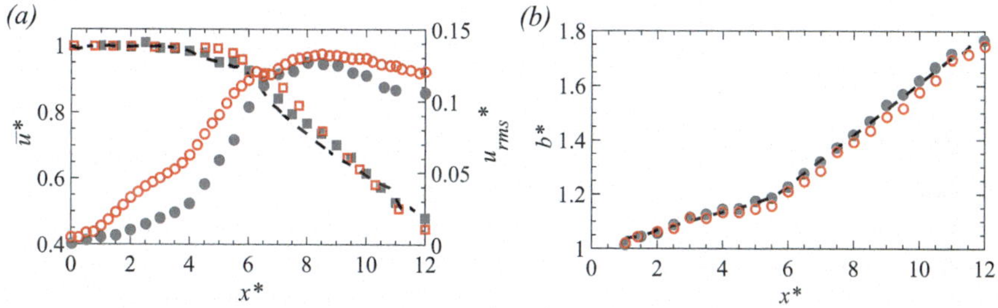

Fig. 6 **a** Centreline mean velocity decay of the Mach 0.3 jet: □, present results; ■, Iqbal and Thomas (2007); (– –), Crow and Champagne (1971) and the evolution of streamwise turbulent intensities: ○, present results; ●, Iqbal and Thomas (2007). **b** Centreline variation of jet half-width of the Mach 0.3 jet: ○, present results; ●, Iqbal and Thomas (2007); (– –), Crow and Champagne (1971)

largely disappears (see Fig. 5d). This trend is unsurprising, as several factors, such as surface roughness between the two facilities, may be at play in a developing jet just downstream of the nozzle exit. As one can see, the u^*_{rms} in the shear layer for the present case is larger than reported in Iqbal and Thomas (2007).

For a natural, axisymmetric jet with an initial top-hat and self-similar profiles, the mean centreline velocity profile is given by

$$\frac{U_o}{\bar{U}_c} = \frac{1}{B_u}\left[\frac{x}{D} - \frac{x_o}{D}\right], \tag{1}$$

where B_u is the decay rate constant and x_o is the streamwise location of the virtual origin of the jet. In the present investigation, $x_o{}^*$ is 2.5 with a characteristic decay constant range (B_u) of 4.96 to 5.6 which lies within the universal decay range (Crow and Champagne 1971). Figure 6 summarizes the streamwise evolution of jet width, mean centreline velocity, and the streamwise fluctuating velocity component.

The black dotted line represents the variation of $\bar{u}^*$ along the jet centreline for a jet at $Re = 1.03 \times 10^5$ from the experiments reported by Crow and Champagne (1971). Figure 6a also shows good agreement with Iqbal and Thomas (2007), Crow and Champagne (1971) with the jet potential core closing at $x^* \approx 3.5$. On the other hand, the variation of u^*_{rms} in the jet core over $1 \lesssim x^* \lesssim 6$ shows some deviation with Iqbal and Thomas (2007), which beyond $x^* \gtrsim 6$ starts showing good agreement, consistent with observations in Figs. 5c, d. In Fig. 6b, b^* is the jet half-width defined as the radial location where $\bar{u}^*$ is half of $\bar{U}_c$. Its magnitude is initially small but linearly grows beyond $x^* \approx 6$. Radial profiles in the self-similar region collapse onto a single curve when non-dimensionalized by these values of b^* (Abramovich et al. 1963). Therefore, these flow-field analyses reveal excellent agreement of jet turbulence and its mean data with the results obtained in other axisymmetric jet facilities in reference studies mentioned above. Further, the following section outlines the acoustic measurement performed at Mach 0.8 to establish the high precision of jet noise spectra.

4.2 *Acoustic Far-Field Validation via Mach 0.8 Jet*

Figure 4*b* shows the polar microphone array arrangement, aligned with the centre point of the nozzle at the same height as the jet centerline, about 3.77 ft (1.15 m) above the ground. Jet noise can be quantified using an arc-shaped microphone array spanning over 0 – 150° coverage with 0° at the nozzle exit plane coinciding with the origin. All the microphones and holders were wrapped with foam material during the experiments to avoid undesired reflections. The dynamic range of the noise measurement system was adequately set to span the range of overall SPL and lowest frequency of interest (Viswanathan 2003). Acoustic data with a bin spacing of 12.5 Hz up to a maximum frequency of 100 kHz were acquired and synthesized to produce the narrow-band spectra. Microphone data was analyzed for spectral response based on advanced signal processing techniques. Note that the polar array in Fig. 4(*b*) used for far-field measurements can be modified for azimuthal measurements in case of non-axisymmetric or controlled jet experiments (Viswanathan 2003).

Since Iqbal and Thomas (2007) does not contain any far-field data; separate measurements with a Mach 0.8 jet were made using 6 microphones mounted on a polar array at $r^* = 100$ away from the nozzle exit. Uncertainty in the calibration of microphones was 0.1 dB at 251.2 Hz, and the flow was maintained within a 0.5% tolerance of the jet exit Mach number. Here, signals were acquired at 100 kHz for 20 s based on the statistical convergence tests. All the data was processed in narrowband spectra using Welch's method to estimate the power spectral density (PSD). Time-domain data was averaged into 512 segments, which canceled any spurious noise. Further,

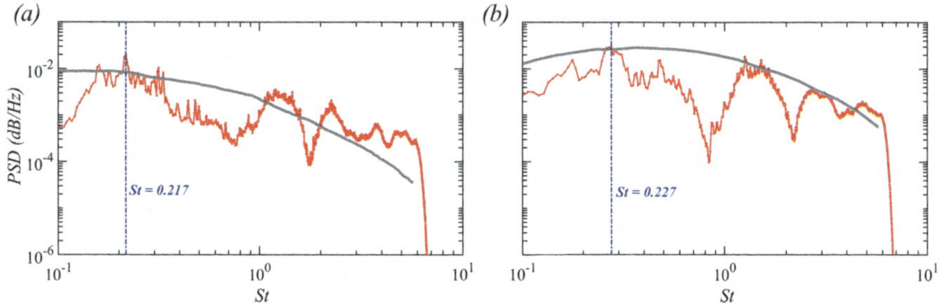

Fig. 7 Power spectral density as a function of Strouhal no. for Mach 0.8 jet at (*a*) 40° and (*b*) 90° at $r^* = 100$. Present data (●); data from Vishwanathan (Viswanathan 2006) are shown as solid line —

a bin size of 12.5 Hz was chosen to resolve the spectra at high frequencies, which provided a ΔSt of 0.002. The cut-off Strouhal number selected here was 7.0, sufficiently high to measure the dominant peaks in the jet noise data (Crow and Champagne 1971). Further, the Hanning window with a 50% overlap and 2048 number of Discrete Fourier Transform points (NFFT) were used for the present spectral analysis. Figure 7a, b represents the PSD of the pressure fluctuations acquired at 40° (in the upstream direction) and 90° (sideline direction) from the upstream jet axis at $r^* = 100$ from the jet centreline. The discrepancy occurring at high frequency was due to the variation in the present sampling rate compared with Vishwanathan's reference data (Viswanathan 2006). This constrained the analyses of high-frequency data at $St \gtrsim 1$. However, since this St value is well outside of the dominant jet noise radiation range, which primarily consisted of radiation from fine-scale jet turbulence (Tam et al. 1996), the poorly resolved data at high frequencies is not expected to provide any meaningful physical insight into the jet noise mechanisms. The spectra in Fig. 7 peaked at St of 0.217 and 0.227 (shown by the blue vertical line) at 40° and 90° respectively, which lie within the range of preferred radiation mode frequency (Crow and Champagne 1971). These measured peaks are fundamentally in good agreement with the reference results. Interestingly, the PSD values for 40° are 0.2 dB/Hz lower than at the 90° location, which indicates sound pressure levels to be 10 dB lower on the decibel scale at the former angle. Note that the 40° direction lies at the upstream side where mostly fine-scale turbulence of jet noise radiates which are at a lower amplitude than at 90°. These results thus indicate good agreement between the present investigation with the reference data for the microphone measurements.

4.3 *Sound Directivity of Mach 0.5 and Mach 0.8 Jets*

Figure 8 shows the overall sound pressure levels (OASPL) expressed for Mach 0.5 and Mach 0.8 jets as a function of different observation angles in the interval of 30° to 150° at an increment of 5°. These measurements provide physical insight into the

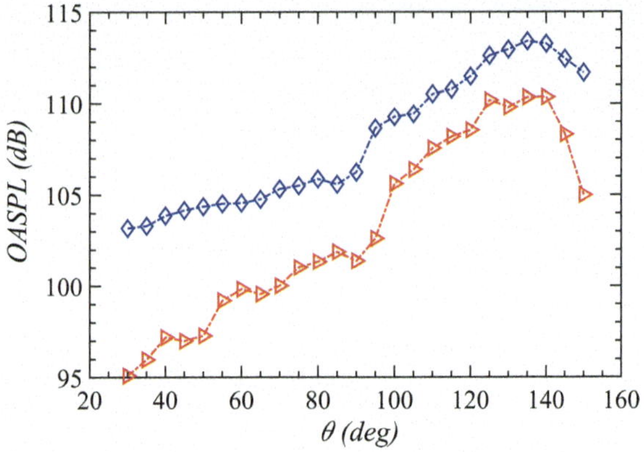

Fig. 8 Dependence of overall sound pressure level on jet exit M = 0.5 (△) and M = 0.8 (◇) as a function of observation angles at $r^* = 72$

sound directivity for high subsonic jets. Both the jets peak around $\theta = 135°$, however, the Mach 0.8 jet is louder than the Mach 0.5 jet by about 3.0 dB. The differences are at 4.8 dB at the sideline (90°) and 6.7 dB at the downstream direction (40°). These results indicate that the $\Delta OASPL$ diminishes on traversing from smaller to higher angles. Perhaps, this indicates similar radiations of large-scale structures along the downstream direction for both the jets. Both these jets individually exhibit an OASPL variation of approximately 10 dB between their extreme directions, and this observation is also consistent with Tam et al. (1996). The results indicate that JAFC lab's anechoic chamber is reliable for investigating uncontrolled and controlled jet noise generation and radiation mechanisms, which can help mitigate the undesired noise caused by jet turbulence.

5 Conclusions

The anechoic chamber commissioned at Jet-Acoustics and Flow Control (JAFC) Laboratory, Department of Aerospace Engineering, IIT Kanpur, was calibrated to determine the cut-off frequency, which revealed the anechoic chamber to attenuate noise attenuation for frequencies above 500 Hz. Additionally, the benchmark tests performed on a Mach 0.3 jet for the flow-field and a Mach 0.8 jet for the acoustic characteristics show good agreement with the reference data. This now ensures the commissioned anechoic chamber to be utilized for aeroacoustic studies of jets at varying Mach numbers.

References

Abramovich, Genrikh Naumovich and Girshovich, TA and Krasheninnikov, S Iu and Sekundov, AN and Smirnova, IP: The theory of turbulent jets. Cambridge, Mass., M.I.T. Press, (1963).

Ahuja, KK, (2003), Designing clean jet-noise facilities and making accurate jet-noise measurements in *International Journal of Aeroacoustics*, 2, pp. 371–412.

Beranek, Leo L. and Sleeper, Harvey P., Jr. (2005),The Design and Construction of Anechoic Sound Chambers, *J. Acoust. Soc. Am.*, 18, 140–150.

Crow, S C, and Champagne, FH (1971), Orderly structure in jet turbulence in *J. Fluid Mech.*, 48, pp. 547–591.

Cunefare KA, Biesel VB, Tran J, Rye R, Graf A, Holdhusen M, Albanese AM. (2003), Anechoic chamber qualification: Traverse method, inverse square law analysis method, and nature of test signal, *J. Acoust. Soc. Am.*, 113, 881–892.

Iqbal, M. O., and Thomas, F. O., (2007), Coherent structure in a turbulent jet via a vector implementation of the proper orthogonal decomposition, *J. Fluid Mech.*, 571, pp. 281–326.

Perumal, A. K., and Zhou, Y. (2018), Parametric study and scaling of jet manipulation using an unsteady minijet, *J. Fluid Mech.*, 848, 592–630.

Tam, Christopher KW and Viswanathan, K and Ahuja, KK and Panda, J (1996), The sources of jet noise: experimental evidence, Aeroacoustics of Flight Vehicles in *Journal of Fluid Mechanics*, 615, pp. 253–292.

Viswanathan, K., (2003), Jet aeroacoustic testing: issues and implications, *AIAA J.*, 41, pp. 1674–1689.

Viswanathan, K., (2006), Instrumentation considerations for accurate jet noise measurements, *AIAA J.*, 44, pp. 1137–1149.

Numerical Simulation of Lateral Migration of Healthy RBC and Malaria Infected RBC in Poiseuille Flow Using Smoothed Particle Hydrodynamics

Justin Antony and Ranjith Maniyeri

Abstract Several diseases, such as sickle cell anaemia, malaria, cancer, etc., are caused by pathological action that alters the mechanical properties of healthy red blood cells (RBC). Understanding the behaviour of healthy and diseased cells will help to develop diagnostic devices for the early detection of diseases. Lateral migration is a cell separation technique in which the cells can be separated from different lateral positions based on their size and deformability resulting from the migration of cells across streamlines under the action of hydrodynamic forces. In the present work, a numerical model based on smoothed particle hydrodynamics has been developed to study complex fluid-structure interaction of RBC and parallelized in GPU, leading to a seventeen-fold reduction in computational cost. RBC is assumed to be circular, and a spring network model is used to model the RBC membrane in the present work. The developed numerical model is used to investigate lateral migration and deformation of malaria infected RBC in Poiseuille flow and compared with that of healthy RBC.

Keywords Lateral migration · RBC dynamics · Smoothed particle hydrodynamics

1 Introduction

Several diseases, such as malaria, sickle cell anaemia, cardiovascular diseases and neurological disorders, are associated with red blood cell (RBC) deformation (Mokken et al. 1992). Understanding the behaviour of healthy and diseased cells will help to develop diagnostic devices for the early detection of diseases. The diseased cells can be separated from the sample using external forces or hydrodynamic forces. The external forces may further alter the properties of cells. Hence, separation using

J. Antony · R. Maniyeri (✉)
Biophysics Laboratory, Department of Mechanical Engineering, National Institute of Technology Karnataka, Surathkal, Mangalore, India
e-mail: mranji1@nitk.edu.in

G. Rajesh et al. (eds.), *Proceedings of the 15th International Symposium on Experimental and Computational Aerothermodynamics of Internal Flows*,
https://doi.org/10.1007/978-981-96-4082-9_12

hydrodynamic forces is preferred as it does not alter the properties of the cells, and it is a label-free method. Lateral migration is a method in which the cells are migrated across the streamlines under hydrodynamic forces to different lateral positions based on size and deformability. In Poiseuille flow, a rigid particle migrates to lateral position of 0.6 times the radius of the channel under the action of wall-repulsion lift force and shear-gradient lift force (Segre and Silberberg 1962). The former is due to the bulk shear flow between the rigid particle and wall, whereas the latter is due to the velocity gradient of the Poiseuille flow. However, a deformable particle such as red blood cell (RBC) migrates to the centre of the channel, and the cell undergoes tank-treading motion due to the velocity gradient (Goldsmith 1971). In the present work, a meshless particle method known as smoothed particle hydrodynamics (SPH) is used to develop a GPU-parallelized numerical model using CUDA Fortran to simulate the complex fluid-structure interaction of RBC. The parallelized model is able to achieve seventeen-fold speedup using Nvidia GeForce GTX 1650 Ti GPU (4GB, 1024 threads per block). The lateral migration, migration time and deformation of malaria infected RBC in a Poiseuille flow are analyzed using the developed model and compared with that of healthy RBC.

2 Methodology

The SPH is a Lagrangian particle-based method in which the field variables and derivatives at nodes are evaluated using a smoothing kernel function. The formulation of conventional SPH is given in Eqs. 1 and 2.

$$f(\mathbf{x}) = \int_{\Omega} f(\mathbf{x}')W(\mathbf{x} - \mathbf{x}, h)d\mathbf{x}' \tag{1}$$

$$\nabla f(\mathbf{x}) = -\int_{\Omega} f(x')\nabla W(\mathbf{x} - \mathbf{x}, h)d\mathbf{x}' \tag{2}$$

where m is mass, ρ is density and $\mathbf{x}$ is the position vector of the node. W is the smoothing kernel function. Indices i and j represent the node and its neighbours, respectively. Zhang and Liu (2018) improvised conventional SPH by introducing additional terms in the denominator to improve the accuracy by reducing inconsistencies arising from non-uniform particle distribution and kernel inconsistency. This improved SPH is known as the decoupled finite particle method (DFPM), and its formulations are given in Eqs. 3 and 4.

$$f(\mathbf{x}_i) = \frac{\sum_{j=1}^{N} \frac{m_j}{\rho_j} f(\mathbf{x}_j) W_{ij}}{\sum_{j=1}^{N} \frac{m_j}{\rho_j} W_{ij}} \tag{3}$$

$$\nabla f(\mathbf{x}_i) = \frac{\sum_{j=1}^{N} \frac{m_j}{\rho_j} f(\mathbf{x}_j) \cdot \nabla W_{ij}}{\sum_{j=1}^{N} \frac{m_j}{\rho_j} (\mathbf{x}_j - \mathbf{x}_i) \cdot W_{ij}} \tag{4}$$

Huang et al. (2016) developed a computationally efficient and accurate scheme for evaluating the Laplacian of the field function (Eq. 5)

$$\nabla^2 f(\mathbf{x}_i) = 2 \frac{\sum_{j=1}^{N} \frac{m_j}{\rho_j} (f(\mathbf{x}_j) - f(\mathbf{x}_i)) W_{ij}}{\chi \psi} \tag{5}$$

where $\chi = 15/7\pi h^2$ and $\psi = 31h^4\pi/210$.

3 Governing Equations

The equations that govern fluid flow are continuity, momentum and artificial equation of state, which are given in Eqs. 6–8, respectively in SPH formalization.

$$\frac{D\rho_i}{Dt} = -\rho_i \frac{\sum_{j=1}^{N} \frac{m_j}{\rho_j} (\mathbf{u}_j^{\beta} - \mathbf{u}_i^{\beta}) \cdot \frac{\partial W_{ij}}{\partial \mathbf{x}_i^{\beta}}}{\sum_{j=1}^{N} \frac{m_j}{\rho_j} (\mathbf{x}_j^{\beta} - \mathbf{x}_i^{\beta}) \cdot \frac{\partial W_{ij}}{\partial \mathbf{x}_i^{\beta}}} \tag{6}$$

$$\frac{D\mathbf{u}_i^{\alpha}}{Dt} = -\frac{1}{\rho_i} \frac{\sum_{j=1}^{N} \frac{m_j}{\rho_j} (p_j - p_i) \frac{\partial W_{ij}}{\partial \mathbf{x}_i^{\alpha}}}{\sum_{j=1}^{N} \frac{m_j}{\rho_j} (\mathbf{x}_j^{\alpha} - \mathbf{x}_i^{\alpha}) \cdot \frac{\partial W_{ij}}{\partial \mathbf{x}_i^{\alpha}}} + 2 \frac{\mu_i}{\rho_i} \frac{\sum_{j=1}^{N} \frac{m_j}{\rho_j} (\mathbf{u}_j^{\alpha} - \mathbf{u}_i^{\alpha}) \frac{\partial W_{ij}}{\partial \mathbf{x}_i^{\alpha}}}{\chi \psi} + F_i \tag{7}$$

$$p_i = c^2(\rho_i - \rho_0) \tag{8}$$

where, **u** is the velocity vector, p is the pressure, μ is the viscosity and **F** is the acceleration due to external body force. α and β denotes the components of the vector in x and y directions. c is a constant which can be taken as ten times the maximum velocity in the domain (Maziar et al. 2017).

For simplicity, the RBC is assumed to be of circular shape, and the plasma and cytoplasm are considered the same fluid in the present study. The RBC membrane is modelled as a spring network with stretching and bending moduli, as presented in Hosseini and Feng (2009). The stretching force of a spring connected between two nodes is calculated using Eq. 9 (Barthes-Biesel et al. 2002).

$$T = \frac{E_S}{2} \lambda_1 (\lambda_1^2 - 1) \sqrt{\frac{1 + C\lambda_1^4}{1 + C\lambda_1^2}} \left[1 + \frac{C(1 + C\lambda_1^2)}{(1 + C\lambda_1^4)^2} \right] \tag{9}$$

where E_S is the shear modulus, λ_1 is the ratio of deformed length to resting length, and C is the ratio between dilation (E_D) and shear modulus. The bending forces at nodes are evaluated from bending moments at the nodes, and this is calculated using Eq. 10 (Hosseini and Feng 2009).

$$m = E_B(\kappa - \kappa_0) \tag{10}$$

where E_B is the bending modulus and κ and κ_0 are local curvature in deformed and resting states.

4 Solution Algorithm

The flow chart describing the solution algorithm is given in Fig. 1. The problem domain is discretized with uniformly distributed nodes, and its neighbours in the support domain are evaluated. The present work uses a cubic spline smoothing kernel function (Liu and Liu 2003) whose support domain is a circle of radius twice the smoothing length. The smoothing length can be taken as 1.2 times the initial particle spacing (Maziar et al. 2017). An artificial particle displacement (Shadloo et al. 2012) is used in each timestep to maintain a uniform particle distribution. Some nodes, called virtual nodes, are created outside the domain to impose boundary conditions. The properties of these nodes are evaluated from their mirror image in the domain. For the no-slip boundary condition, the values of velocity obtained from mirror images are reversed and given to the corresponding virtual nodes. The pressure of all nodes is updated using the artificial equation of state. The required gradients and Laplacian to solve the governing equations are evaluated using Eqs. 3–5. The forces on RBC membranes are evaluated using Eq. 9, and the bending moment is determined using Eq. 10 and converted into forces. Finally, the forward Euler method is used to update the velocity, position, and density at each node from momentum and continuity equations. A density-diffusive term (Marrone et al. 2013) is added to the right-hand side of the continuity equation to reduce spurious oscillations arising from the use of an artificial equation of state. The procedure is continued till a specified time step is reached. The stability criteria presented by Morris et al. (1997) is used for determining the time step size which is given in Eq. 11.

$$\Delta t = 0.125\frac{\rho h^2}{\mu} \tag{11}$$

5 Numerical Simulation

In the present study, the RBC is assumed to be of circular shape with diameter 5.5 μm. A two-dimensional channel of dimension 24 μm × 10 μm is considered, and the RBC is placed at a distance of 2 μm from the centre of the channel towards the

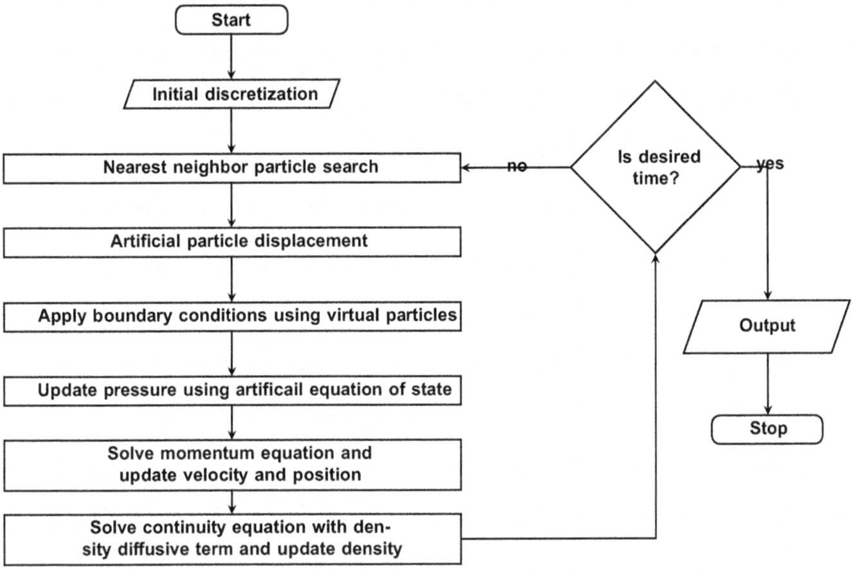

Fig. 1 Flow chart of the solution algorithm

Fig. 2 Initial distribution of nodes in the domain

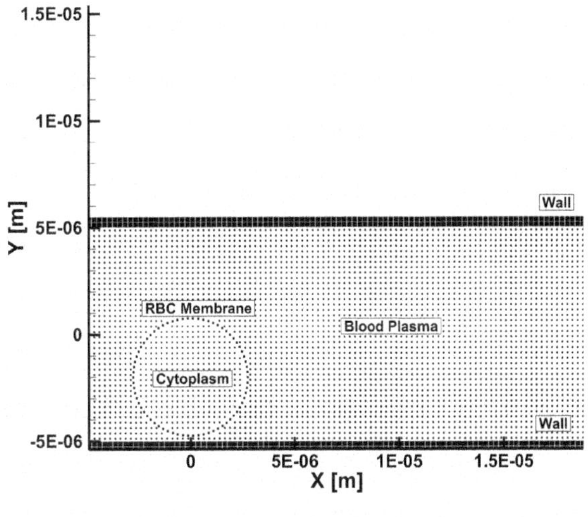

bottom wall, as shown in Fig. 2. The density is taken as $1080\,\text{kg/m}^3$ and viscosity is taken as 1.2×10^{-3} Pas. The shear and bending modulus is taken as $5 \times 10^{-6}\,\text{N/m}$ and 2×10^{-19} Nm respectively for healthy RBC (Hosseini and Feng 2009), and $15 \times 10^{-6}\,\text{N/m}$ and 6×10^{-19} Nm respectively for infected RBC (Hosseini and Feng 2012). An acceleration is applied in the x-direction to initiate the fluid flow. Periodic boundary conditions are applied at left and right boundaries. The number of particles and time step size is 96×40 and 4×10^{-10} s. The simulations are carried out for a

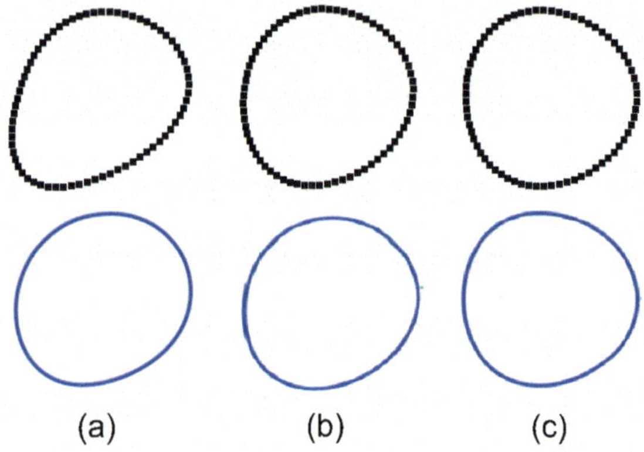

Fig. 3 Shape of RBC in Poiseuille flow of present study (top) and Shi et al. (2012) (bottom) at time **a** 0.8 ms, **b** 1.5 ms, **c** 7.5 ms

time period of 0.01 s. The developed numerical model is validated for the simulation of lateral migration of RBC in Poiseuille flow and compared with the results of Shi et al. (2012) (Fig. 3).

6 Results and Discussion

The motion and deformation of healthy and malaria infected RBC are investigated separately using the developed numerical model. The RBC placed initially at an offset of 2 μm from the centre towards the bottom wall in both cases. It is observed that the RBC migrates towards the centre in both cases and undergoes tank-treading motion. The tank treading frequency of infected RBC is observed to be more than that of healthy RBC. The final shape of the healthy RBC is a bullet shape. At the same time, that of the infected RBC remains more or less circular with very small deformation, as shown in Fig. 4. The path traced by the centre of RBC in both cases is presented in Fig. 5. It can be observed that the healthy RBC reaches the center of the channel much faster than infected RBC. The lateral migration time is approximately 0.004 s for healthy RBC and 0.009 s for infected RBC. The migration time for the infected RBC is considerably longer than that of healthy RBC. The shear gradient lift force which forces the RBC towards the channel wall is small in the case of healthy RBC due to its higher deformation than that of infected RBC. Since the migration time is higher for infected RBCs, in a mixture of healthy and infected RBCs, the healthy RBCs may force the infected RBCs to move towards the wall and accumulate at a different lateral position other than the centre of the channel. Then, it will be possible to separate the cells based on their lateral position. However, further investigations, considering their mutual interactions, must be considered to shed light on developing efficient techniques for separating healthy and infected RBCs.

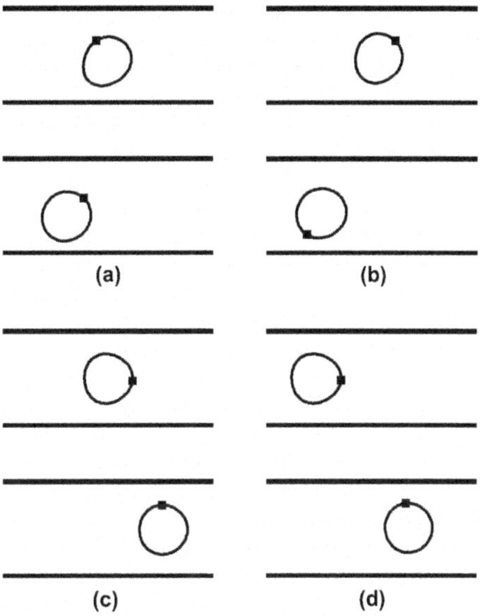

Fig. 4 Motion and deformation of healthy RBC (top) and infected RBC (bottom) at time **a** 0.9 ms, **b** 1.6 ms, **c** 7.5 ms and **d** 9 ms

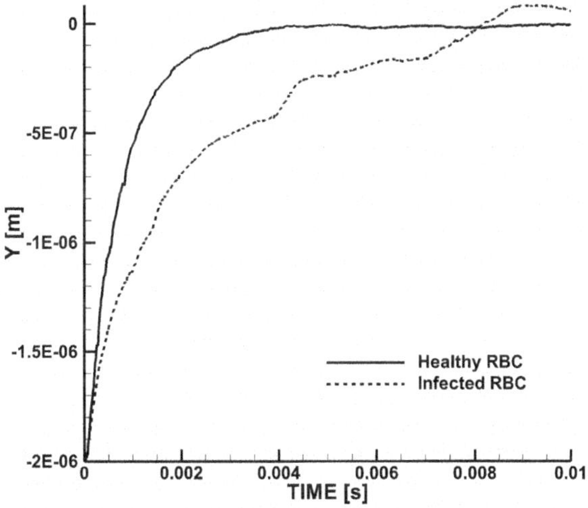

Fig. 5 Comparison of lateral migration of healthy and infected RBC in Poiseuille flow

7 Conclusion

The developed numerical model captured the dynamics of the healthy and infected RBC very well. It is found that there is a significant difference in the migration time of infected RBC due to its reduced deformability. The higher migration time for infected RBC can be utilized to separate infected cells from healthy cells. The healthy RBC due to its lower migration time reaches the center of the channel faster, while the infected RBC is only approaching the center. Hence, the interaction between the healthy and infected RBC will force the infected RBC towards the wall and can be separated easily. In the present study, both cells are considered separately, and the developed numerical model will be implemented to investigate the mutual interaction of healthy and infected RBCs in the future.

References

Barthes-Biesel D., Diaz A. and Dhenin E.: Effect of constitutive laws for two-dimensional membranes on flow-induced capsule deformation. Journal of Fluid Mechanics, 460, 211–222 (2002).

Goldsmith H. L.: Red cell motions and wall interactions in tube flow. Federation Proceedings, 30, 1578–1590 (1971).

Hosseini S. M., Feng J. J.: A particle-based model for the transport of erythrocytes in capillaries. Chemical Engineering Science, 64(22), 4488–4497 (2009).

Hosseini S. M. and Feng J. J.: How malaria parasites reduce the deformability of infected red blood cells. Biophysical Journal, 103(1), 1–10 (2012).

Huang C., Lei J. M., Liu M. B., Peng X. Y.:An improved KGF-SPH with a novel discrete scheme of Laplacian operator for viscous incompressible fluid flows. International Journal for Numerical Methods in Fluids, 81(6), 377–396 (2016).

Liu G. R., Liu M. B.: Smoothed Particle Hydrodynamics A Meshfree Particle Method. World Scientific Publishing Co. Pte. Ltd. (2003).

Marrone S., Colagrossi A., Antuono M., Colicchio G., Graziani G.: An accurate SPH modeling of viscous flows around bodies at low and moderate Reynolds numbers. Journal of Computational Physics, 245, 456–475 (2013).

Maziar G. K., Sergio A. G., Alexander S., David J. W.: Parametric study on smoothed particle hydrodynamics for accurate determination of drag coefficient for a circular cylinder. Water Science and Engineering, 10, 143–153 (2017).

Mokken F. C., Kedaria M., Henny C. P., Hardeman M. R., Gelb A. W.: The clinical importance of erythrocyte deformability, a hemorrheological parameter. Annals of Hematology, 64(3), 113–122 (1992).

Morris J. P., Fox P. J., Zhu Y.: Modeling low Reynolds number incompressible flows with curved boundaries using SPH. Journal of Computational Physics, 136, 214–226 (1997).

Segre G. and Silberberg: Behaviour of macroscopic rigid spheres in Poiseuille flow. Journal of Fluid Mechanics, 14, 136–157 (1962).

Shadloo M. S., Zainali A., Yildiz M., Suleman A.: A robust weakly compressible SPH method and its comparison with an incompressible SPH. International Journal for Numerical Methods in Engineering, 89(8), 939–956 (2012).

Shi L., Pan T. W., Glowinski R.: Numerical simulation of lateral migration of red blood cells in Poiseuille flows. International Journal for Numerical Methods in Fluids, 68(11), 1393–1408 (2012).

Zhang Z. L., Liu M. B.: A decoupled finite particle method for modeling incompressible flows with free surfaces. Applied Mathematical Modelling, 60, 606–633 (2018).

Effect of Slip on the Dynamics of Vortex Shedding Behind a Rotating Cylinder

T. Arunjith and K. S. Santhosh

Abstract Slip surface can significantly reduce the drag experienced on a stationary cylinder surface. The present numerical study is about the effect of slip on the rotating cylinder surface on vortex shedding within the laminar region for different Reynolds numbers. A domain of 50D × 100D with structured quadrilateral grids which has 0.002D thickness of first layer around the cylinder considered for analysis in COMSOL Multiphysics 5.6 software by finite element method. Implicit scheme with backward difference formula used for the computation. Overall, the study highlights the influence of slip and rotation on the flow characteristics around a rotating cylinder, providing insights into the drag, lift, and vortex shedding behavior. The results indicate that slip has a positive effect on reducing drag, and the rotation of the cylinder further decreases drag. On the other hand, the presence of slip distribution leads to a decrease in lift despite an increase observed with rotation. The combined effect of rotation and slip distribution results in the re-appearance of vortex shedding at high rotation rates and increased slip distribution.

Keywords Rotating cylinder · Slip · Vortex shedding

1 Introduction

Many research has been conducted on the flow past a rotating cylinder, which refers to the scenario where a cylinder is subjected to an oncoming fluid stream and undergoes rotation. This configuration is encountered in various engineering and scientific applications, such as rotating machinery, wind turbines, underwater drones etc. In the wake of a rotating cylinder, vortex shedding takes place similar to a stationary cylinder. However, the rotation of the cylinder can interact with the shed vortices, causing changes in their size, position, and stability. The shedding frequency and strength of the vortices are influenced by the rotational speed of the cylinder. Slip describes

T. Arunjith (✉) · K. S. Santhosh
College of Engineering Trivandrum, Thiruvananthapuram, Kerala, India
e-mail: arunjitht1997@gmail.com

G. Rajesh et al. (eds.), *Proceedings of the 15th International Symposium on Experimental and Computational Aerothermodynamics of Internal Flows*,
https://doi.org/10.1007/978-981-96-4082-9_13

the motion that occurs for fluid relative to cylinder's surface near the surface. When slip happens, there is a velocity difference between the fluid at the cylinder's surface and free stream fluid velocity. Slip length is particularly relevant in situations where the fluid flows in very close proximity to a solid boundary, such as micro channels, nanofluidics, or when dealing with superhydrophobic surfaces.

Prandtl (1925) conducted one of the earliest experiments on the flow past a rotating cylinder. In his work, he proposed that the maximum lift produced by a cylinder spinning in a uniform flow is constrained to 4π. Measuring the lift on a rotating cylinder presents challenges due to the inherent limitations associated with the cylinder's rotation. Tokumaru et al. (1991) devised a method to estimate the mean lift acting on a rotating cylinder in uniform flow. The research conducted by them for a Reynolds number of 3.8×10^3 shows that the lift coefficient limit proposed by Prandtl can be surpassed. In the experimental investigation, Coutanceau et al. (1985) observed that only one vortex shed when the rotation rate (α) surpasses a threshold value. They noted that this critical rotation rate is roughly equal to 2 and it is relatively unaffected by Reynolds number in the laminar shedding range. Their view is supported by computations of Badr et al. (1985). Mittal et al. (2003) done computational study for Reynolds number of 200 with various rotation rates ($0 \leq \alpha \leq 5$). They discovered that a von Karman street is visible over a range of 0–1.9. The flow remains steady when the rotation rate is between 1.91 and 4.34. However, it becomes unstable when the rotation rate exceeds 4.35. When the rotation rate surpasses 4.8, the flow stabilizes once again. Present range of rotation is within $0 \leq \alpha \leq 2$ with increment of 0.5.

In their study, Legendre et al. (2009) used numerical simulations to examine how the slip boundary condition affected the wake dynamics within a two-dimensional fluid flow around a cylinder for Re = 50, 100, 200, 500 and 800. Their research showed that slip causes a delay in the start of vortex shedding and recirculation in the wake. Additionally, slip was observed to reduce both drag and lift forces, decrease vorticity intensity in the immediate wake region, and increase the shedding frequency. To create a partial slip boundary condition, Daniello et al. (2013) implemented a superhydrophobic surface coating on the circular cylinder through roughening of PTFE. Through their study, they observed that this implementation led to an increase in both length and width of the recirculation region. Simultaneously, they noted a reduction in the lift coefficient and intensity of shed vortices. Li et al. (2014) conducted a numerical study to examine the impact of different slip configurations on the flow characteristics over a circular cylinder. No slip ($\theta_s = 0$), fore-side slip, flank slip, aft-side slip, and all slip($\theta_s = 180$) were the five distinct types of slip conditions that they examined (see Fig. 1). By comparing different types of slip distribution, it was found that implementing slip on the flanks of the cylinder yields the best results in terms of reducing drag and lift.

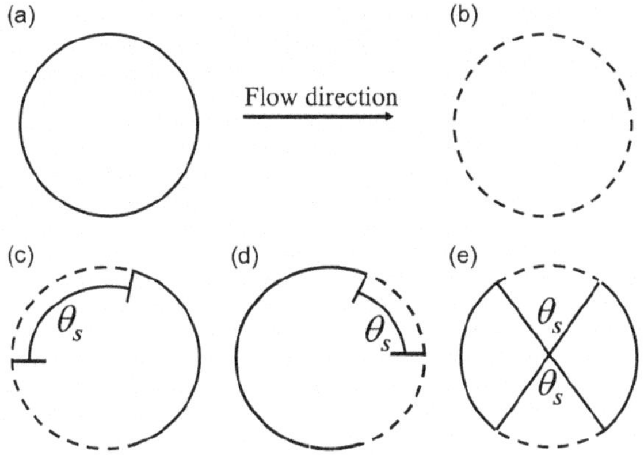

Fig. 1 Five slip boundary distributions (spanwise distributed): **a** no slip; **b** all slip; **c** fore-side slip; **d** aft-side slip; and **e** flank slip (Li et al. 2014)

2 Numerical Solution Procedure

2.1 Governing Equations and Boundary Conditions

The governing equations for two dimensional unsteady, viscous and incompressible flow are:

Continuity equation

$$\frac{\partial(\mathbf{u})}{\partial \mathbf{x}} + \frac{\partial(\mathbf{v})}{\partial \mathbf{y}} = \mathbf{0} \tag{1}$$

Momentum equation

X-momentum equation

$$\rho\frac{\partial(\mathbf{u})}{\partial \mathbf{t}} + \rho\mathbf{u}\frac{\partial(\mathbf{u})}{\partial \mathbf{x}} + \rho\mathbf{v}\frac{\partial(\mathbf{u})}{\partial \mathbf{y}} = \frac{-\partial \mathbf{p}}{\partial \mathbf{x}} + \mu\left[\frac{\partial^2 \mathbf{u}}{\partial \mathbf{x}^2} + \frac{\partial^2 \mathbf{u}}{\partial \mathbf{y}^2}\right] \tag{2}$$

Y-momentum equation

$$\rho\frac{\partial(\mathbf{v})}{\partial \mathbf{t}} + \rho\mathbf{u}\frac{\partial(\mathbf{v})}{\partial \mathbf{x}} + \rho\mathbf{v}\frac{\partial(\mathbf{v})}{\partial \mathbf{y}} = \frac{-\partial \mathbf{p}}{\partial \mathbf{y}} + \mu\left[\frac{\partial^2 \mathbf{v}}{\partial \mathbf{x}^2} + \frac{\partial^2 \mathbf{v}}{\partial \mathbf{y}^2}\right] \tag{3}$$

where ρ is the fluid density, μ is dynamic viscosity, p is the pressure, u and v are velocity along X and Y directions respectively. The boundary conditions for slip and impermeability on the cylinder surface

$$\begin{gathered}\mathbf{n.u} = \mathbf{0}, \\ \mathbf{n} \times \mathbf{u} = \mathbf{2L_s n} \times (\mathbf{S.n})\end{gathered} \tag{4}$$

where n is the unit normal vector, $\mathbf{S} = (\Delta\mathbf{u} + (\Delta\mathbf{u})^T)/2$ and $\mathbf{L_s}$ is the slip length.

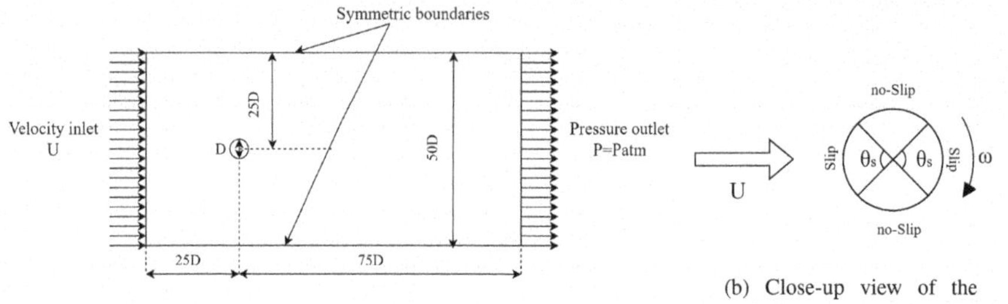

(a) Flow setup for computation

(b) Close-up view of the cylinder

Fig. 2 Schematic diagrams showing domain for computation and boundary conditions

Boundary conditions: The cylinder is positioned within a rectangular domain of 50D×100D as suggested by Ren et al. (2018) (see Fig. 2a), and the flow velocity is set based on the specified Reynolds number at the surface of the cylinder. The rotation of the cylinder follows a clockwise direction. The upstream boundary is assigned the value of the free-stream velocity, while a pressure boundary condition is specified at the downstream boundary. For the lateral boundaries, a zero value is prescribed for the component of velocity perpendicular to the boundary as well as the component of stress vector along the boundary. The surface of the cylinder is subjected to slip, which is characterized by slip span (θ_s) (see Fig. 2b)

There are three dimensionless parameters involved in the study: Reynolds number, $Re = \frac{\rho U D}{\mu}$:the nondimensional rotation rate, $\alpha = \frac{D\omega}{2U}$; and nondimensionalized slip length (0.02), $\beta = \frac{L_s}{D}$. Here, L_s is the slip length (0.2D), D is the cylinder's diameter, and U is the far field velocity. We will also use following definitions: Coefficient of drag, $C_D = \frac{2F_D}{\rho U^2 A}$; Coefficient of lift, $C_L = \frac{2F_L}{\rho U^2 A}$; and the Strouhal number, $St = \frac{fD}{U}$ where f is the vortex shedding frequency. Integrating pressure and viscous stresses around the cylinder's perimeter required for computing force coefficients.

2.2 *Numerical Method*

In this study, numerical simulation is conducted using the commercial software COMSOL Multiphysics 5.6. COMSOL utilizes the finite element method to solve the given problem. To validate the numerical results, comparisons are made with results from other researchers in the field. The numerical solution is obtained using an implicit scheme with the Backward Difference Formula (BDF). For solving the system of equations, the direct solver named PARDISO is employed.

Finite element mesh The domain is of 50D × 100D size. There are 53020 elements and 53610 mesh vertices in the mesh that was used for all calculations. The number of divisions around the cylinder is represented by C_t. R_t refers to the first layer of mesh element immediately surrounding the cylinder. Figure 3a displays typical

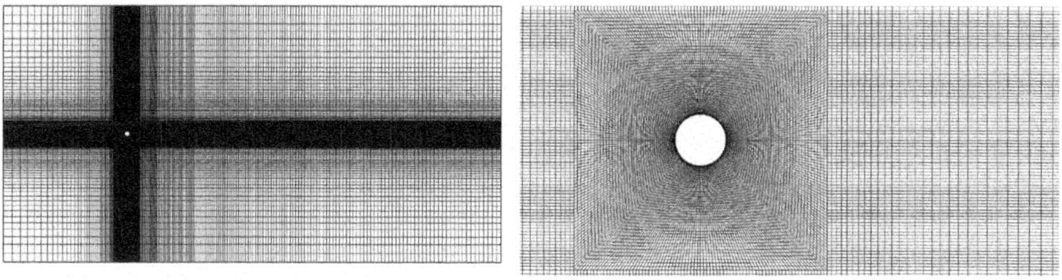

(a) Finite element mesh with 53020 elements and 53610 mesh vertices

(b) Finite element mesh near the cylinder surface: Close-up view

Fig. 3 Mesh configuration

mesh configuration. It is evident that the mesh near the cylinder is extremely fine, consisting of smaller elements. The element sizes gradually grow greater as you get further away from the cylinder. Figure 3b displays a close-up of a typical mesh. During the grid independent study, it was determined that for a fixed radial thickness (R_t) of 0.002D, the optimal number of points in the circumferential direction (C_t) was found to be 320 in order to obtain the best results (see Fig. 4).

Validation and comparison The values for drag and lift coefficients compared for Re = 100 and 200 with former researchers. Also effect of rotation and slip distribution were also validated.

Lift and drag comparison: Table 5 shows the comparison of C_D, C_L and Strouhal number with other researchers. The numerical results of this investigation are more closely aligned with the numerical values obtained by Ren et al. (2018) and Mittal (2005) and compared to the experimental results by Clift et al. (2005) (see Table 1).

Validating rotation of cylinder: The validation of the change in the Strouhal number with the nondimensional rotation rate is performed by comparing the results with the

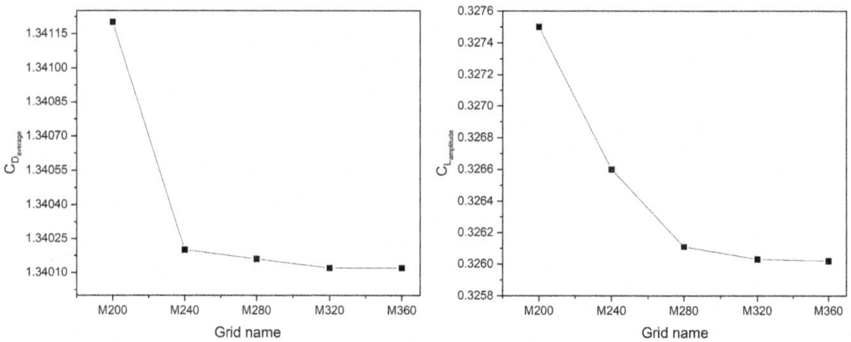

(a) (b)

Fig. 4 Grid independent study result for **a** average coefficient of drag and **b** maximum coefficient of lift

Table 1 Drag and lift coefficients comparison

Author(s)	Method	C_D	$C_{L_{amplitude}}$	$C_{L_{rms}}$	St
Ren et al. (2018)	Numerical	1.34	0.32	–	0.17
Mittal (2005)	Numerical	1.322	–	0.226	0.1644
Clift et al. (2005)	Experimental	1.24	0.27	–	–
Present	Numerical	1.3401	0.326	0.2306	0.165

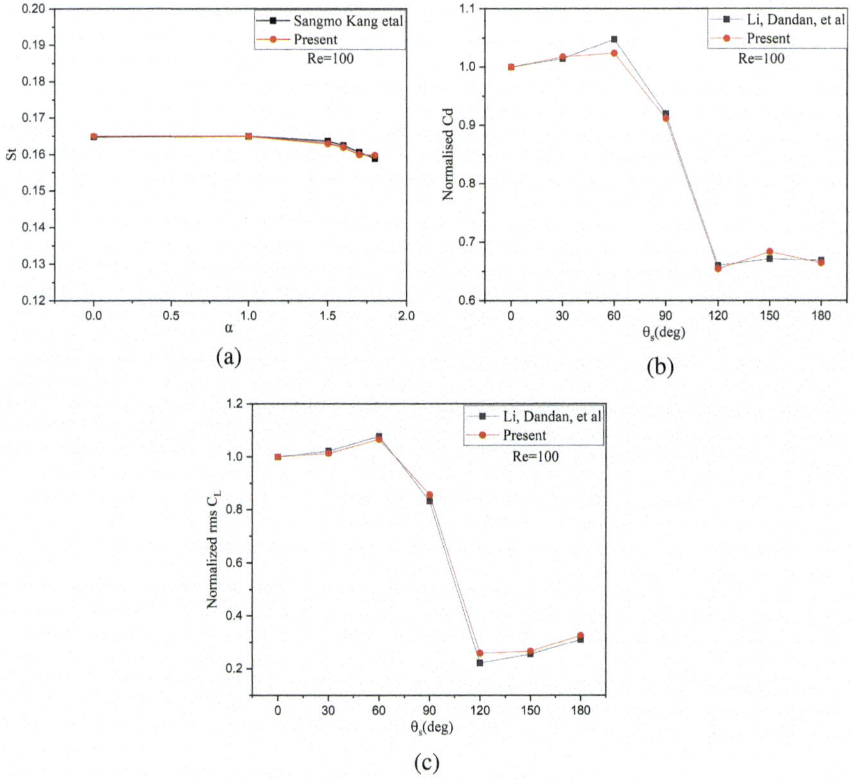

Fig. 5 Validation of **a** rotation, **b** drag and **c** lift for Re = 100

findings of Kang et al. (1999). As rotation speed is increasing, the Strouhal number value reduces (see Fig. 5a).

Slip distribution: The slip length fixed as 0.2D where D is the diameter of cylinder. Half of the total azimuth that is symmetrical about the central axis and along the flow direction is designated as the slip span and is denoted by the symbol θ_s.

Li et al. (2014) considered 3 types of distributions as fore-side slip, flank slip and aft-side slip. Among them, fore-side slip was validated as shown in Fig. 5b and c.

3 Computational Results

3.1 Effect of Various Rotation Rates

The effect of the cylinder's slip distribution on the drag and lift coefficients was investigated in the study for Reynolds numbers of 100 and 200 at five rotation rates. Figure 6 shows variation in drag and lift coefficients with slip azimuth for different rotation rate and Reynolds numbers.

Drag coefficient: As the rotation rates increase, the effective total drag on the cylinder surface is reduced. This reduction in drag can be attributed to the interaction between the rotating cylinder and the flow of the surrounding fluid. On a stationary cylinder, up to $\theta_s = 120°$, flow seperation point remains within no-slip region and for $\theta_s = 150°$ and $\theta_s = 180°$, it is in slip region. Near wall velocity at no slip surface decreases the surface vortices. For values of α ranging from 0.5 to 1.5, the drag coefficient initially decreases then with an increase in slip azimuth angle (θ_s) from 60° to 150°, drag increases. For 180°, the drag coefficient reaching the lowest value for the corresponding α. Initial decrease of drag is by more mixing at downstream due to rotation of cylinder, and then increased turbulence effects create pressure drag moves to high values. The work of Pralits et al. (2001) has demonstrated the presence of two distinct regions of instability, referred to as Mode-I and Mode-II, in the (Re, α) plane. The Mode-I instability, as identified by Pralits et al., exhibits instability through a supercritical Hopf bifurcation. This mode is observed within the range of $0 \leq \alpha < 2$. So, $\alpha = 2$ condition behaves differently from others. Vortex shedding is suppressed at $\alpha = 2$ when there is a no-slip condition on the surface. However, if the slip azimuth angle (θ_s) reaches 145° for Reynolds number (Re) 100 and 95° for Re 200, there is a re-appearing of vortex shedding. This indicates that the slip has an effect on the onset of vortex shedding.

Lift coefficient: The findings indicate that as the rotation rate increases, the lift coefficient also increases for a given slip azimuth. This can be attributed to the rotation creating a pressure gradient between the top and bottom of the cylinder due to the differential velocities. A larger pressure gradient in the vertical direction leads to an increased effective lift force. It is evident that the lift coefficient exhibits a decreasing trend with increasing slip distribution. The lowest value of the lift coefficient ($C_{L_{rms}}$) is achieved when the surface is fully slip with $\alpha = 0$ and maximum when surface is fully no slip with $\alpha = 2$. The lift coefficient rises with increasing rotation rate but falls with increasing θ_s. Because the strength of the shedded vortices decreases, lift force diminishes as slip increases.

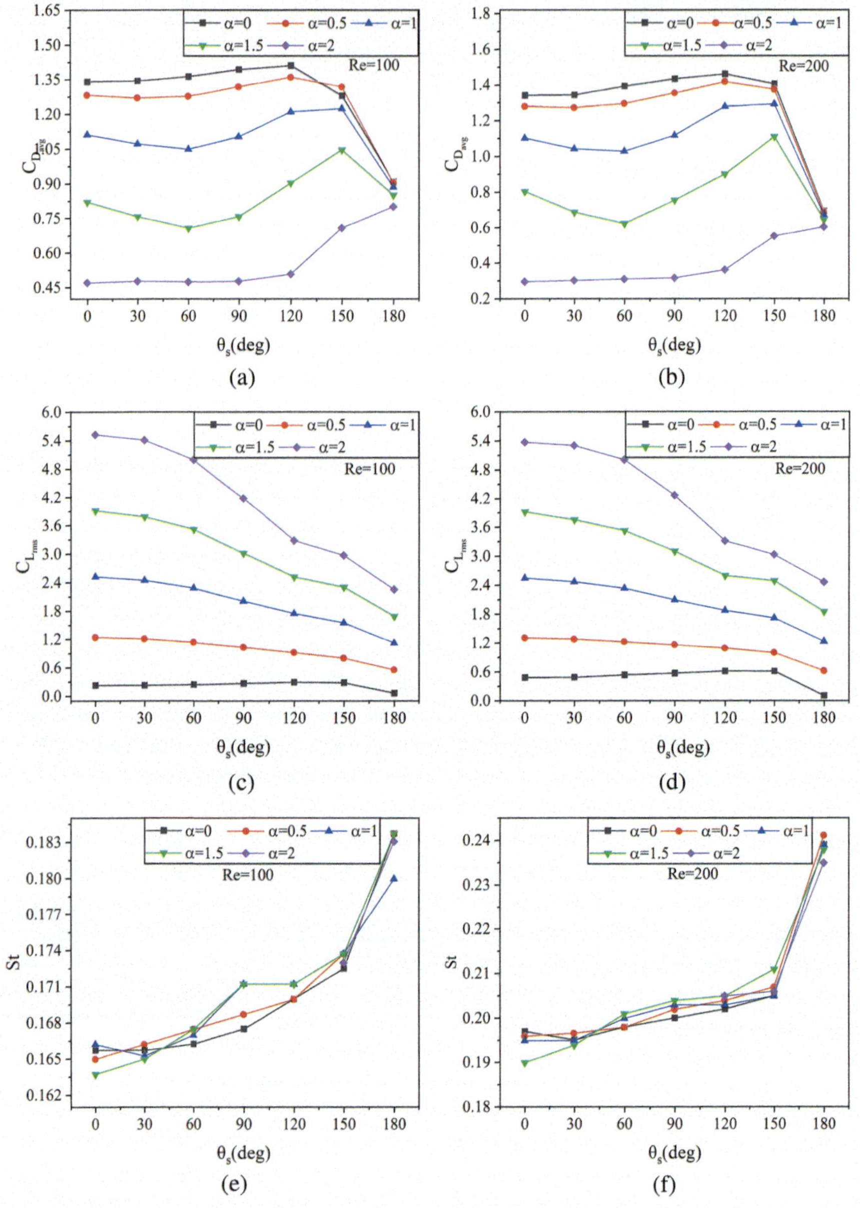

Fig. 6 Variation in drag and lift and Strouhal number with slip distribution for different rotation rates for Re = 100 and Re = 200

Strouhal number: The nondimensional frequency of shedding is what the Strouhal number actually is. As rotation rate is increasing, the frequency of shedding is increasing. For a value of α, maximum shedding frequency is recorded for fully slip condition. Both advection along the cylinder's surface and diffusion through the boundary layer contribute to the rate of change of the vortices there (Legendre et al. 2009). Strength of advection increases with tangential velocity at the cylinder surface. As discussed before, for $\alpha = 2$, corresponding to Re $= 100$, $0 \leq \theta_s \leq 140°$ and for Re $= 200$, $0 \leq \theta_s \leq 90°$ are no-shedding regions. That is why there is no graph plot for $\alpha = 2$.

3.2 *Comparison of Reynolds Numbers*

Figure 7 depicts the comparison graph between two different Reynolds numbers.

Drag coefficient: For $\alpha = 0$, there is a high drag coefficient observed at a Reynolds number of 200, and this behavior shifts as the rotation rate increases to 2. Since friction drag is of the same scale of order as the pressure drag, the total drag has a rather strong reliance on the Reynolds number for high rotation speeds (Kang et al. 1999). Near wall vorticity effect increases with increasing effective near wall fluid velocity. For both Re, when $\alpha = 0$, maximum C_D is corresponding to $\theta_s = 120°$. It moves to 150° as α changes to 1, and then to 180° as $\alpha = 2$. Corresponding to $\alpha = 2$, the average drag coefficient exhibits an increasing trend from no slip to all slip conditions due to turbulence and as a results, pressure drag increases (Fig. 7i). Re-appearing of vortex shedding further increases drag force on the cylinder surface.

Lift coefficient: For a stationary cylinder, there is a peak for rms lift coefficient at $\theta_s = 120^0$ (Fig. 7b). According to Li et al. (2014), the boundary layer seperation angle moves downstream as slip distribution increases. When the seperation point enter into slip region, overall lift decreases. Up to $\theta_s = 120°$, lift coefficient increases and later seperation point is at slip zone. When the cylinder is rotating, the rms lift coefficient decreases with the slip distribution from a maximum value (Fig. 7d–j). Also for a rotating cylinder, the slip distribution over the surface has the potential to reduce the lift coefficient. The effect of these two Reynolds numbers on lift diminishes at high rotation rates (Fig. 7j) (Kang et al. 1999). This suggests that slip can significantly affect a rotating cylinder's lift characteristics, resulting in less lift generation than in situations where slip is absent or where slip is distributed differently.

3.3 *Reappearing Vortex Shedding*

For both Reynolds numbers, $\alpha = 2$ is steady flow regime up to some θ_s. Re $= 100$ have steady flow region for $\theta_s \leq 140°$ and Re $= 200$ have steady flow region for $\theta_s \leq 90°$. On some critical slip azimuth angle (θ_{s_c}), vortex shedding re-appears and

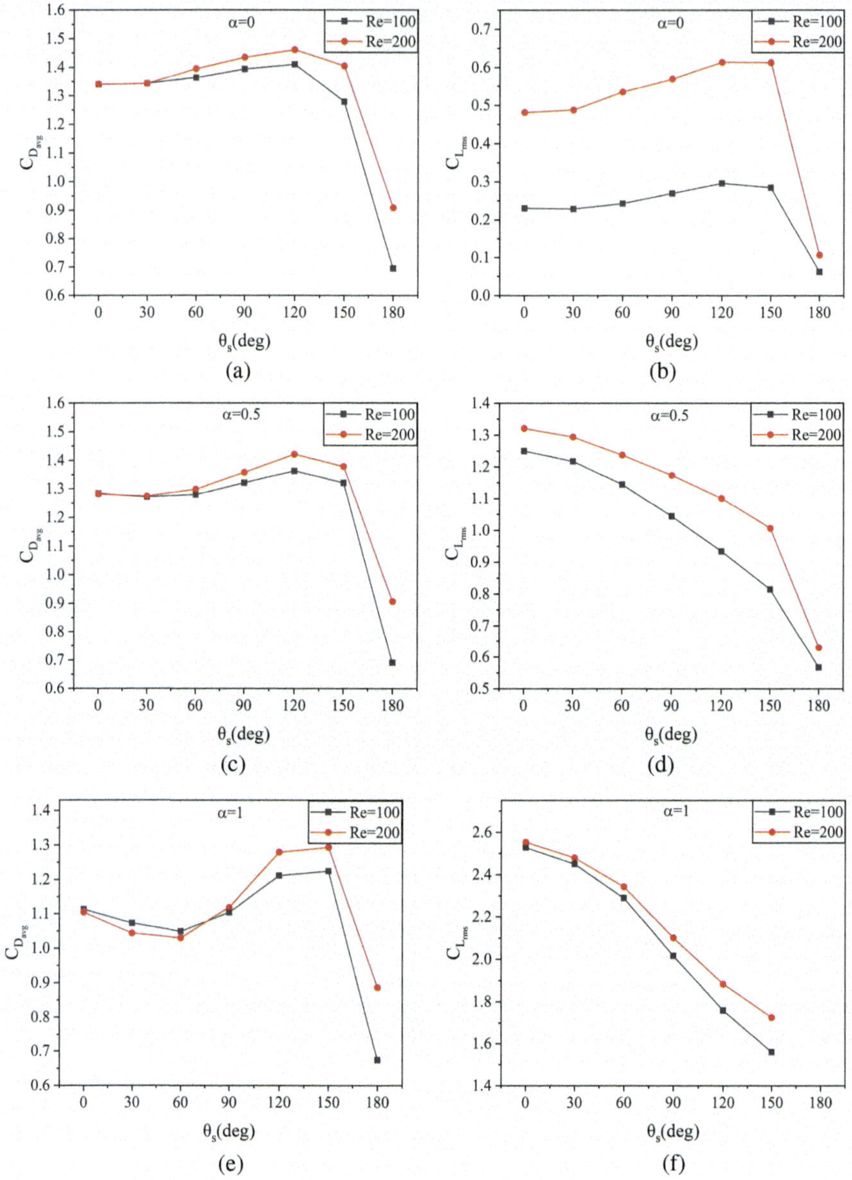

Fig. 7 Lift and drag comparison among different Reynolds numbers

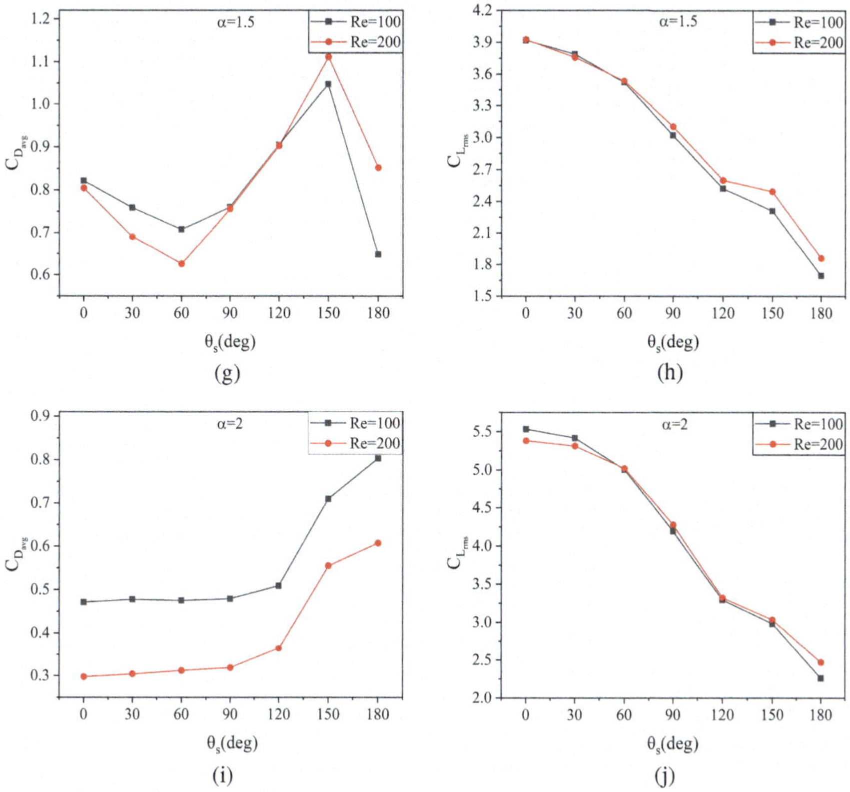

Fig. 7 (continued)

flow becomes unsteady with asymmetric vortices on the downstream. Generated vortices at critical angle are weak and disappears fast. In Figs. 8a and 9a, the total drag divided into pressure drag and skin friction drag and plotted versus the slip distribution (F_D represents total drag, F_P represents Pressure drag and F_V is Skin friction drag). The lines of pressure drag and skin friction drag crosses each other at a slip azimuth angle of 145° which is regarded as critical angle. The same effect observed for Re = 200 but critical angle observed at 95°. At the critical slip azimuth angle, vortex shedding appeared and strength of vortices increases up to full slip.

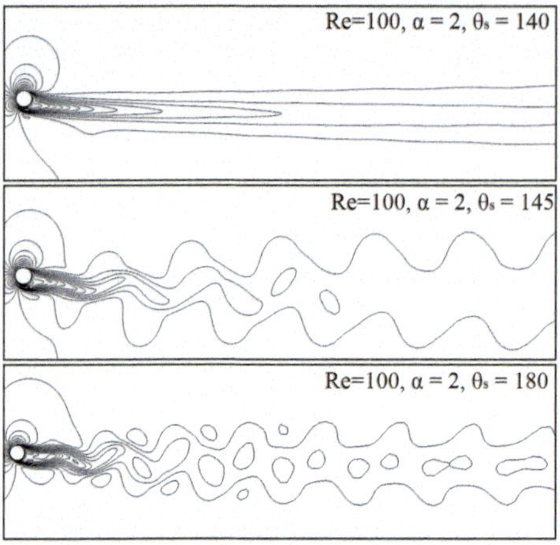

(a) Velocity field for rotating cylinder ($\alpha = 2$) at a time instant when the Coefficient of lift is at its highest point for (a) $\theta_s = 140°$ and (b) $\theta_s = 145°$ and (c) $\theta_s = 180°$

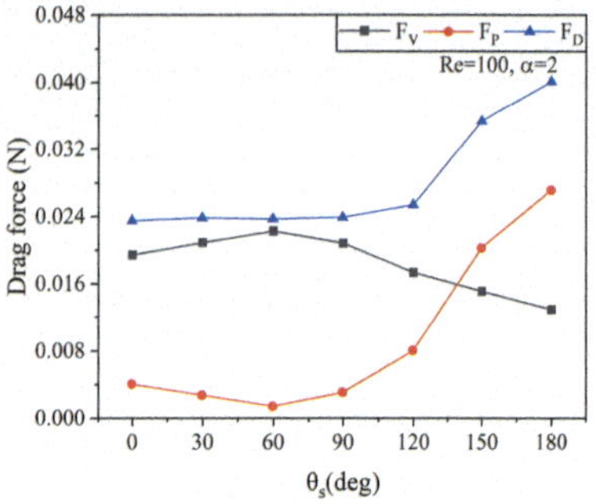

(b) Contributors to total drag for $\alpha = 2$.

Fig. 8 Re = 100: onset of vortex shedding-drag force relationship

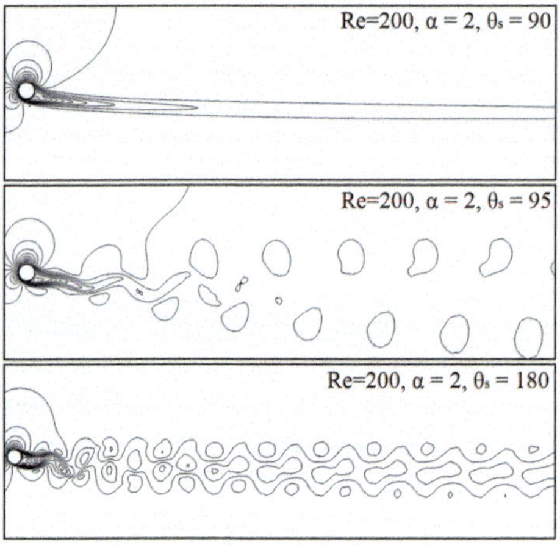

(a) Velocity field for rotating cylinder ($\alpha = 2$) at a time instant when the Coefficient of lift is at its highest point for (a) $\theta_s = 90°$ and (b) $\theta_s = 95°$ and (c) $\theta_s = 180°$

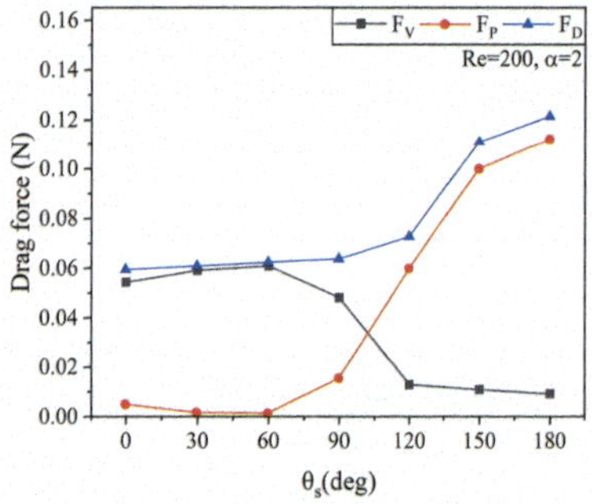

(b) Contributors to total drag for $\alpha = 2$.

Fig. 9 Re = 200: onset of vortex shedding-drag force relationship

4 Conclusions

The study focused on investigating the impact of slip distribution over the cylinder on the drag and lift coefficients. The analysis was conducted for Reynolds numbers 100 and 200, considering five different rotation rates. The study's primary goal was to look at how slip distribution and rotation together affected the cylinder's drag and lift characteristics. The results revealed that the introduction of slip can effectively reduce the overall drag experienced on the cylinder surface within the range of rotation rates $0 \leq \alpha_s \leq 1.5$. However, for a rotation rate of $\alpha = 2$, the presence of slip actually increases the drag experienced by the cylinder. Additionally, the slip distribution was found to reduce the lift experienced by the cylinder, while rotation was observed to increase the lift. The occurrence of vortex shedding was observed, which was absent until a slip azimuth angle (θ_s) of 140° for Re = 100 and 90° for Re = 200, specifically for a rotation rate of $\alpha = 2$. The critical slip azimuth angles, which mark the onset of vortex shedding, are identified as 145° degrees for Reynolds number 100 and 95° degrees for Re = 200. These findings highlight the complex interplay between slip, rotation, drag, and lift, and emphasize the importance of considering both factors in analyzing and predicting the aerodynamic behavior of rotating cylinders.

References

Badr, H. & Dennis, S. Time-dependent viscous flow past an impulsively started rotating and translating circular cylinder. *Journal Of Fluid Mechanics*. **158** pp. 447–488 (1985)

Clift, R., Grace, J. & Weber, M. Bubbles, drops, and particles. (Courier Corporation, 2005)

Coutanceau, M. & Menard, C. Influence of rotation on the near-wake development behind an impulsively started circular cylinder. *Journal Of Fluid Mechanics*. **158** pp. 399–446 (1985)

Daniello, R., Muralidhar, P., Carron, N., Greene, M. & Rothstein, J. Influence of slip on vortex-induced motion of a superhydrophobic cylinder. *Journal Of Fluids And Structures*. **42** pp. 358–368 (2013)

Kang, S., Choi, H. & Lee, S. Laminar flow past a rotating circular cylinder. *Physics Of Fluids*. **11**, 3312–3321 (1999)

Legendre, D., Lauga, E. & Magnaudet, J. Influence of slip on the dynamics of two-dimensional wakes. *Journal Of Fluid Mechanics*. **633** pp. 437–447 (2009)

Li, D., Li, S., Xue, Y., Yang, Y., Su, W., Xia, Z., Shi, Y., Lin, H. & Duan, H. The effect of slip distribution on flow past a circular cylinder. *Journal Of Fluids And Structures*. **51** pp. 211–224 (2014)

Mittal, S. & Kumar, B. Flow past a rotating cylinder. *Journal Of Fluid Mechanics*. **476** pp. 303–334 (2003)

Mittal, S. Excitation of shear layer instability in flow past a cylinder at low Reynolds number. *International Journal For Numerical Methods In Fluids*. **49**, 1147–1167 (2005)

Pralits, J., Brandt, L. & Giannetti, F. Instability and sensitivity of the flow around a rotating circular cylinder. *Journal Of Fluid Mechanics*. **650** pp. 513–536 (2010)

Prandtl, L. & Others The Magnus effect and windpowered ships. *Naturwissenschaften*. **13**, 93–108 (1925)

Ren, Q., Xiong, Y., Yang, D. & Duan, J. Flow past a rotating circular cylinder with superhydrophobic surfaces. *Acta Mechanica*. **229** pp. 3613–3627 (2018)

Tokumaru, P. & Dimotakis, P. Rotary oscillation control of a cylinder wake. *Journal Of Fluid Mechanics*. **224** pp. 77–90 (1991)

Mixing Characteristics of a Subsonic Twin Jet Configuration

Chitimada Narendrakumar and K. P. Sinhamahapatra

Abstract Turbulent jets have numerous applications in aerospace and several other industries. In particular, studying turbulent flows issuing from multi-jets is vital in designing the optimal aircraft for thrust vectoring and stealth applications. This paper investigates the basic multi-jet, a twin jet issuing from circular orifices, with a spacing-to-diameter ratio of 2 at a jet exit Mach number of 0.8 and was simulated using a RANS-based two equation turbulence model SST $k - \omega$. To study the effect of twin jets on mixing characteristics, the potential core length, turbulent intensity, decay and spread rates, and mean velocity profiles in the transverse direction are studied and compared with the free jet case.

Keywords Twin jet · Free jet · Mixing · Decay rate · Half-width

1 Introduction

Turbulent jets are important and widely studied flow phenomenon in fluid dynamics due to their prevalence in various engineering and industrial applications. These flows are characterized by the rapid mixing of a high-speed fluid jet with a surrounding ambient fluid, which results in the formation of complex and highly turbulent structures, which enhances the mixing. In the aerospace field, turbulent jets are commonly encountered in aircraft exhaust nozzles, orifices, pipes, and other similar devices. The proper understanding and control of these flows are crucial for optimizing processes such as noise reduction in aircraft exhaust, combustion chambers, cooling, mixing, and chemical reactions. In recent years, various experimental and numerical studies have been conducted on turbulent jets and it has been discovered that the flow's mixing characteristics and turbulence transport are delicate to initial conditions such as Reynolds number at the exit, nozzle type, nozzle geometry, and spacing. Of particular interest have been turbulent jets from non-circular nozzles, as these offer a

C. Narendrakumar (✉) · K. P. Sinhamahapatra
Department of Aerospace Engineering, IIT Kharagpur, West Bengal, India
e-mail: narendrakumar.iitkgpian@gmail.com

G. Rajesh et al. (eds.), *Proceedings of the 15th International Symposium on Experimental and Computational Aerothermodynamics of Internal Flows*,
https://doi.org/10.1007/978-981-96-4082-9_14

passive way to shape the turbulence structure and transport to meet specific design needs. Previous research, including work with elliptic, rectangular, star-shaped, and triangular nozzles (Quinn 2007; Mi and Nathan 2010) has observed the occurrence of the axis-switching phenomenon and confirmed that it's the primary reason for improved mixing in non-circular jets as compared to circular jets.

Along with non-circular jets, jet mixing can also be increased by employing the multi-jet concept (Morris et al. 2020), practically seen in fighter aircraft and injection systems. In the multi-jet flows, a twin jet is fundamental, and understanding the flow field of twin jets helps comprehend the flow behaviour of multi-jets (Aleyasin and Tachie 2019; Narendra and Sinhamahapatra 2023). Twin jets consist of two characteristic points: a merging point and a combined point. These two points divide the flow of the twin jet into three regions: converging region, merging region, and combined region, respectively (Narendra et al. 2023; Narendra and Sinhamahapatra 2023). The near-field flow dynamics of two-dimensional parallel twin jets was experimentally investigated by Tanaka (1970) and found that two jets attract each other due to the interaction of individual jets. Also, in another study by the same author (Tanaka 1974), the twin jet flow field in the combined region was investigated and reported that the velocity profiles in this region are similar to the single jet, but turbulent intensities show some discrepancy with respect to the single jet. Laban et al. (2019) investigated the effects of nozzle spacing on twin jets mean velocity and turbulent quantities using planar particle image velocimetry at a Reynolds number of 10,000. The results showed that, with increasing spacing ratio, the interaction between jets was delayed, indicated by enhanced velocity decay rate. The Kumar and Sinhamahapatra (2023) of the current study also studied the effects of spacing ratio on non-circular twin jets in a similar manner as Laban et al. (2019), but the spacing distance was much smaller and observed that the reduction in spacing ratio leads to an intense rapid interaction between the jets, verified by reduced velocity decay rate.

Since 1970s the computational fluid dynamics (CFD) methods are largely used in aerospace and mechanical applications. In particular, CFD simulations can provide valuable insights into the behavior of turbulent jets, including the development and mixing of the jet with the surrounding fluid, and the decay of the jet. In a recent study, Miller et al. (1995) used DNS to investigate non-circular jets, providing a detailed analysis of the entrainment rate and axis-switching phenomenon. Holdo and Simpson (2002) utilized LES to analyze a rectangular turbulent free jet with a 10:1 aspect ratio and found good agreement with experimental results only in the far-field region. While DNS is highly accurate at low Re and LES can be used at high Re, both methods require significant computational costs and time. As a result, RANS-based solvers are frequently used in industrial and engineering applications due to their ability to simulate high Re with low computational cost and time. Additionally, RANS-based solvers have been shown to be effective in solving jet flows with minimal loss of accuracy. The flow field of two-dimensional twin parallel jets was investigated experimentally and numerically by Anderson and Spall (2001) using an x-type hot-wire probe and RANS-based turbulence models. They reported that numerical results predicted the merge and combined points with good accuracy and were consistent with experimental results. In their study, Durve et al. (2012) utilized the RANS-based Reynolds stress model to numerically analyze the mixing properties of twin and triple jets with both equal and unequal velocity ratios. The findings indicated that the triple jet scenario resulted in a more rapid decay of velocity and spread of the jet than both twin and single jet configurations. In addition, compared to the triple jet, the twin jet flow exhibited smaller turbulent fluctuations due to the deflection of recirculation vortices towards the jet's symmetry line.

Studies on single and twin jets have been conducted for many years with the aim of understanding the complex flow field phenomena and its effect on jet mixing properties. However, most previous experimental and numerical studies were only limited to low subsonic and subsonic Mach numbers and direct comparison of twin jet flow field with single jet flow field is not yet discussed significantly. So, the present investigation is focused on analysing the twin jet flow field emerging from two circular orifices using the RANS-based SST $k - \omega$ turbulence model at high subsonic Mach number of 0.8. The single circular jet was also simulated with the same testing conditions to compare it with the flow field of twin jet. For the present investigation, the twin jets are separated by spacing to diameter ratio (S/D) of 2 where S is the spacing provided between the twin jets and D is the diameter of the orifice. The chosen model has an exit area of 176.625 mm^2 and mass flow rate of 0.06 kg/s.

2 Computational Methodology

In this study, the authors employed the finite volume based ANSYS Fluent software (Anderson and Wendt 1995) package with SST $k - \omega$ turbulence model to simulate the steady-state compressible free and twin jet flows. The simulations were conducted using a pressure-based coupled solver for steady-state RANS simulations. The jet fluid was modeled as air, an ideal gas, and viscosity was calculated using the Sutherland three-coefficient method. The under-relaxation factors for momentum and pressure were set to 1, and the convection terms were discretized using the bounded QUICK scheme (Leonard 1979). Viscous terms were solved using a second-order central differencing scheme, and the pressure-velocity coupling was implemented using the SIMPLEC procedure with a skewness correction of 1 (Patankar 2018). The simulations were performed with a double-precision solver setting to minimize computer-generated round-off errors. The iterative solution procedure continued until convergence of the results was achieved, considering the residuals converged when the absolute values of pressure, temperature, and velocity components were below 10^{-6}. The turbulence model employed in the present numerical study is the SST $k - \omega$ model (Menter 1994), where turbulent kinetic energy is denoted by k and specific rate of dissipation is denoted by ω. The SST $k - \omega$ model is advantageous as it switches from $k - \omega$ in the near fields to the $K - \varepsilon$ model in the far fields. This model is particularly suitable for flows involving adverse pressure gradients, shock waves, and transonic flows.

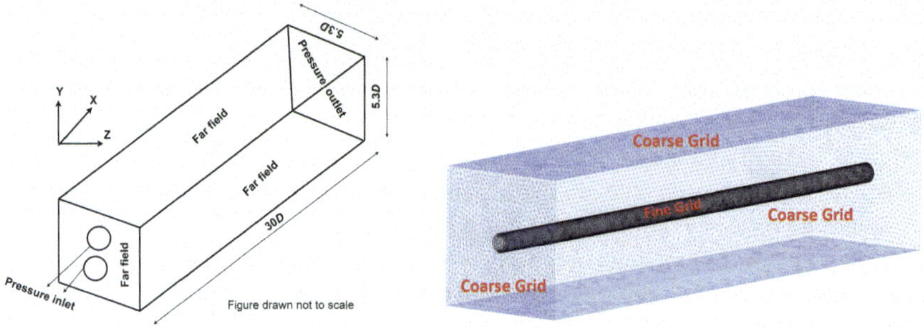

(a) (b)

Fig. 1 **a** Computational domain with boundary conditions, **b** Coarse mesh of computational domain

The numerical simulations of the free jet and twin jet were conducted within a rectangular computational domain measuring 30D × 5.3D × 5.3D, as depicted in Fig. 1a. To generate the computational mesh, an unstructured approach utilizing triangular elements was employed with different mesh methods available in the ANSYS Workbench grid generation tool. The mesh near the jet boundary was refined to ensure accurate prediction of the shear layer and recirculation region. Figure 1b illustrates the coarse unstructured mesh around the circular orifice near the exit of the computational domain. Additionally, Fig. 1a presents the applied boundary conditions for the computational domain. The pressure and temperature within the computational domain were set as ambient conditions, specifically 101,325 Pa and 300 K, respectively. At the inlet of the domain, a stagnation pressure of 154,453.80 Pa and a stagnation temperature of 300 K were fixed to simulate an adiabatic jet with a Mach number of 0.8. The outlet conditions were set as atmospheric, with a static pressure of 101,325 Pa and a temperature of 300 K. Far field conditions were defined with ambient pressure and temperature of 101,325 Pa and 300 K, respectively.

3 Grid Independence and Validation

To assess the grid dependency of the free and twin jet simulations, three different meshes consisting of 1.6, 2.7, and 3.7 million cells were employed. It was found that meshes with 2.7 million or more elements produced nearly identical results for the current study. Therefore, a mesh size of 2.7 million cells was chosen for the present investigation. Although a finer mesh size of 3.7 million cells was tested, it did not yield significant improvements in the predictions of jet velocity. Consequently, for computational convenience, a grid size of 2.7 million cells was adopted for all cases of free and twin jet flow.

To evaluate the performance of the selected SST $k - \omega$ turbulence model, the obtained results from the simulations of free and twin jets were compared with

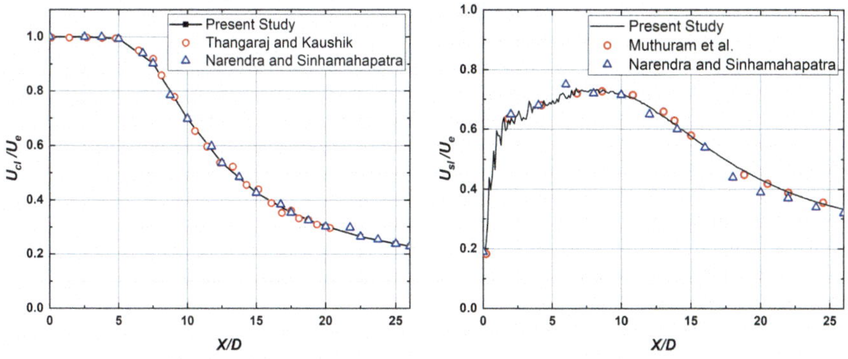

(a) Free jet mean center line velocity

(b) Twin jet mean symmetry line velocity

Fig. 2 Validation of free and twin jet mean velocity

existing experimental (Thangaraj et al. 2022; Muthuram et al. 2020) and numerical (Narendra and Sinhamahapatra 2023) data from the literature. Figure 2a presents the decay of centerline velocity (U_{cl}) for an M 0.8 jet in the case of the free jet. The results obtained from the SST $k - \omega$ turbulence model closely resemble the findings of Kumar and Sinhamahapatra (2023), Thangaraj et al. (2022), demonstrating good agreement up to a distance of 10 jet diameters. Beyond this point, the predicted jet decay slightly deviates. In Fig. 2b, a comparison of the mean symmetry line velocity (U_{sl}) for the twin jet is shown, along with experimental and numerical results. The mean velocity decay obtained through the present numerical solution reasonably matches the findings of Muthuram et al. (2020) and the previous work by the authors (Narendra and Sinhamahapatra 2023). The twin-jet profile exhibits good agreement in all flow regions, aligning well with the experimental and numerical data.

4 Results and Discussion

4.1 Evolution of Mean Velocity and Turbulent Intensity in the Streamwise Direction

The analysis of turbulent jet flows commonly involves examining the jet profile in both the streamwise and transverse directions to understand their mixing characteristics. Figure 3a displays the normalized velocity contours for both free and twin jet configurations. In these contours, the jet velocity and streamwise distance are normalized with the jet's exit velocity and orifice diameter, respectively. In the case of the free jet, the profile exhibits a potential core region where the jet velocity remains constant up to a certain axial distance. Beyond this region, a transition phase occurs due to rapid mixing between the jet and the surrounding fluid. In this phase, the

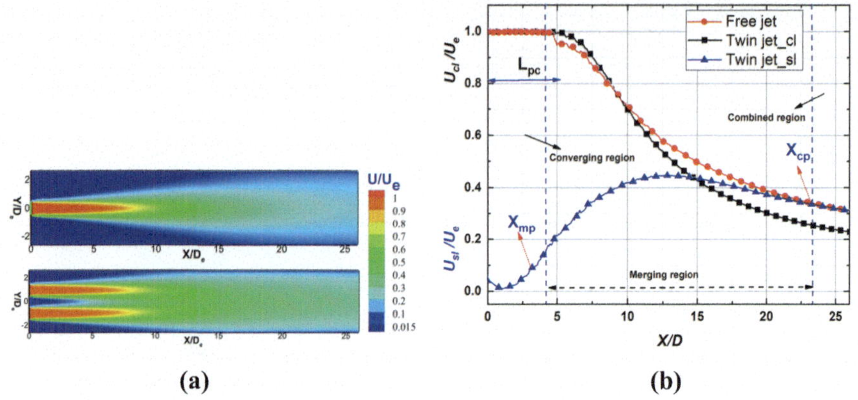

Fig. 3 **a** Mean velocity contours and **b** Streamwise velocity decay of free and twin jets

streamwise velocity decays and linearly varies with the downstream distance. Following the transition region, the free jet profile becomes self-similar. Figure 3a also illustrates the velocity contours for a twin jet with two jets separated by a spacing ratio of 2. Due to the interaction between the individual jets, the flow properties of twin jets differ significantly from those of free jets. A twin jet consists of converging, merging, and combined regions, which help identify the locations where the twin jets merge and combine. As the flow progresses downstream, the centerline velocity, defined along the jet centerline, shifts towards the jet's symmetry line ($Y/D = 0$). As a result, the twin jets merge into each other and eventually combine into a single or free jet.

One approach to comprehending the mixing characteristics of jet flow is to analyze the potential core region (L_{pc}) of each individual jet. L_{pc} is defined as the region where the ratio of centerline velocity (U_{cl}) to exit velocity (U_e) equals 0.99 (Morris et al. 2020; Tanaka 1974). Figure 3b depicts the normalized decay of centerline velocity for both free and twin jets, as well as the decay of velocity along the symmetry line in the streamwise direction. In the case of the twin jet, since the upper and lower jets exhibit statistical similarity about the jet's symmetry line, the estimation of L_{pc} is based on the upper jet. The figure reveals that both the free jet and twin jets maintain an L_{pc} at $X/D = 5$, indicating that the twin jet configuration does not significantly impact the L_{pc} in this region. However, beyond the L_{pc}, the neighboring jets of the twin jet interact and spread inwards and outwards. Consequently, the velocity decays much more rapidly in the twin jet scenario compared to the free jet scenario. Up to $X/D = 9.5$, the twin and free jets exhibit similar mixing behavior. However, beyond this point, the growth of the inner and outer shear layers in the twin jets enhances the efficiency of mixing the jet fluid with the surrounding fluid. This phenomenon becomes evident in the plot after $X/D = 10$.

Another way to assess the turbulent mixing characteristics of the twin jets is by examining their converging, merging, and combined regions. These regions can be identified by analyzing the decay of streamwise symmetry line velocity along the X/D ratio. Along this line, the velocity first rises from almost zero to a maximum value and then remains approximately constant in the flow direction until it merges with the centerline velocity at the combined point. Figure 3b also illustrates the decay of symmetry line velocity in the twin jet configuration, which also provides insights into the locations where the twin jets merge and combine. At the nozzle exit, the twin jets interact due to the presence of a sub-atmospheric region between them, resulting from the entrainment of fluid by the turbulent jet. The deflection of the jet axis causes the two jets to be attracted to each other and eventually merge. The merging point (X_{mp}) is defined as the streamwise location where the recirculation region between the two jets ends (Morris et al. 2020; Aleyasin and Tachie 2019). The combined point (X_{cp}) is the point where the twin jets fully merge into a single jet (Morris et al. 2020; Aleyasin and Tachie 2019). According to the given definitions, the X_{mp} and X_{cp} locations were found to be at 3.52 and 25, respectively and this locations are good in agreement with the previous experimental (Morris et al. 2020; Aleyasin and Tachie 2019) and numerical (Morris et al. 2020; Aleyasin and Tachie 2019) data.

Achieving high turbulent intensity in turbulent jets contributes to greater flow mixing, and there are various methods available to achieve this, including the implementation of the multi-jet concept. Figure 4a presents streamwise contours of turbulent intensity for the simulated free and twin jets. The contours indicate lower levels of turbulent intensity in the core region, while the highest turbulence levels are observed within the shear layer near half-velocity width. In the case of the twin jet, both the outer and inner shear layers play a significant role in mixing the jet fluid with the surrounding ambient fluid. On the other hand, the free jet exhibits only an outer shear layer that expands outward and interacts with the surrounding or atmospheric fluid. The elevated turbulence levels observed in the outer shear layers are a result of vortical structures formed due to the Kelvin-Helmholtz instability. It is widely recognized that the potential core endpoint represents the location where

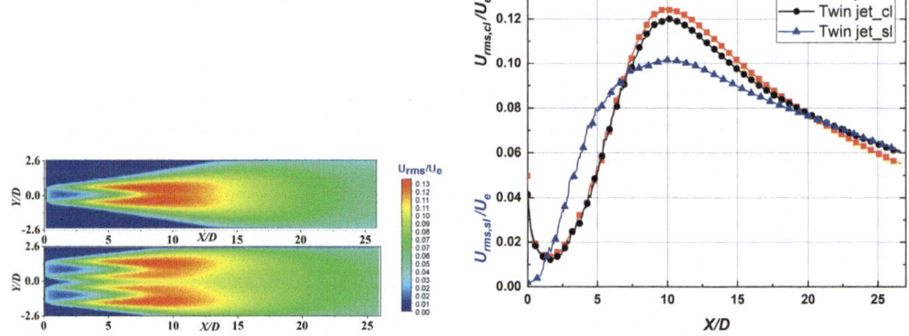

(a) (b)

Fig. 4 **a** Turbulent intensity contours and **b** Distribution of turbulent intensity for free and twin jet maximum mixing is expected to occur along the jet centerline. As the jet develops, the maximum turbulent intensity gradually shifts from the shear layers towards the centerline in the case of the twin jet configuration.

The development of turbulent intensity along the jet centerline and symmetry line is often considered an indicator of mixing capability in the converging and merging regions. Greater growth in turbulent intensity signifies enhanced mixing in turbulent jets. The distributions of streamwise turbulent intensity along the jet centerline ($U_{rms,cl}/U_e$) and symmetry line ($U_{rms,sl}/U_e$) are illustrated in Fig. 4b. Along the centerline, streamwise turbulent intensities exhibit rapid increments and reach their maximum values for both the free jet and twin jet configurations. However, the free jet achieves a higher peak intensity compared to the twin jet due to the deflection of the jet towards the jet symmetry line, resulting in a lower peak value in the twin jet. After reaching the peak, both the free jet and twin jet encounter a sharp decrease in turbulence intensity. A similar pattern is observed along the symmetry line, but the peak values are relatively lower and located further downstream as compared to the centerline. The diffusion of vortices from the inner shear layer towards the centerline and symmetry line (in the case of twin jet) leads to a sudden surge in turbulent intensities. Turbulence intensities are higher on the centerline because these structures diffuse more rapidly towards it compared to the symmetry line, and their positions are much closer to the orifice exit than those on the symmetry line.

4.2 Jet Decay and Spread Rates

The jet emerging from the orifice exit enters into a quiescent ambience, mixes with the atmospheric fluid, and spreads, leading to the velocity decay with the increasing streamwise distance. Along with the L_{pc}, jet centerline velocity decay rate and spread rate are typically used to understand the jet mixing with the ambient fluid. The centerline velocity decay rate is calculated by fitting the following Eq. 1 to the streamwise velocity decay on the jet centerline.

$$\frac{U_e}{U_{cl}} = K_u \left(\frac{X}{D} - \frac{X_s}{D} \right) \tag{1}$$

Here K_u stands for the decay rate, and X_s stands for the kinematic virtual origin of the jet. The Eq. 1 is strictly valid in the far-field region, but many researchers (Morris et al. 2020; Aleyasin and Tachie 2019; Narendra and Sinhamahapatra 2023; Narendra et al. 2023) also used it in the transition region to estimate the decay rate. The velocity decay rate on the centerline of the free and twin jets is shown in Fig. 5a. The values of the decay rates for free and twin jets were calculated within the streamwise range of X/D of 6–26. Based on the Eq. 1, the K_u values of free and twin jets are 0.12 and 0.18, respectively. The greater value of the twin jet decay rate indicates that the mixing of jet fluid with the ambient fluid is more prominent in twin jet case as

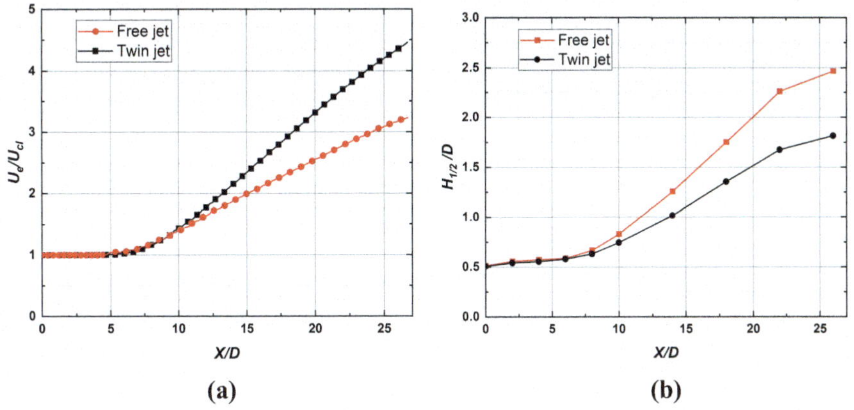

Fig. 5 **a** Mean velocity decay rate and **b** Jet half-width profile for free and twin jet

compared to free jet case. This observation is also verified by the higher decay of centerline velocity in the transition region in Fig. 3b.

The spreading behaviour of jets was investigated by using the half-velocity width plot. The half-velocity width ($H_{1/2}$) represents the distance from the centerline of the jet to the location where the mean centerline velocity is half of the maximum centerline velocity (Morris et al. 2020; Aleyasin and Tachie 2019; Narendra and Sinhamahapatra 2023; Narendra et al. 2023). Figure 5b displays the evolution of the half-velocity width of the jet in the X-Y plane, starting from the origin. From Fig. 5b, it is observed that the half-velocity width is nearly flat near the jet exit for free and twin jets in the potential core region. Beyond L_{pc}, the half-velocity widths of free and twin jets monotonically increase with axial distance.

For the present study, the quantification of the spread rate is achieved by fitting the following Eq. 2 to the half-velocity width profile presented in Fig. 5b.

$$\frac{H_{1/2}}{D} = K_s \left(\frac{X - X_v}{D} \right) \tag{2}$$

Here, K_s and X_v are the rate of spread and geometric virtual origin, respectively. The spread rates are calculated within the stream wise range of 5–26. The K_s value free jet is 0.096, while it is 0.065 for twin jet. The decay and spread rates of the present studied free and twin jet are consistent with previous work of Morris et al. (2020), Aleyasin and Tachie (2019) and current authors (Narendra and Sinhamahapatra 2023).

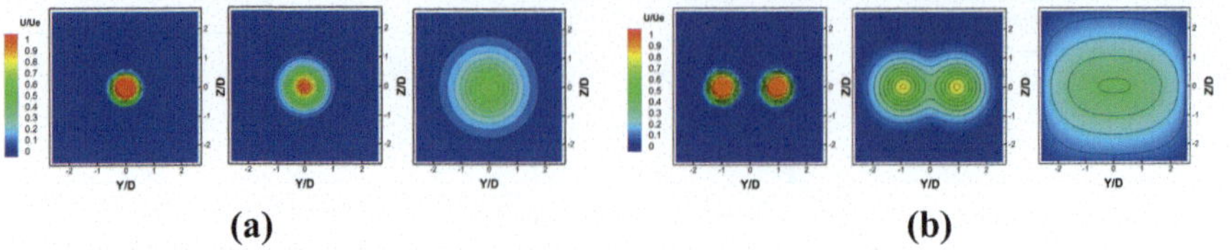

Fig. 6 Transverse velocity contours of **a** free jet and **b** twin jet at various streamwise locations

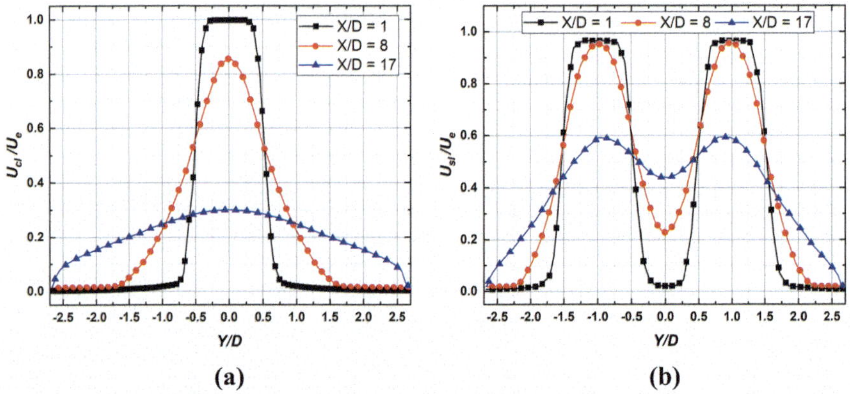

Fig. 7 Transverse velocity profiles of **a** free jet and **b** twin jet

4.3 *Mean Velocity Profiles of Free and Twin Jets*

In order to gain a comprehensive understanding of the behavior of both free and twin jets, it is essential to examine mixing measurements in the transverse direction. Figure 6a illustrates the mean velocity contours of the free jet at specific locations denoted by X/D (= 1, 8 and 17). It can be observed that the mean velocity profiles exhibit a considerable degree of symmetry around the $Y/D = 0$ axis in the case of the free jet. Initially, at $X/D = 1$, the contours reflect the characteristic jet shape of the orifice and are closely spaced, indicating limited mixing at this particular location. Also at $X/D = 1$, it becomes evident that the L_{pc} remains unchanged. As the jet progresses, it rapidly decays towards an asymptotic value beyond the potential core region, due to diffusion and energy dissipation. This behavior is observed until $Y/D = 1$, as depicted in Fig. 7a. Following the transition region, the jet gradually loses its distinct characteristics and eventually behaves similarly to the surrounding ambient fluid. At $X/D = 8$, the absence of a potential core region is notable, and the jet begins to decay from $Y/D = 0$ itself, which can be observed in the velocity contours at these locations. Further downstream, at $X/D = 17$, the contours have lost their initial effects and resemble those of the ambient fluid, as confirmed by Fig. 7a. Moreover, the spacing between contour levels has increased at this location, indicating self-similarity.

The twin jets under consideration are positioned parallel to each other. Figures 6b and 7b display the mean velocity contours and profiles of a twin jet configuration with an S/D ratio of 2. The selected streamwise locations, namely $X/D = 1$, 8, and 17, correspond to the converging, merging, and combined regions, respectively. The velocity profile at $X/D = 1$ represents the distribution typically observed at the jet exit plane, and the presence of the neighboring jet does not significantly impact the jet behavior. At the nozzle exit ($X/D = 1$) along the centerline direction, the mean velocity profile exhibits a small region with no influence from either jet, followed by a rapid increase in flow velocity. Subsequently, the flow velocity remains relatively constant before decreasing towards the outer edge of the flow. At this streamwise location, a wake region is present in the twin jet, remaining unaffected until $Y/D = 0.25$. The formation of this wake region is attributed to the lack of interaction between the two individual jets due to the spacing between them. Moving to $X/D = 8$, the mean velocity profiles correspond to the merging region of the twin jets. No consistent velocity profile is observed at this location, but the wake region persists, and the velocity peak is observed at approximately $Y/D = 0.95$. Finally, at $X/D = 17$, the mean velocity profile represents the combined region of the twin jet. The wake region still exists, and the peak velocity is observed around $Y/D = 1$. The combined point refers to the specific streamwise location where a single peak pattern emerges in the velocity profile of the merged jet ($X/D \geq 17$). Downstream of the combined point, the characteristics of the mean velocity profile with a single peak become increasingly similar to those of a self-similar single jet. Consequently, after $X/D = 17$, the twin jet gradually loses its effect and behaves akin to a free jet.

5 Conclusions

The numerical simulations utilized the RANS-based SST $k - \omega$ turbulence model to investigate the flow field of a twin jet operating at a Mach number of 0.8. Several mixing characteristics, including the potential core length, turbulent intensity profiles, decay rate, spread rate, and transverse mean velocity profiles, were examined and compared to those of a free jet. The findings indicated that the velocity decay in the transition region was more rapid for the twin jet compared to the free jet. The turbulent intensity profiles showed that the highest peaks of turbulence occurred at the end of the potential core length for both the free and twin jets. The decay rate was higher for twin jets, while the spread rate was greater for the free jet. The transverse velocity profiles demonstrated the presence of a wake region in the twin jets due to the spacing, which could be a potential explanation for their superior mixing capabilities compared to the free jet.

References

Aleyasin, S. S., Tachie, M. F.: Statistical properties and structural analysis of three-dimensional twin round jets due to variation in Reynolds number. International Journal of Heat and Fluid Flow, 76, 215–230 (2019).

Anderson, JD., Wendt, J.: Computational fluid dynamics, Vol 206. Springer (1995).

Anderson, E. A., Spall. R. E.: Experimental and numerical investigation of two-dimensional parallel jets. J. Fluids Eng., 123(2), 401–406 (2001).

Durve, A. P., Banarjee. I., Padmakumar. G., Vaidyanathan. G.: Numerical investigation of mixing in parallel jets. Nuclear engineering and design, 242, 78–90 (2012).

Holdo, A., Simpson, B.: Simulations of high aspect-ratio jets. International Journal for Numerical Methods in Fluids, 39(4), 343–359 (2002).

Kumar, Narendra, Chandiramani, P. N., Sinhamahapatra K. P., Pranaykumar S.: A CFD Study on Mixing Characteristics of Non-Circular Twin Orifice Jets. In AIAA SCITECH 2023 Forum, pp. 1616 (2023).

Kumar, Narendra, Sinhamahapatra K. P.: Mixing Characteristics of Circular and Non-Circular Twin Jets at a Transonic Mach number 0.8. In Fluid Mechanics and Fluid Power (Vol. 3) Select Proceedings of FMFP 2021, pp. 363–368. Singapore: Springer Nature Singapore, (2023).

Kumar, Narendra, Sinhamahapatra K. P.: The effects of spacing to diameter ratio on mixing characteristics of circular and elliptical twin jets. Proceedings of the Institution of Mechanical Engineers, Part G: Journal of Aerospace Engineering, 09544100231155692, (2023).

Laban, A., Aleyasin, S. S., Tachie, M. F., Koupriyanov, M.: Experimental investigation of nozzle spacing effects on characteristics of round twin free jets. Journal of Fluids Engineering, 141(7) (2019).

Leonard, BP.: A stable and accurate convective modelling procedure based on quadratic upstream interpolation. Computer methods in applied mechanics and engineering, 19(1), 59–98 (1979).

Menter, FR.: Two-equation eddy-viscosity turbulence models for engineering applications. AIAA journal, 32(8), 1598–1605 (1994).

Mi, J., Nathan, G J.: Statistical properties of turbulent free jets issuing from nine differently-shaped nozzles. Flow, turbulence and combustion, 84, 583–606 (2010).

Miller, R. Madnia, C., Givi P.: Numerical simulation of non-circular jets. Computers and Fluids, 24(1), 1–25 (1995).

Morris, E. M., Aleyasin, S. S., Biswas N., Tachie M. F.: Turbulent Properties of Triple Elliptic Free Jets With Various Nozzle Orientation. Journal of Fluids Engineering, 142(3) (2020).

Muthuram, A., Thanigaiarasu, S., Divvela. R.: Nozzle geometry effect on twin jet flow characteristics. International Journal of Turbo & Jet-Engines, 39, 525–532 2020.

Patankar, SV.: Numerical heat transfer and fluid flow. CRC press (2018).

Quinn, WR.: Experimental study of the near field and transition region of a free jet issuing from a sharp-edged elliptic orifice plate. European Journal of Mechanics-B/Fluids, 26(4), 583–614 (2007).

Tanaka, E.: The interference of two-dimensional parallel jets: 1st report, experiments on dual jet. Bulletin of JSME, 13(56), 272–280 (1970).

Tanaka, E.: The interference of two-dimensional parallel jets: 2nd report, experiments on the combined flow of dual jet. Bulletin of JSME, 17(109), 920–927 (1974).

Thangaraj, T., Kaushik, M., Thangaiarasu, S.: Effects of circular and non-circular nozzle exit geometries on subsonic and supersonic jet propagation's. Proceedings of the Institution of Mechanical Engineers, Part G: Journal of Aerospace Engineering, 237, 209–229 (2022).

Experimental and Numerical Investigation of Negative Buoyancy Jets at Laminar Conditions

P. V. Sai Hemant, N. Aditya, Ayan Sarkar, K. Jishad, and Chockalingam Prathap

Abstract This study focused on the analysis of laminar negative buoyant Boussinesq fountains generated by circular nozzles. The behavior of these fountains was thoroughly investigated using experiments and simulations. High-speed shadowgraphy and particle Image Velocimetry techniques were employed to study the jet dynamics. The high-speed images revealed that the jet initially ascended to a larger length, known as initial height, subsequently, it stabilised at a slightly shorter height known as steady state penetration height. The steady-state penetration height was estimated from the shadowgraphy data and validate with the Philippe et al.'s correlation. Additionally, three-dimensional simulations were performed using a commercial software to predict the jet development and the penetration height. Moreover, the transient flow behaviour was analysed both experimentally and numerically. Non-dimensionalised plots were plotted between $Ri^{0.5}.H_s/D$ and Reynolds number which showed good agreement between the experimental, numerical and literature data.

Keywords Negative buoyant · Boussinesq · Penetration height · PIV

1 Introduction

A buoyant jet refers to a jet of fluid having a certain density being injected into another fluid having a different density and provided that both the fluids are in the same phase. These types of flows occur in many practical applications such as room ventilation, jet mixing (the process of combining a fresh fluid with other contents in a tank), waste disposal systems, seawater desalination plants, leakages of pipeline in water-bodies like oceans, and natural phenomena such as explosive volcanic jets, plume and cloud motion in the atmosphere, and so on. Buoyancy jets are categorized

P. V. Sai Hemant · N. Aditya · A. Sarkar · K. Jishad · C. Prathap (✉)
Department of Aerospace Engineering, Indian Institute of Space Science and Technology, Thiruvananthapuram, Kerala, India
e-mail: prathapc@iist.ac.in
URL: http://www.iist.ac.in

G. Rajesh et al. (eds.), *Proceedings of the 15th International Symposium on Experimental and Computational Aerothermodynamics of Internal Flows*,
https://doi.org/10.1007/978-981-96-4082-9_15

Table 1 Validity of Boussinesq approximation as per different literature

References	Valid for	Proposed approximation
Ai et al. (2006)	Forced plumes	$\left(\frac{\rho_b - \rho_i}{\rho_b}\right) \approx 0.15$
Baddour and Zhang (2009)	Fountains	$\left(\frac{\rho_b - \rho_i}{\rho_b}\right) \approx 0.003$
Crapper and Baines (1977)	Positive Jets	$\left(\frac{\rho_b - \rho_i}{\rho_b}\right) \approx 0.05$

into two types based on the orientation of the buoyancy force and the direction in which the liquid jet propagates. They are:

1. Positive Buoyant Jet.
2. Negative Buoyant Jet.

A positive buoyant jet refers to a jet where the buoyancy force and the momentum of the injected jet are aligned in the same direction. Conversely, when the buoyancy force and jet propagation were oriented in opposite directions, the jets were classified as negative buoyant jets. Negative buoyant jets were generated by injecting a lighter fluid vertically downward into a denser fluid or by injecting a denser fluid vertically upward into a lighter fluid. Positive buoyant jets were the exact opposite of the same. These jet classifications involve the participation and mixing of two fluids, namely the injected fluid and the bath fluid. The properties of these two fluids determine the categorization of buoyancy jets into either miscible or immiscible, as well as Boussinesq or non-Boussinesq. Different researchers proposed different ranges of relative density ratio to be eligible for Boussinesq approximation as listed in Table 1.

An extensive research has been conducted in the domain of buoyancy jets with the primary objective of comprehending the influence of non-dimensional parameters, such as Reynolds number, Froude number, Archimedes number, Richardson number, and others, on the penetration height. Different terminologies of penetration heights were reported in the literature, such as, initial height, steady state penetration height, laminar length, and others. Turner and Baines et al. (1990) were one among the first to investigate the upward Boussinesq fountains. They measured the initial height of the jet and formulated a non-dimensional expression to find the same as: $H/D \propto Ri^{-1/2}$. They investigated the effects of buoyancy flux by varying the density difference between the injected and the bath fluid and the momentum flux by varying the jet exit velocity on the jet penetration height. Lin and Armfield (2003) numerically studied a weak laminar (i.e. $Fr < 2.1$) Boussinesq circular fountains and obtained a non-dimensional correlation of penetration height as:

$$\frac{H}{D} = Fr \qquad \text{if } 0.14 \lesssim Fr \lesssim 0.71 \text{ and } Re = 400$$

$$\frac{H}{D} = Fr \cdot Re^{-1/2} \text{ if } Fr = 0.71 \text{ and } 10 \lesssim Re \lesssim 400$$

Pantzlaff and Lueptow (1999) performed experiments on both positive and negative buoyant turbulent round jets. They used an acrylic cylindrical tank, with bath fluid as aqueous potassium chloride (KCL) solution and the injected fluid as water in case of positive buoyancy case and the vice versa for the negatively buoyant case. Their study highlighted the importance of momentum and buoyancy fluxes in determining the development of the injected jet. Graphs depicting the penetration height of the jet over time were presented in the research paper, offering fundamental insights into the behavior of buoyancy-assisted jet flow.

Philippe et al. (2005) conducted an experimental investigation on the penetration height of negatively buoyant Boussinesq jets. They developed a theoretical model based on fundamental conservation equations to predict the transient behavior of the jet, and the model's results were validated against their own experimental data. Flow visualization techniques, including high-speed photography and Particle Image Velocimetry (PIV), were employed. Furthermore, the researchers established a correlation between the non-dimensional penetration height and non-dimensional parameters such as Reynolds number and Froude number, H/D $= 0.25 Re^{\phi} Ri^{-0.5}$, where the value of ϕ ranges from 0.5 to 0.6.

Vinoth et al. (2014) performed experimental research on non-Boussinesq fountains with various nozzle geometries. Specifically, they used four different nozzles. Their study revealed that at low Reynolds and Froude numbers, the established fountain remained steady. However, as the Reynolds and Froude numbers increased, the fountain becomes unsteady and displayed periodic oscillations. They classified the flow behavior into three regimes, namely the Steady regime, Flapping regime (Anti-Symmetric Oscillations), and Flapping and Bobbing regime (Anti-Symmetric + Stream-wise Oscillations). They introduced a new height scale using the equivalent diameter (D_e), which is the square root of the cross-sectional area. The empirical formula for the penetration height scale and Froude number is as follows:

$$\frac{H_s}{D_e} \approx \begin{cases} 1.2 Fr_{D_e}^{1/2} & \text{for } 0.5 \lesssim Fr_{D_e} \lesssim 1.6 - \text{Weak fountain} \\ 0.9 Fr_{D_e} & \text{for } 1.6 \lesssim Fr_{De} \lesssim 9.6 - \text{Forced fountain} \end{cases}$$

Hassanzadeh et al. (2021) has investigated positive buoyancy jets generated at the exit of a cylindrical injector. The study investigated various jet features, including jet width, laminar length, spread angle, and others. They presented a non-dimensional correlation for laminar length estimation as:

$$L_m^* \approx \begin{cases} 0, & \text{if } \mathrm{Re} \geqslant 3000 + 0.05\mathrm{Ar} \\ 324.4\left\{\left[2\left(\frac{\mathrm{Re}}{\mathrm{Fr}}\right)\right]^{0.29} + 38.6\right\}\mathrm{Fr}^{-1}\mathrm{Re}^{-1}, & \text{if } 7.4\sqrt{\mathrm{Ar}} < \mathrm{Re} < 3000 + 0.05\mathrm{Ar} \\ 23.7\mathrm{Fr}^{-1.35}, & \text{if } \mathrm{Re} \leqslant 7.4\sqrt{\mathrm{Ar}} \end{cases}$$

The current research focuses on the investigation of negative buoyancy jets using advanced flow visualization techniques, specifically high-speed shadowgraphy and Particle Image Velocimetry (PIV). Moreover, a numerical model was developed to simulate the negative buoyancy jets and was rigorously validated with the present experimental and literature data.

2 Experimental Rig and Procedure

Present experimental rig comprised of a transparent cuboidal bath, and a digital infusion pump to investigate the characteristics of negative buoyant jets at 1 bar, 300 K, and laminar conditions. High-speed Z-type shadowgraphy was used for the visualization of transient phase of the jet development. Two-dimensional Particle image visualization also used to quantitatively measure the velocity field of the jet. The transparent bath had four sidewalls made of float glass, and it had a stainless steel bottom plate. The inner dimensions of the cuboidal bath was carefully designed so that the width of the container exceeded four-times the averaged width of the jet and the depth exceeded fifty-times the largest size of the injector used as suggested by Papanicolaou et al. (2008) to prevent the wall effects. Injectors of different diameters, 1.19, 2.146 and 3.988 mm were meticulously designed to account for the entry length of the jet, ensuring the establishment of fully-developed flow conditions. Throughout the conducted experiments, clean water was employed as the denser fluid, while a mixture of iso-propyl alcohol and water was utilized as the lighter fluid. Iso-propyl alcohol (99% purity), procured from NICE Chemical has a density of 786 kg/m^3 at a temperature of 20 °C. The density of alcoholic solution was altered by changing the concentration of iso-propyl alcohol in the solution. For the attainment of precise and accurate flow rates of the injected jet, a HOLMARC HO-SPLF-2 syringe pump was employed. The syringe pump was operated by a electronic programmed controller, which allowed to set the desired volumetric flow rate of the jet, and it also maintained the same quite consistently. To initiate or halt the flow, a micro-needle valve was securely connected to the output of the syringe pump. The density measurements of all samples were done using the conventional method of weighing a known volume of fluid. To ensure accurate volume measurements, a Hamilton 5 ml syringe was employed. Each sample, comprising 5 ml of fluid, was carefully drawn into the syringe and subsequently transferred to a measuring cylinder. The weight of the fluid was then precisely measured to obtain the density value. High-speed shadowgraphy and Particle Image Velocimetry (PIV) were employed in the current experiments. The high-speed shadowgraphy rig consisted of a LED light source, a pair of 8" parabolic reflectors having 80" and 60" focal lengths, a high-speed camera

(Phantom-V1210) mounted with a NIKON AF Zoom-Nikkor 80–200 mm f/2.8D ED Telephoto Zoom Lens to capture the shadowgraph images of the negative buoyant jets. The camera was operated at a frame rate of 200 fps and 768 pixels × 768 pixels, ensuring the acquisition of high-resolution images exceptional temporal resolution. The shadowgraph images were post-processed using an in-house MATLAB/IMAGEJ program to obtain the jet width and the penetration length.

In 2D-PIV, glass spheres of density 1.1 g/ml and average particle diameter ranging from 9 to 12 μm was utilised as seeding particles. It was customary to estimate the Stokes number based on the seeding particle to ensure that it was faithfully following the flow. The Stokes number (St) is defined as the ratio of the particle response time to the fluid characteristic time scale. It can be expressed mathematically as (see Fig. 1):

$$\boxed{St = \frac{\tau_p}{\tau_f}} \quad (1)$$

where τ_p is the particle response time, τ_f is the fluid characteristic time scale. τ_p and τ_f are mathematically expressed as shown below:

$$\boxed{\tau_p = \frac{\rho_p d_p^2}{18\mu_g}} \quad (2)$$

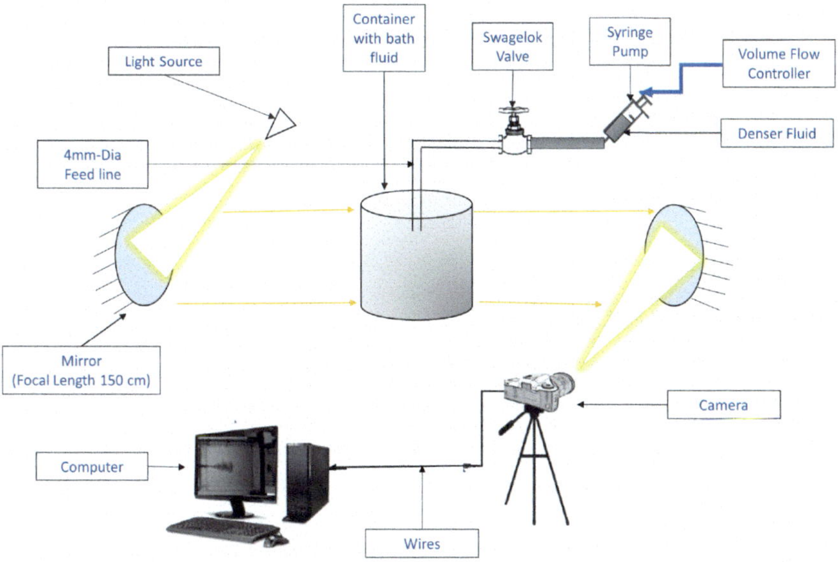

Fig. 1 Schematic of Z-type shadowgraphy setup

where ρ_p is the density of the particle, d_p is the average diameter of the particle, and μ_g is the dynamic viscosity of the fluid.

$$\boxed{\tau_f = \frac{L_o}{V}} \tag{3}$$

where L_o is the characteristic length (here diameter of the injector), V is the average speed at the exit of the injector.

The calculated Stokes number for all the present experiments was within the range of 1e–3, which is significantly smaller than 1 indicating that the employed glass spheres serve as highly appropriate seeding particles for the present experimental setup. A double pulsed Nd-YAG laser (Quantel-Evergeen 200) was utilised in this experiment. The laser sheet with a thickness of approximately 1 mm was created using Lavision's divergent sheet optics. The laser system operated at an average energy of 15 mJ/pulse, with a wavelength of 532 nm and a frequency of 5 Hz. A PCO 2000 camera was utilized to capture the Mie scattering images, operating at a frequency of 5 Hz. A synchronisation unit was employed to trigger the camera and laser according to the experimental requirement. The delay time was calculated based on the formula (Raffel 2002) given:

$$\boxed{dt = \frac{PS \times 1000}{V \times MF}} \tag{4}$$

Here, PS represents the pixel shift, V denotes the average speed of the flow, and MF signifies the magnification factor. Traditionally, a pixel shift of 4–5 pixels has been commonly used to estimate the delay time. In the present experiments, the calculated delay time ranges from 1 to 5 milliseconds. The post-processing was done in a commercial DAVIS software. A cross-correlation scheme was used while post-processing the Mie scattered images.

3 Numerical Model

3.1 Computational Method

The numerical modeling was performed in Ansys Fluent V231. The pressure-velocity coupling was done by employing SIMPLE technique. Considering the laminar nature of the flow regime, the Laminar turbulence model was appropriately selected. Since both the bath and injected fluids were miscible, Species Transport model was employed. With the aforementioned models and settings, a transient simulation was performed, by setting all the residuals to 1e–6. The governing equations were listed below:

Continuity Equation:

$$\frac{\partial \rho}{\partial t} + \nabla.(\rho U) = 0 \tag{5}$$

Momentum Equation:

$$\frac{\partial(\rho U)}{\partial t} + \nabla.(\rho U U) = -\nabla p + \nabla.\tau + \rho g \tag{6}$$

Species Transport Equation:

$$\frac{\partial(\rho Y_i)}{\partial t} + \nabla.(\rho U Y_i) = -\nabla J_i + R_i + S_i \tag{7}$$

3.2 Computational Domain and Boundary Conditions

A 3D cuboidal domain with a cylindrical inlet was created in Ansys Spaceclaim. The geometrical domain was meshed in Ansys Fluent. To achieve a fine mesh in the region of interest, cylinders with 3 different diameters (10, 20, and 40 mm) and a height of 200 mm were separately created inside the cuboidal outer domain. Body of Influence (BOI) was assigned to the cylinders whereas Body Sizing was assigned to the rest of the domain. The outer domain was intentionally made coarser since no injected fluid was anticipated to flow through this region, allowing for computational efficiency without compromising the accuracy of the simulation. The figure shown in Fig. 2 illustrates the boundary conditions applied to the current numerical setup. The inlet boundary was assigned a velocity-inlet condition, allowing for the specification of the flow velocity. On the other hand, the outlet boundary employed a pressure-outlet condition, with the outlet pressure set to 1 atm. The average exit velocity served as the prescribed velocity for the velocity inlet boundary. The wall boundaries, including the nozzle walls, were assigned no-slip conditions, ensuring that the fluid adheres to the solid surfaces without slipping. Adiabatic boundary conditions were implemented to accurately simulate the absence of heat transfer across these boundaries.

3.3 Grid Independence Study

The accuracy and convergence of the simulation results depend upon the grid. Thus, the influence of density of the grid had been investigated. A poly-hex core mesh was generated in Ansys Fluent with Fluent Meshing. Three simulations were performed for ϕ-2 mm injector at Reynolds number 80 and density difference of 4 kg/m^3 with three different mesh sizes. In the first case, the number of elements generated was 1.5 million, in the second case, the number of elements generated was 4.15 million

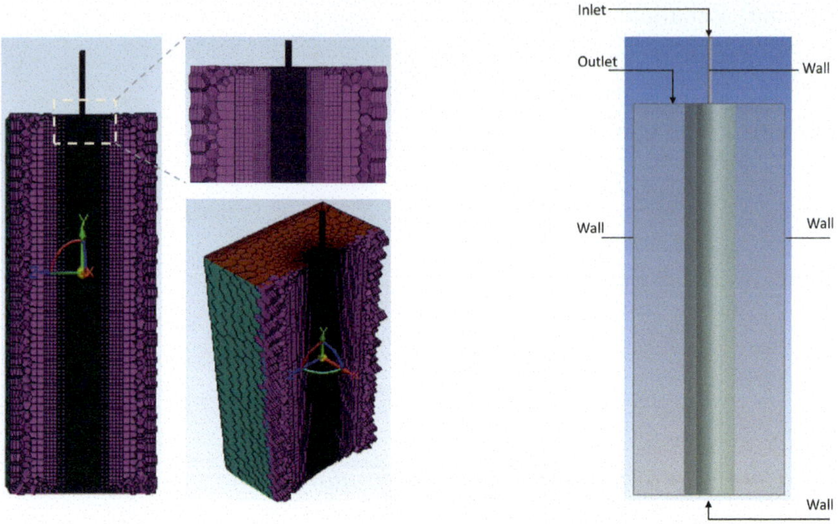

Fig. 2 Meshed domain (left) with assigned boundary conditions (right) for the simulation

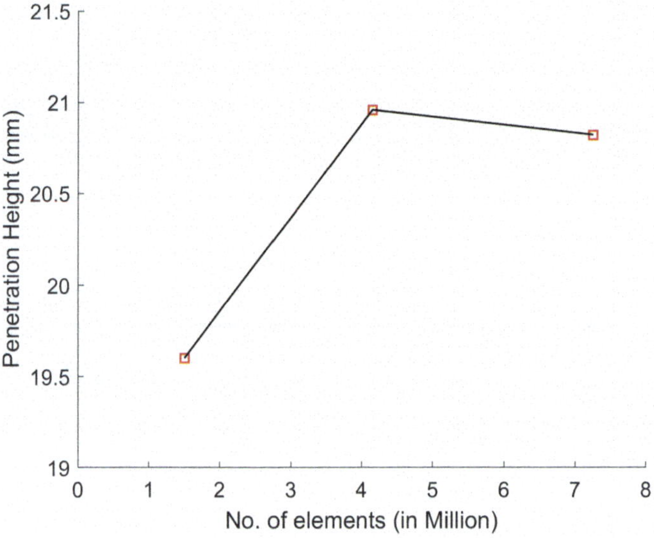

Fig. 3 Grid independence study

and in the third case, the number of elements generated was 7.25 million. Penetration height obtained from all of the simulations were analysed to get an understanding on dependence of grid (see Fig. 3).

Since the variation between second and third data point is less than 10%, the element sizes were chosen such that the number of elements generated were 4.15 million.

4 Results and Discussion

All the experiments were performed in the laminar regime at a Reynolds number 40, 60, 80 and 100 for three nozzles of diameters 1.19, 2.1 and 3.97 mm for various relative density ratios. The injector sizes were kept as small as possible because as the jet diameter increased, the penetration height decreased, making it difficult to estimate the penetration height accurately. Since the emphasis is on Boussinesq. Fountains, the relative density ratio is less than 0.05, as suggested by Baddour et al. (2009). Repeatability experiments were regularly performed at few operating conditions in order to avoid any system errors. Iso-propyl alcohol was used instead of ethanol (as used by Philippe et al.) due to the high volatility of ethanol. In order to get jet morphology, shadowgraphy technique was utilised, eliminating the requirement of dye as used by Philippe et al. The post-processed shadowgraph images were shown in Fig. 4. Figure 4 shows the fully developed fountain at various Reynolds number, and these images were post processed to find the steady-state penetration height. To enhance clarity, an in-house MATLAB code for edge detection was employed to detect and highlight all the edges in the shadowgraph images, wherein the color scheme was inverted.

A series of experiments were conducted under various flow conditions to investigate the steady-state penetration height. The obtained data was then used to create

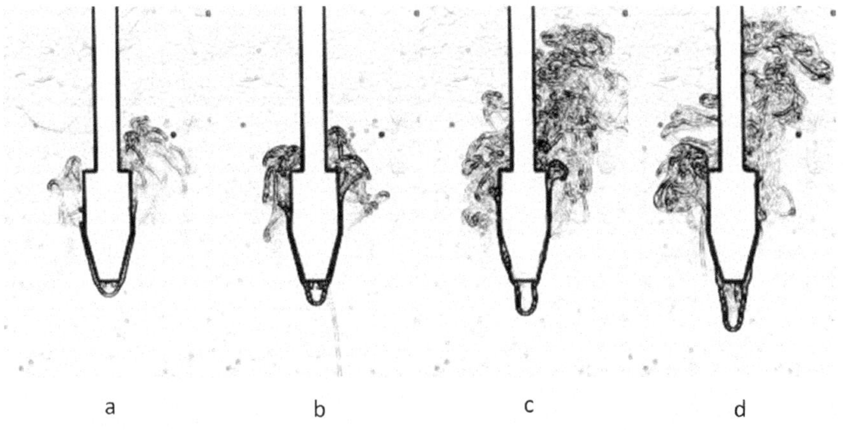

Fig. 4 Experimental steady state heights for a 2 mm diameter injector at Re = 40 (**a**), Re = 60 (**b**), Re = 80 (**c**) and Re = 100 (versus) for $\Delta\rho = 8\,\text{kg/m}^3$ (relative density ratio 0.008) at T = 300 K, P = 1 bar

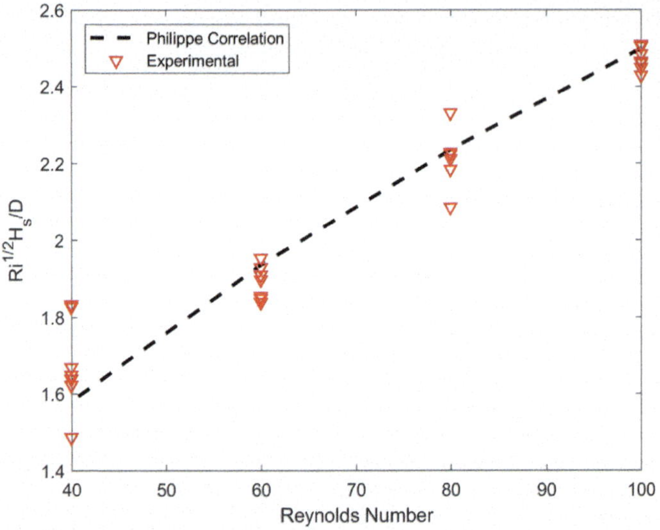

Fig. 5 Plot of $Ri^{1/2}.H_s/D$ versus the Reynolds number Re for all the experimental data and Philippe's correlation

a non-dimensional plot of $Ri^{0.5}.H_s/D$ against Re. The experimental results were subsequently compared to the correlation data established by Philippe, allowing for a comprehensive evaluation and analysis.

From the Figs. 4 and 5, it was observed that the penetration height increased as the Reynolds number was hiked at a constant relative density ratio. This was expected as the inertial forces was only increased without any change in the buoyancy forces. Apart from the fully developed steady fountains, transient behaviour of the jet was also analysed. As shown in Fig. 6, the jet first ascends to a specific height, commonly known as the initial height, and subsequently experiences a reversal in its direction, ultimately stabilizing at a constant height that was shorter than the initial height. The back-flow of the jet was caused by the negative buoyancy effect, which reduced the momentum of the downward motion till the axial velocity reaches zero, but the buoyancy force still persists, leading to a change in the flow direction. However, because the buoyancy force is exactly opposite to the direction of the injected flow, the back-flow pushed the injected flow, resulting in the formation of an annular/cap shape.

After obtaining the penetration height, PIV experiments were performed to measure the velocity field of non-buoyant and buoyant jets. The objective of the PIV experiments was to validate present simulations. Initially, water was injected into a stagnant pool of water, a non-buoyant jet configuration, and the velocity field was measured. Later, lighter density mixture of water and iso-propyl alcohol was injected into a bath of water, a negative-buoyant configuration, and the velocity field of the jet

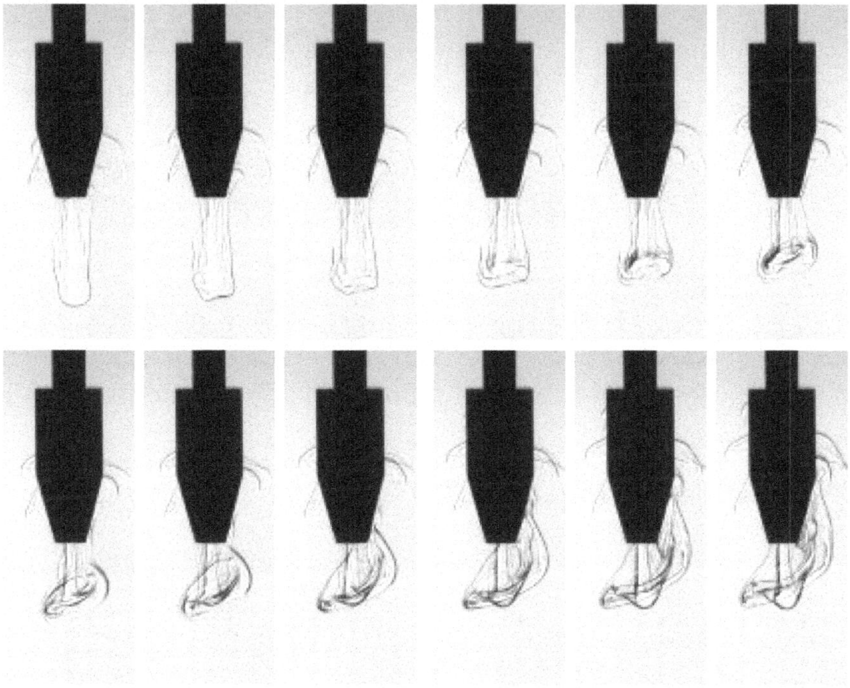

Fig. 6 Jet transient behaviour obtained from experiments for a span of every 2 milli-seconds

was also measured. The results obtained from the both the configuration were compared to clearly demarcate the effects of negative buoyancy. To validate the present PIV measurements, the radial profile of axial velocity, obtained at a distance of 8 mm from the jet exit, was integrated to determine the mass flow rate. Subsequently, this flow rate was compared to the flow rate obtained from the syringe pump. The observed deviation between them was $\pm 9\%$, which was well within limits of experimental uncertainties. The identical procedure was carried out for the negative buoyant jet experiment too, and there also a good agreement was obtained between the flow rates estimated from velocity field and recorded using syringe pump. The axial and radial velocity contours obtained from PIV experiments under steady-state conditions were compared for both negative buoyant and non-buoyant jets, as depicted in Figs. 7 and 8. In the case of the non-buoyant jet, the injected fluid, driven by its higher momentum, penetrated the surrounding quiescent fluid, thereby initiating momentum exchange. As the jet continued to propagate downstream, it eventually fully mixed with the bath fluid. Whereas in case of negative buoyant jets, the jet decelerated by means of momentum dissipation at the shear layers, and also due to buoyancy forces. After a certain height, the buoyancy forces dominated well above the inertial forces, and the jet reversed its direction. The distance between the point of flow reversal at the jet axis and the jet exit at steady state was called steady state

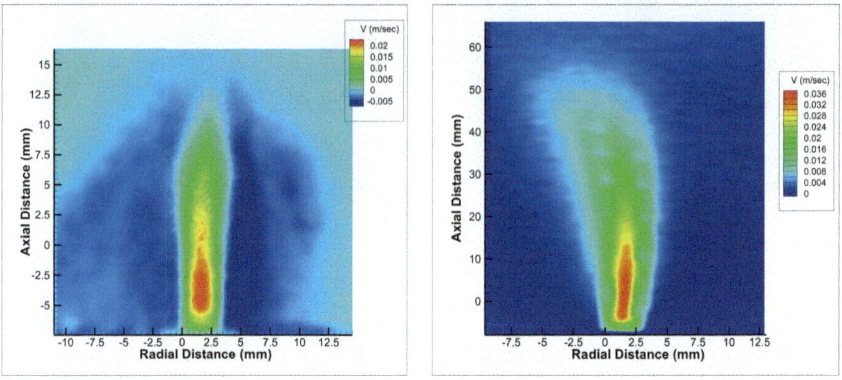

a b

Fig. 7 Axial velocity profiles (obtained from PIV) for negative buoyant jet at Re 100 and density difference 7 kg/m^3 (**a**) and non-buoyant jet at Re 100 (**b**)

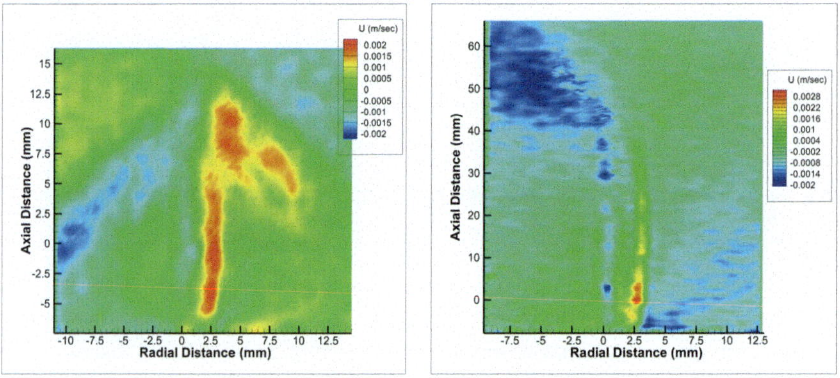

a b

Fig. 8 Radial velocity profiles (obtained from PIV) for negative buoyant jet at Re 100 and density difference 7 kg/m^3 (**a**) and non-buoyant jet at Re 100 (**b**)

penetration height. The radial velocities were quite smaller (almost zero) in both the cases as shown in Fig. 8.

From the contour plots discussed above, it was quite clear that the peak velocity was observed at the axis. To understand the effect of negative buoyancy, the axial velocity at the axis at different axial heights from the jet exit is extracted and plotted in Fig. 9 at Re = 100. It was evident that centre-line axial velocity decreased in both non-buoyant jets and negative buoyant jets, although the reduction rate was comparatively smaller in non-buoyant jets. As depicted in Fig. 7, it was apparent that the potential core length in non-buoyant jets was longer compared to negative buoyant jets. This phenomenon can be elucidated based on the previous explanation. In the case of

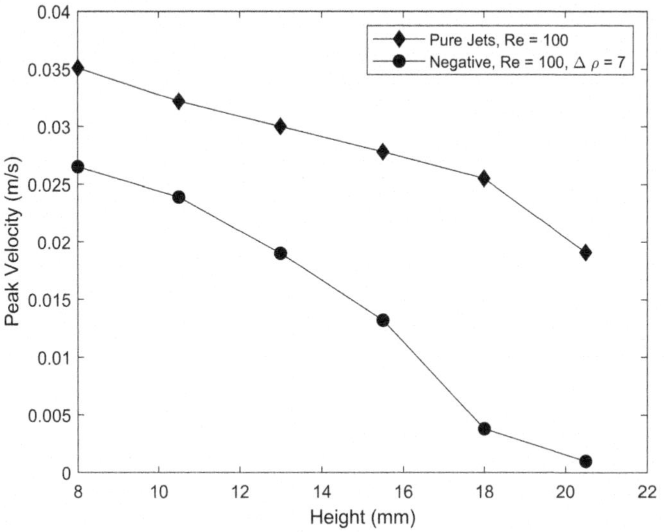

Fig. 9 Comparison of peak velocities (obtained from PIV) of non-buoyant and negative buoyant jets at various heights

negative buoyant jets, the presence of buoyancy introduced an additional dissipating force that operated in conjunction with the existing forces observed in non-buoyant jets, and as a consequence, the potential core was shorter for negative buoyant jets as compared to the non-buoyant scenario. The axial velocity profiles obtained from simulation and experimental at 8 mm from the injector exit were plotted together in Fig. 10. It was observed that the peak velocities obtained from simulations and experiments had a close match, whereas the width of the profiles vary slightly.

The steady state penetration height obtained from the numerical simulations were compared with the Philippe's correlation data. The determination of the penetration height for each simulation case involved analyzing the axial density variation. The distance between the exit of the injector to that point where the density equals the density of the ambient fluid is considered as the penetration height. A non-dimensionalised plot was plotted (similar to the experimental case) between $Ri^{0.5}.H_s/D$ against Reynolds number for the experimental data and simulation data as shown in Fig. 11.

Additionally, the transient behavior of the jet was examined based on the simulation results. The observed jet behavior closely resembled that observed in the experimental cases. Notably, the formation of an annular shape as the flow reverses can be observed in Fig. 12, consistent with the experimental findings.

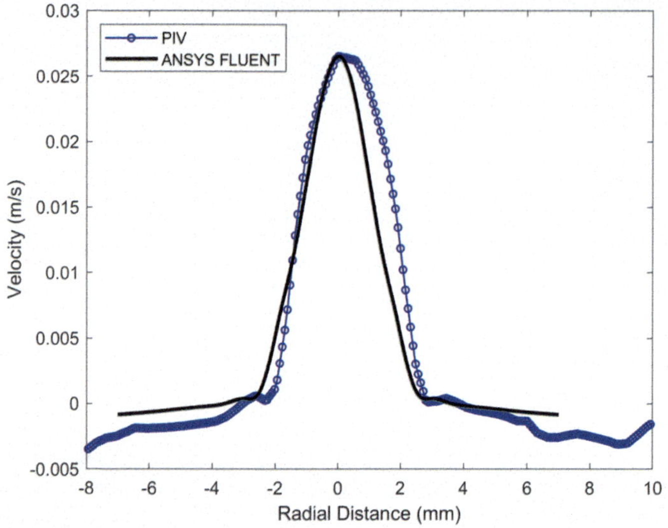

Fig. 10 Comparison of velocity profiles at a distance of 8 mm from the injector exit: PIV data (blue) versus FLUENT simulation (black)

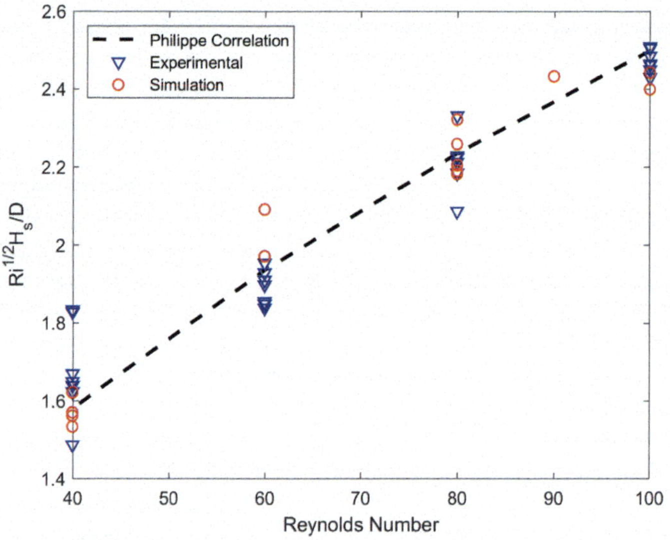

Fig. 11 Plot of $Ri^{1/2}.H_s/D$ versus the Reynolds number Re for all the experimental, simulation data and Philippe's correlation

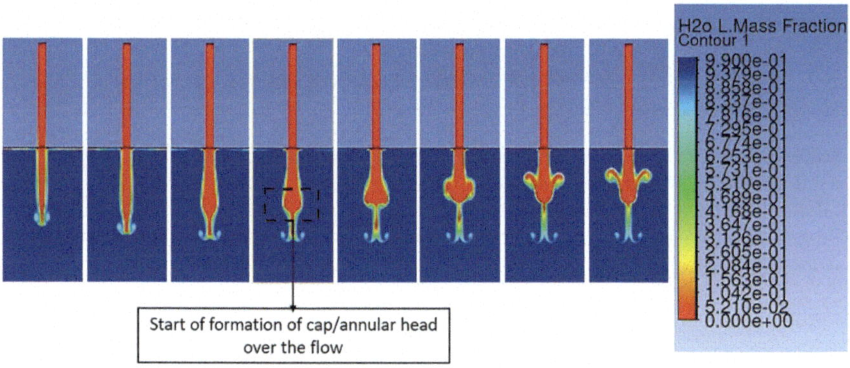

Fig. 12 Transient behaviour of the jet obtained from numerical simulations

5 Conclusion

In this work, an experimental and numerical investigation were performed to study the behaviour of negative buoyancy jets. Three different injectors were used to investigate the jet development in the laminar regime at 1 bar and 300K. The flow characteristics such as penetration height and velocity field were measured using shadowgraphy and Particle Image Velocimetry methods. To validate the obtained results, a thorough comparison was made with existing literature, particularly Philippe et al.'s correlation, revealing a high level of agreement. Furthermore, the shadowgraph data and flow field data were then utilised to validate the numerical simulations and it was found that they were in good agreement. Consequently, the numerical model used in this study was effectively validated, enhancing its reliability and applicability. Hence, the simulation shall be extended for elevated thermodynamic conditions where experiments become quite complicated.

References

J. Ai, A. W.-K. Law, and S. Yu, "On boussinesq and non-boussinesq starting forced plumes," *Journal of Fluid Mechanics*, vol. 558, pp. 357–386, 2006.

R. E. Baddour and H. Zhang, "Density effect on round turbulent hypersaline fountain," *Journal of Hydraulic Engineering*, vol. 135, no. 1, pp. 57–59, 2009.

W. Baines, J. Turner, and I. Campbell, "Turbulent fountains in an open chamber," *Journal of Fluid Mechanics*, vol. 212, pp. 557–592, 1990.

P. Crapper and W. Baines, "Non boussinesq forced plumes," *Atmospheric Environment (1967)*, vol. 11, no. 5, pp. 415–420, 1977.

H. Hassanzadeh, A. Eslami, and S. Taghavi, "Positively buoyant jets: semiturbulent to fully turbulent regimes," *Physical Review Fluids*, vol. 6, no. 5, p. 054501, 2021.

W. Lin and S. Armfield, "The reynolds and prandtl number dependence of weak fountains," *Computational Mechanics*, vol. 31, no. 5, pp. 379–389, 2003.

L. Pantzlaff and R. M. Lueptow, "Transient positively and negatively buoyant turbulent round jets," *Experiments in fluids*, vol. 27, no. 2, pp. 117–125, 1999.

P. N. Papanicolaou and T. J. Kokkalis, "Vertical buoyancy preserving and non-preserving fountains, in a homogeneous calm ambient," *International Journal of Heat and Mass Transfer*, vol. 51, no. 15-16, pp. 4109–4120, 2008.

P. Philippe, C. Raufaste, P. Kurowski, and P. Petitjeans, "Penetration of a negatively buoyant jet in a miscible liquid," *Physics of Fluids*, vol. 17, no. 5, p. 053601, 2005.

M. Raffel, "Particle image velocimetry: A practical guide," *null*, 2002.

B. Vinoth and P. Panigrahi, "Characteristics of low reynolds number non-boussinesq fountains from non-circular sources," *Physics of Fluids*, vol. 26, no. 1, p. 014106, 2014.

Study of Flow Physics During Transient Operation of a Zero Secondary Flow Ejector

Martin Raju, S. L. N. Desikan, and Aravind Vaidyanathan

Abstract Zero secondary flow ejectors/vacuum ejectors are primarily used to produce low back pressure condition in a high altitude test facility where high area ratio nozzles are tested at ground level condition. Here we performed experiments in the subscale model of the vacuum ejector to study the flow physics involved during the starting as well as stopping transient operation. Simultaneous acquisition of unsteady pressures and time resolved schlieren helped in establishing a relation between the jet dynamics to the unsteady pressure fluctuation in the secondary chamber of the vacuum ejector. The degradation of vacuum for pressures higher than the starting jet chamber pressure was studied. The pressure evolution curve of the vacuum ejector operation during the started regime showed the degradation of vacuum level for pressure ratio values above and below the optimum line of operation. The modal analysis of the time resolved schlieren images revealed the influence of longitudinal and lateral oscillations of the jet on the unsteadiness measured in the secondary chamber. The frequency of the longitudinal oscillation and lateral oscillation of the jet were found for different cases of unstarted ($P_0 = 2$, 3 bar) and started condition ($P_0 = 3.8$ bar). It was found that the frequency of oscillation of lateral modes were higher compared to the longitudinal mode of oscillation. The jet oscillation mode responsible for causing frequency peak at started mode of operation ($P_0 = 3.8$ bar) was also identified.

Keywords Vacuum ejector · Modal analysis · Unsteady pressure

M. Raju (✉) · A. Vaidyanathan
Indian Institute of Space Science and Technology, Thiruvananthapuram, Kerala, India
e-mail: martinraju4360@gmail.com

M. Raju · S. L. N. Desikan
Vikram Sarabhai Space Centre, Thiruvananthapuram, Kerala, India

G. Rajesh et al. (eds.), *Proceedings of the 15th International Symposium on Experimental and Computational Aerothermodynamics of Internal Flows*,
https://doi.org/10.1007/978-981-96-4082-9_16

1 Introduction

An ejector works on the principle of momentum transfer from the primary jet to the surrounding secondary fluid. The entrained secondary fluid mixes with the primary fluid through shear layer formed between them (Fabri and Siestrunck 1958). Ejectors with limited volume of secondary region is called zero secondary flow ejectors and they are typically used in high altitude test (HAT) facilities (Goethert 1962), supersonic wind tunnels and hypersonic wind tunnels (Gu and Olivier 2020) to produce low back pressures which brings down the required stagnation pressure to run the nozzle in full flow condition. A vacuum ejector consists of a primary nozzle through which the flow expands and attaches to the diffuser wall around it and entrains the secondary fluid from the secondary chamber located on its upstream section causing its evacuation (Massier et al. 1962). This is known as started mode of operation. The condition where the jet boundary is not attached to the diffuser wall is called unstarted mode of operation.

The literatures related to zero secondary flow ejectors were predominantly related to improving its performance by changing the diffuser and primary nozzle design parameters (Sivo et al. 1960; German et al. 1966). Recent studies have mainly concentrated in describing the flow physics during the transient operation of the vacuum ejectors which have explained the presence of pseudo shocks in the diffuser section, generation of recirculation bubble and also its effect on the evacuation of secondary chamber (Arun Kumar and Rajesh 2018; Desevaux and Lanzetta 2004). The presence of this complex pseudo shock system and its interaction with the diffuser wall is an area less explored. The findings of Annamalai et al. (1998) shows the presence of large amplitude pressure oscillations in the secondary chamber during the operation of the vacuum ejector. The cause of the pressure oscillation and their frequencies were not investigated or explored.

The present study is continuation of the previous study done by the current authors (Raju et al. 2022, 2023) who had reported the frequencies found in the secondary chamber due to the pressure oscillations and also explained the influence of the dynamics of the primary supersonic jet on the observed frequencies. It showed the presence of longitudinal and lateral oscillations in the jet during unstarted mode of operation. The reduction in the level of unsteadiness for short startup duration and the presence of frequencies even for reduced startup duration was reported in the study. The study was done for a configuration with axisymmetric primary nozzle and planar diffuser (Raju et al. 2022). The current study is done for planar nozzle and planar diffuser but with same throat to diffuser area ratio of the previous model for the purpose of comparing the change in flow physics. This study also looks into the off-design operation of the vacuum ejector during started mode of operation and also study its effect on degradation of vacuum in the secondary chamber.

2 Experimental Setup

Experiments are conducted in the Open Jet Facility (OJF) at the Vikram Sarabhai Space Center (VSSC), Trivandrum. This facility has three pressure reservoir vessels with a total capacity of 35 m^3. This facility can achieve a stagnation pressure/primary jet chamber pressure of up to 15 bar, and a maximum mass flow rate of 5 Kg/s. The primary jet chamber pressure is adjusted using a pressure regulator. Fig. 1 shows a schematic representation of the vacuum ejector on which experiments were conducted. It consists of a planar geometry primary nozzle, diffuser section and a top and bottom secondary region. The length (L) and height (D) of the diffuser is 240 and 72 mm respectively which gives an L/D = 3.36. The width (W) of the diffuser is 30 mm. The area ratio of the diffuser section and primary nozzle is 3.6 which is similar to the previous study with the axisymmetric nozzle and planar diffuser (Raju et al. 2022). P_1 and P_2 are the locations of the unsteady pressure sensors in the top and bottom secondary regions respectively kept at the middle of the fore-wall. Air is used as the working fluid to conduct the experiments. During steady state experiments, the primary jet chamber pressures, P_0 are 2 bar (unstarted mode), 3 bar (unstarted mode) and 3.8 bar (started mode). For the transient study of ejector operation, the primary jet chamber pressure is increased from ambient to 7 bar. The mass flow rate varies from 0.14 to 1 Kg/s.

The secondary chamber pressure measurements, P_1 and P_2 are done using two 25 psia range Kulite make piezoresistive unsteady pressure sensors. The sensitivity of the pressure sensors are 1.99 and 2.01 mv/psi respectively. The full-scale error (FSE) for both the sensors is 0.02%. The unsteady pressure is acquired at a sampling rate of 100 KHz and a 25 KHz filter is used to overcome the aliasing effect. The primary jet chamber pressure is measured using a 10 bar range pressure sensor of GP-50 make. The measured data was recorded using a 16-channel data acquisition system of HBM make.

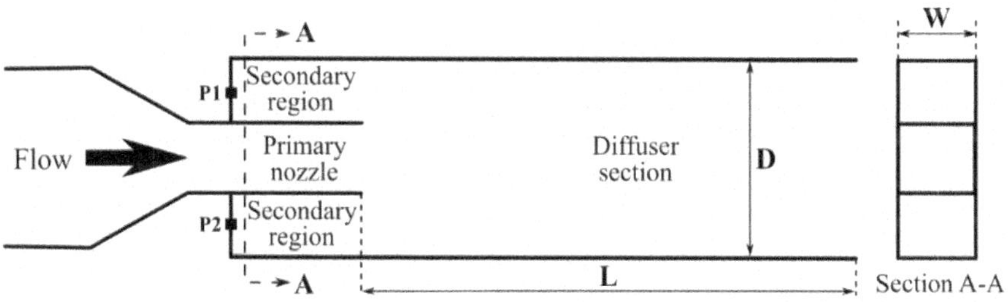

Fig. 1 Schematic of the experimental setup

3 Results

3.1 *Pressure Evolution During the Transient Operation from Startup to Shutdown*

The change in secondary chamber pressure P_1 and P_2 with change in primary jet chamber pressure P_0 from startup to shutdown is shown in Fig. 2a. Their corresponding schlieren images are also shown in Fig. 3a–f. The started mode is marked in Fig. 2a from t_o+2.4s to t_o+25.4s. The region before started mode is the starting transient and the region after started mode is the stopping transient. The vacuum ejector starts at P_0= 3.8 bar and the secondary chamber pressures P_1 and P_2 are 0.36 bar. This is the maximum evacuation possible for this configuration. Further increase in P_0 to 7 bar degrades the vacuum level in the secondary chambers to 0.58 bar as shown in Fig. 2a in the started regime of operation. Further decrease in P_0 from 7 bar to 3.8 bar brings back the evacuation level to the optimum point of 0.36 bar. The vacuum ejector stays in started regime as the primary jet chamber pressure decreases to P_0= 3 bar which is dissimilar to the starting pressure of P_0= 3.8 bar. During this period a degradation in evacuation is also observed. This observation has not yet been reported in previous studies. When the primary jet chamber pressure decreases beyond 3 bar the jet detaches from the diffuser wall and marks the start of stopping transient. The dissimilar evacuation path taken by P_1 and P_2 during the starting transient as well as stopping transient is due to the jet attachment to one side of the diffuser wall which has been already explained in the previous study (Raju et al. 2022). On examining the primary jet chamber pressure (P_0) to secondary chamber pressure (P_1) ratio shown in Fig. 2b, the pressure ratio P_0/P_1 increased from 1 to 10.5 to reach started mode as P_0 increased from 1 bar to 3.8 bar (started mode). Whereas the increase in P_0 from 3.8 bar to 7 bar during the started regime only saw a small increase in P_0/P_1 to 12. This is due to the vacuum degradation which causes increase in secondary chamber pressure P_1 as P_0 increases beyond the minimum starting pressure of 3.8 bar during the started regime. During the latter phase of started regime, we observe a reduction in P_0/P_1 as P_0 decreases from 3.8 bar to 3 bar. This is also due to vacuum degradation as observed in Fig. 2a. From these observations we could conclude that P_0/P_1 values above or below the optimum line during the started regime will have vacuum degradation effect.

The schlieren images in Fig. 3a–f. shows the shape and shock structure of the primary jet at different instances from startup to shutdown of the vacuum ejector corresponding to the pressure evolution shown in Fig. 2. The deflection of jet to one side of diffuser wall during starting transient is shown in Fig. 3a. The jet boundary expands and attaches to both diffuser walls at P_0= 3.8 bar which marks the beginning of started mode of operation (Fig. 3c). The shock pattern changes from Mach Reflection (MR) to Regular Reflection (RR) as P_0 increases to 7 bar from 3.8 bar in started regime of operation. The jet boundary detaches after P_0 decreases beyond 3 bar which marks the end of started regime and beginning of stopping transient.

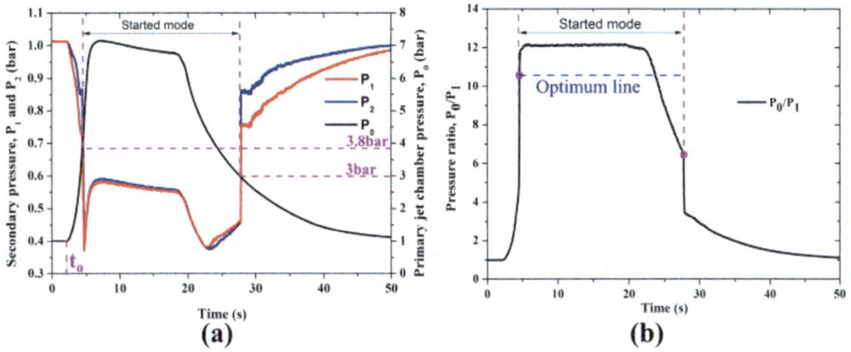

Fig. 2 **a** Pressure evolution of secondary chamber pressures P_1, P_2 and primary jet chamber pressure P_0 during the transient operation. **b** Primary jet chamber pressure to secondary chamber pressure ratio (P_0/P_1) with time

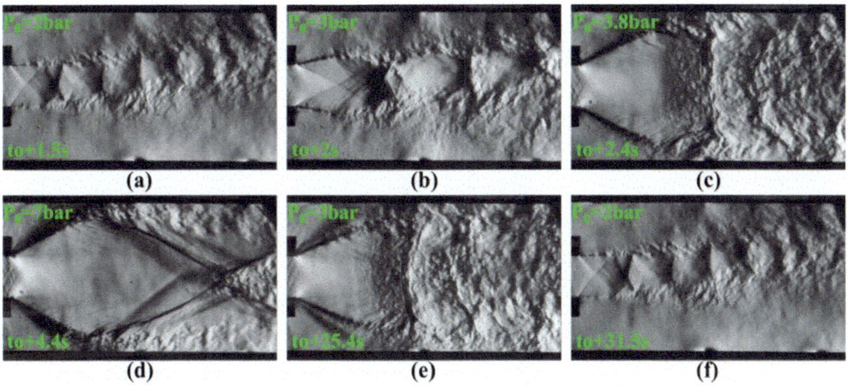

Fig. 3 **a–f** Schlieren image at different instances during the transient operation from startup to shutdown

3.2 *Study of Jet Oscillations Causing Perturbations in Secondary Chamber*

At P_0= 2 bar (unstarted mode):

The influence of primary jet dynamics during unstarted mode on the pressure perturbations observed in the secondary chamber P_1 is investigated by comparing the POD (proper orthogonal decomposition) mode frequencies from time resolved schlieren images to the frequencies from the PSD (power spectral density) of secondary chamber pressure signal P_1. This experiment was conducted at a steady jet chamber pressure of P_0=2 bar. Normalized PSD vs frequency is plotted for P_1 and first 9 POD modes as shown in Fig. 4. The energy content extracted from POD mode is much lesser than energy content from the pressure sensor P_1. Therefore, the normalization is done by dividing with the peak PSD value of P_1 and POD mode 1 separately. Figure 5a–i shows the POD mode shapes. The frequency tones from P_1 signal are 219, 512 and 976 Hz. The combination of tones from POD mode 1 and 4 matches with the tones from P_1 signal. Their corresponding mode shapes are seen in Fig. 5a and d. Both these modes exhibit longitudinal mode of oscillation as the density variation is alternating in the longitudinal axis. Previous study with an axisymmetric nozzle showed jet attachment to one side of the diffuser wall during the latter stage of starting transient (Raju et al. 2022). But in the present study with a planar nozzle we see that the jet is attached to one side of the diffuser wall at an early stage of the starting transient. For examining its cause we observed lower energy POD modes from 5 to 9. Higher frequency tones were observed from POD modes 6 to 9 and their corresponding POD modes shapes in Fig. 5f–i reveal that they undergo lateral mode of oscillation. The presence of these lateral modes of oscillation could be reason for jet deflection to either sides of the diffuser walls.

Fig. 4 Comparison of normalized PSD of secondary pressure P_1 and POD modes for $P_0 = 2$ bar

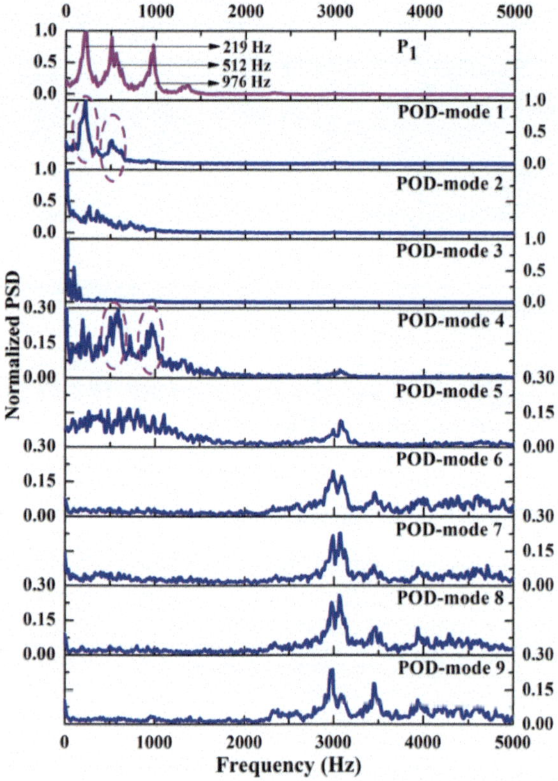

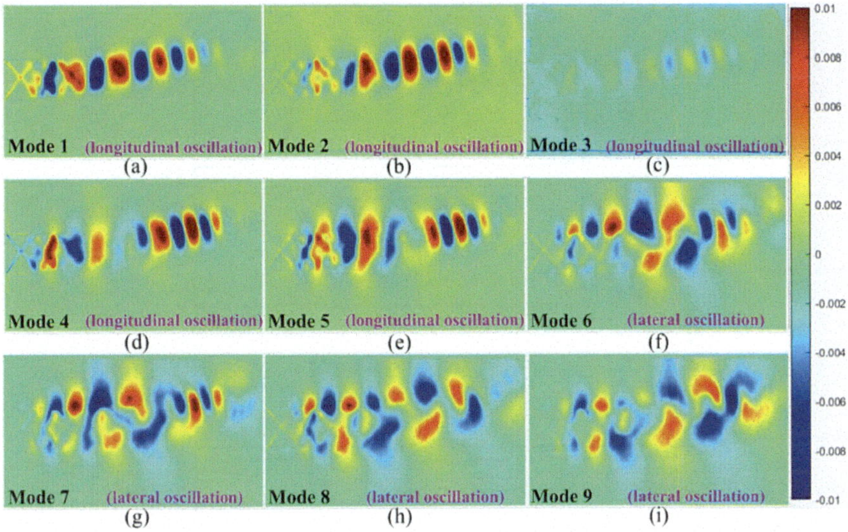

Fig. 5 **a–i** POD modes arranged in decreasing order of their intensity levels for $P_0 = 2$ bar

At P_0= 3 bar (unstarted mode):

Experiments with steady jet chamber pressure of 3bar was done and the tones from the P_1 signal are 231 and 3942 Hz as shown in Fig. 6. POD mode 1 frequency matches with the 231 Hz tone from P_1 signal. The corresponding mode shape from Fig. 7a shows longitudinal mode of oscillation of the jet. POD mode 2 and mode 3 frequency matches close to the 3942 Hz tone from P_1 signal and their corresponding mode shapes are shown in Fig. 8b and c. They exhibit lateral mode of oscillation of the jet. Therefore, it is evident from the above results that the lateral mode of oscillation is of higher frequency compared to the longitudinal mode of oscillation. The lateral modes of oscillation which were present in lower intensity levels (mode 6 to 9) at P_0= 2 bar is now present at higher intensity levels (mode 2 and 3) at P_0= 3 bar. This shows increased contribution of lateral oscillation modes on the unsteadiness in the secondary chamber as P_0 increases in unstarted condition.

At P_0= 3.8 bar (Started mode):

Previous study (Raju et al. 2023) showed the presence of frequency tones in secondary chamber pressure signal even at started mode of operation at minimum starting pressure of P_0= 3.8 bar and showed that as P_0 was increased further no tones were observed. Here we investigate the mode shape that contributes to the frequency tone

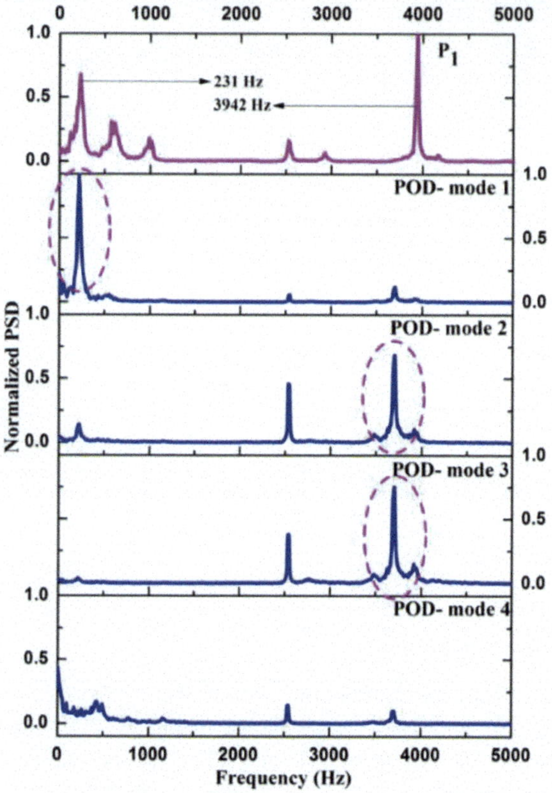

Fig. 6 Comparison of normalized PSD of secondary pressure P_1 and POD modes for $P_0 = 3$ bar

observed at the started mode (P_0= 3.8 bar). From Fig. 8, the frequency tone from pressure signal P_1 is 183 Hz and the POD mode that contributes to this tone is mode 1. The mode shape of POD mode 1 is given in Fig. 9a and it exhibits longitudinal mode of oscillation of the jet. Although the mode shape is different from the longitudinal oscillation at P_0= 2 bar and 3 bar at unstarted condition, their frequencies are in the same range.

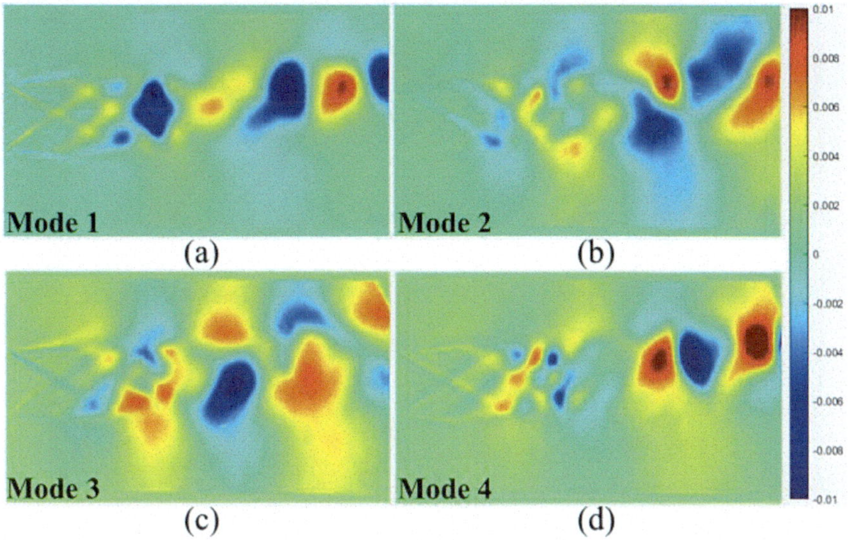

Fig. 7 **a–d** POD modes arranged in decreasing order of their intensity levels for $P_0 = 3$ bar

Fig. 8 Comparison of normalized PSD of secondary pressure P_1 and POD modes for $P_0 = 3.8$ bar

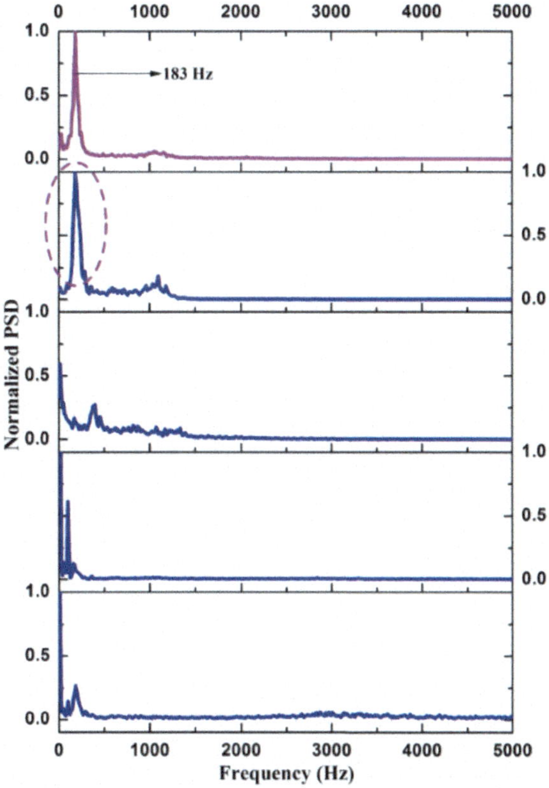

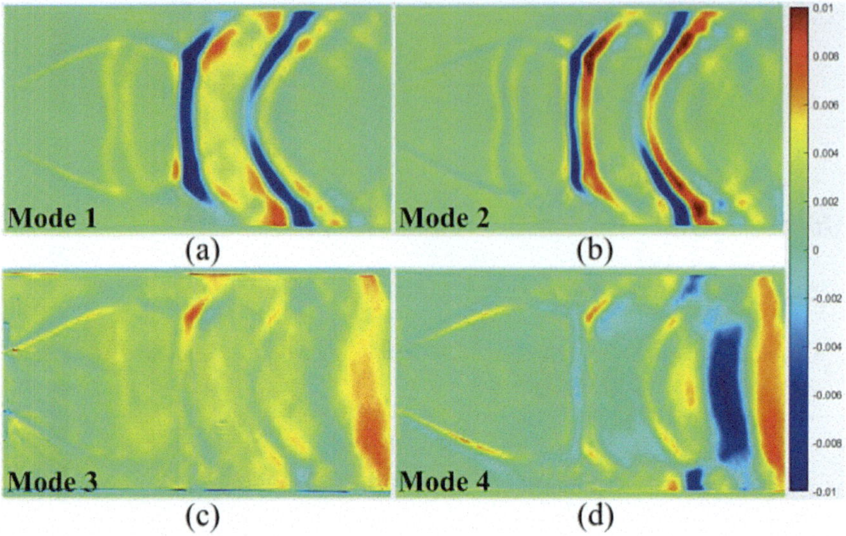

Fig. 9 **a–d** POD modes arranged in decreasing order of their intensity levels for $P_0 = 3.8$ bar

4 Conclusion

Experiments were done to study the flow physics inside a vacuum ejector during its transient as well as steady operating condition. The secondary chamber pressure was analysed both in time domain and frequency domain (PSD). Time resolved schlieren images were also analysed in time and frequency domain (POD). Following observations were made from the present study:

i. The pressure evolution of secondary chamber pressures during started mode of operation revealed the degradation of vacuum level at off design operating condition. An optimum line was introduced in the P_0/P_1 curve during the started regime of operation. The values above and below the optimum line were present and were associated with vacuum degradation effect.
ii. Simultaneously acquired time resolved schlieren images revealed the jet shape and shock structure at different stages of the vacuum ejector operation from startup to shutdown.
iii. The frequency tones from PSD of secondary pressure signal matched with the POD modes for different steady jet chamber pressures at unstarted and started mode of operation. This establishes the influence of jet dynamics on the pressure fluctuations in the secondary chamber.
iv. The POD mode shapes show the presence of longitudinal and lateral mode of jet oscillation and also the ascend of lateral mode of oscillation into higher intensity levels during the starting transient at higher jet chamber pressure. The frequency of lateral mode of oscillation was observed to be higher than longitudinal mode of oscillation.

v. The frequency tone sensed in the secondary chamber at started mode of oscillation was due to the presence of longitudinal mode of jet oscillation.

Additional experiments are planned to study the scaling effect of the vacuum ejector by changing its geometric area ratios. The influence of scaling on the level of unsteadiness and frequency tones in the secondary chamber are planned as part of future studies.

References

Annamalai, K., Visvanathan, K., Sriramulu, V. and Bhaskaran, K. A.: Evaluation of the performance of supersonic exhaust duct using scaled down models, Experimental Thermal and Fluid Science, 17, 217–229 (1998).

Arun Kumar, R. and Rajesh, G.: Physics of vacuum generation in zero-secondary flow ejectors, Physics of Fluids, 30, 066102 (2018).

Desevaux, P. and Lanzetta, F.: Computational fluid dynamic modelling of pseudo shock inside a zero-secondary flow ejector, AIAA Journal, 42(7), 1480–1483 (2004).

Fabri, J. and Siestrunck, R.: Supersonic air ejectors, Advances in applied mechanics, 5, 1 (1958).

German, R. C., Bauer, R. C. and Panesci, J. H.: Methods for determining the performance of ejector-duct systems, Journal of Spacecrafts and Rockets 3(2), 193–200 (1966).

Goethert, B. H.: High altitude and space simulation testing, ARS Journal. 32, 872–882 (1962).

Gu, S. and Olivier, H.: Capabilities and limitations of existing hypersonic facilities, Program in Aerospace Sciences, 113, 100607 (2020).

Massier, P. F. and Roschke, E. J., and Gier, H. L.: Experimental Investigation of Exhaust Ducts for Rocket Engines, JPL Report, 32–210 (1962).

Raju, M., Desikan, S. L. N. and Vaidyanathan, A.: Transient characteristics of a typical vacuum ejector-An experimental study, Physics of Fluids, 34(9), 096108, (2022).

Raju, M., Desikan, S. L. N. and Vaidyanathan, A.: Investigation of the unsteadiness during starting transient operation of a zero secondary flow ejector. In: 25th AIAA International Space Planes and Hypersonic Systems and Technologies Conference, pp. 3063, (2023).

Sivo, J.N., Meyer, C.L. and Peters, D.J.: Experimental evaluation of rocket exhaust diffusers for altitude simulation. National Aeronautics and Space Administration, Vol. 298 (1960).

Numerical Study of Rigid Filament in a Biomagnetic Flow

Dinesh Kumar Ravada and Ranjith Maniyeri

Abstract The current research work proposes two-dimensional computational model using immersed boundary method to simulate biomagnetic fluid flow in a channel with a rigid filament. The fluid flow in the channel under an applied magnetic field is investigated for two orientations of rigid stationary filament (straight and inclined) using the model. By varying Reynolds number (Re) and magnetic number (Mn), we observed the emergence of multiple recirculation zones resulting from the combined influence of flow, magnetic field, and filament. These findings can provide valuable insights into the behaviour of biomagnetic fluid flow under magnetic field in the presence of immersed structures.

Keywords Biomagnetic flow · Immersed boundary method · Magnetic number · Rigid filament

1 Introduction

Biomagnetic fluid flow is a subject of detailed research, which focuses on flow behaviour of biological fluid, particularly blood, which is sensitive to magnetic field. The study of biomagnetic fluid flow have significant application in medicine and health care. By understanding the behaviour of blood in the magnetic field helps in the diagnosis of many cardiovascular diseases. Additionally, manipulating biomagnetic fluid flow helps in targeted drug delivery where the magnetic nano particle with medicine are transported to specified locations in the body. Magnetic resonance imaging (MRI) which is used for imaging of internal organs is also relied on the concept of biomagnetic fluid flow.

Biomagnetic fluid flow also have significant uses in biotechnology and bioengineering. Controlling the movement and flow of biomagnetic fluids with magnetic

D. K. Ravada · R. Maniyeri (✉)
Department of Mechanical Engineering, Biophysics Laboratory, National Institute of Technology Karnataka (NITK), Surathkal, Mangalore, Karnataka, India
e-mail: mranji1@nitk.edu.in

G. Rajesh et al. (eds.), *Proceedings of the 15th International Symposium on Experimental and Computational Aerothermodynamics of Internal Flows*,
https://doi.org/10.1007/978-981-96-4082-9_17

fields helps in handling of fluids in microchannels, which helps in developing optimized microfluidic devices for lab-on-a-chip applications such as mixing, cell sorting, transporting and magnetic separation technique for removing specific particles from biomagnetic fluid. Furthermore, biomagnetic fluid flow has industrial applications such as magnetic levitation and magnetic sensing, which enhanced separation, mixing, and transportation processes.

Numerous numerical simulations were done on the flow behaviour of the biomagnetic fluid. Loukopoulos and Tzirtzilakis (2004) had done numerical simulation of steady fluid flow of conduit under a spatially variable magnetic field. Tzirtzilakis (2008) conducted research on the flow of biomagnetic fluid in a symmetrical narrow passage. They observed the formation of two symmetric vortices downstream of the constriction, solely due to the fluid flow characteristics. When a magnetic field was applied, they found that the attachment point of the flow shifted downstream as the vortex near the magnetic source expanded. Habibi et al. (2012) investigated the impact of a non-uniform magnetic field on blood flow. They discovered that at low Reynolds numbers, blood pressure was highest near the magnetic source and decreased away from it. They also took into consideration the effect of wall shear stress on the biomagnetic fluid and also considered the electrical and magnetic properties of blood. Kenjeres (2008) focused on studying blood flow in arteries exposed to a magnetic field generated by an electric current in a wire. Their findings provided insights into changes in local pressure at the location of the strongest magnetic field. Mousavi et al. (2016) examined the effects of the Lorentz force and magnetic force on biomagnetic fluid under a non-uniform magnetic field. Recently, we simulated the flow of biomagnetic fluid under shear flow condition for non-uniform magnetic field using finite volume method (Ravada and Maniyeri 2023). The current study focuses on the problem of fluid–structure interaction, where a structure (in this case, a stationary rigid filament) is immersed in a biomagnetic fluid flow. To numerically analyse this fluid–structure interaction problem, we employed immersed boundary method. This method allows us to accurately model the mutual influence between the fluid and the filament without relying on a complex mesh.

Peskin first used this method for study of blood flow in heart valves (Peskin 1977). Additionally, Basil et al. (2022) explored the interaction of a rigid filament in oscillatory flow using the immersed boundary method. Lai and Peskin (2000) introduced a second-order accurate immersed boundary method, specifically for studying low viscous fluid flow over a cylinder. Maniyeri and Kang (2012) employed the immersed boundary method to investigate the propulsion of a helical-shaped bacterial flagellum, driven by external torque. Kang and Maniyeri (2014) focused on studying the bundling and tumbling motion of bacterial flagella in viscous fluids. Kanchan and Maniyeri (2019) conducted research on the recuperation dynamics of a flexible filament using Peskin's immersed boundary method. Moreover, Kanchan and Maniyeri (2020) explored the asymmetric movements of the flexible filament in shear flow. The current work presents development of a numerical model to simulate biomagnetic fluid flow in a two-dimensional channel with a rigid stationary filament using immersed boundary finite volume method. The study mainly discusses the effect of magnetic field on the fluid flow under different Reynolds (Re) and magnetic numbers (Mn) for two orientations (straight and inclined) of the filament.

2 Mathematical Formulation

The present work uses non-dimensional form of governing equations which are given below (Mousavi et al. 2016).

Continuity equation:

$$\frac{\partial u}{\partial x}+\frac{\partial v}{\partial y}=0 \tag{1}$$

Momentum equation x-direction:

$$\frac{\partial u}{\partial t}+u\frac{\partial u}{\partial x}+v\frac{\partial u}{\partial y}=-\frac{\partial p}{\partial x}+\frac{1}{Re}\left(\frac{\partial^2 u}{\partial x^2}+\frac{\partial^2 u}{\partial y^2}\right)+MnH\frac{\partial H}{\partial x}+f(x) \tag{2}$$

Momentum equation y-direction:

$$\frac{\partial v}{\partial t}+u\frac{\partial v}{\partial x}+v\frac{\partial v}{\partial y}=-\frac{\partial p}{\partial y}+\frac{1}{Re}\left(\frac{\partial^2 v}{\partial x^2}+\frac{\partial^2 v}{\partial y^2}\right)+MnH\frac{\partial H}{\partial y}+f(y) \tag{3}$$

where magnetic number (Mn) $=\frac{\mu_0^2 H_0^2 \varkappa}{\rho U^2 \mu_0}$, Reynolds number (Re) $=\frac{Uh}{\nu}$, H represents magnetic intensity and $\varkappa$ denotes magnetic susceptibility (Loukopoulos and Tzirtzilakis 2004).where ρ is density, μ_0 is magnetic permeability and $f(\mathbf{X})$ known as Eulerian force functions which is calculated using Lagrangian force $\mathbf{F}(s)$.

$$f(\mathbf{x})=\int \mathbf{F}(s)\delta(\mathbf{x}-\mathbf{X}(s))ds \tag{4}$$

The Lagrangian force $\mathbf{F}(\mathbf{s})$ denote the force exerted by body on the fluid and $\mathbf{x}$ denotes the Eulerian coordinates. The non-dimensional form of intensity due to magnetic effect in directions of x and y are taken from reference (Kenjeres 2008). Since the filament is rigid and stationary the Lagrangian force is calculated as tether force as given below

$$\mathbf{F}(s)=S\left(\mathbf{x}_e(s)-\mathbf{X}(s)\right) \tag{5}$$

where 'S' is the stiffness constant. $\mathbf{x}_e(s)$,$\mathbf{X}(s)$ are the initial and new position of the IB points. To keep the rigid filament in equilibrium and close to its initial position, the stiffness constant must be quite high ($S >> 1$). δ represents Dirac delta function which is given below as

$$\delta = \frac{1}{h^2}\Phi\left(\frac{x}{h}\right)\Phi\left(\frac{y}{h}\right)$$

$$\text{where } \Phi\begin{cases} \frac{1}{8}\left(3 - 2|r| + \sqrt{1 + 4|r| - 4r^2}\right), 0 \le |r| \le 1 \\ \frac{1}{8}\left(5 - 2|r| - \sqrt{-7 + 12|r| - 4r^2}\right), 1 \le |r| \le 2 \\ 0 \text{otherwise} \end{cases} \tag{6}$$

$$\mathbf{U^{n+1}}(\mathbf{X}) = \int \mathbf{u^{n+1}}(\mathbf{x})\delta\left(\mathbf{x} - \mathbf{X}^n(\mathbf{s})\right)\mathrm{dx} \tag{7}$$

$$\mathbf{U^{n+1}}(\mathrm{s}) = \frac{\mathbf{X^{n+1}}(\mathrm{s}) - \mathbf{X^{n}}(\mathrm{s})}{\Delta t} \tag{8}$$

In the above expression the Lagrangian velocity variable is represented as $\mathbf{U}^{n+1}(\mathrm{s})$ and $\mathbf{X^{n+1}}(\mathrm{s})$ denotes new coordinates of Lagrangian variable.

Accordingly, we considered a channel domain of 10 × 1 where no-slip boundary conditions are applied at top and bottom walls. Eulerian grid of 501 × 51 and 51 Lagrangian points are considered for the discretization of the channel and rigid filament. For the numerical simulation a FORTRAN code is developed to solve the non-dimensional form of governing differential equations which are discretised using semi-implicit fractional step method. The discretized momentum equations coupled with Eulerian force density function are solved using Incomplete Cholesky Conjugate Gradient (ICCG) method (Kershaw 1978).

3 Results and Discussion

Firstly, the numerical model without magnetic field and filament is simulated for the case of simple channel flow and validated with well-known analytical results of Plane Poiseuille flow as depicted in Fig. 1.

In the second stage of the study, we consider the influence of magnetic fields and examine the variation of the magnetic number for fixed Reynolds numbers of 10, 30, and 50, with varying magnetic numbers (Mn = 100, 300, and 600). The magnetic source is located at coordinates (3.0, −0.05).

From Fig. 2, we observed that variation in streamline patterns for the fixed Reynolds numbers and different magnetic numbers, indicate the formation of recirculation regions near the location of source. It is noted that as we increase the magnetic number, there is an increase in the size of the recirculation regions. Additionally, we identify the formation of a secondary flow at the top of the wall for a Reynolds number of 50 and Mn = 600, indicating the strong influence of the magnetic field. We also observe that for a constant magnetic number, as the Reynolds number increases, the size of the flow vortex increases. This suggests that as the flow becomes less viscous (increasing Reynolds number), the magnetic forces become more important and affect the flow more.

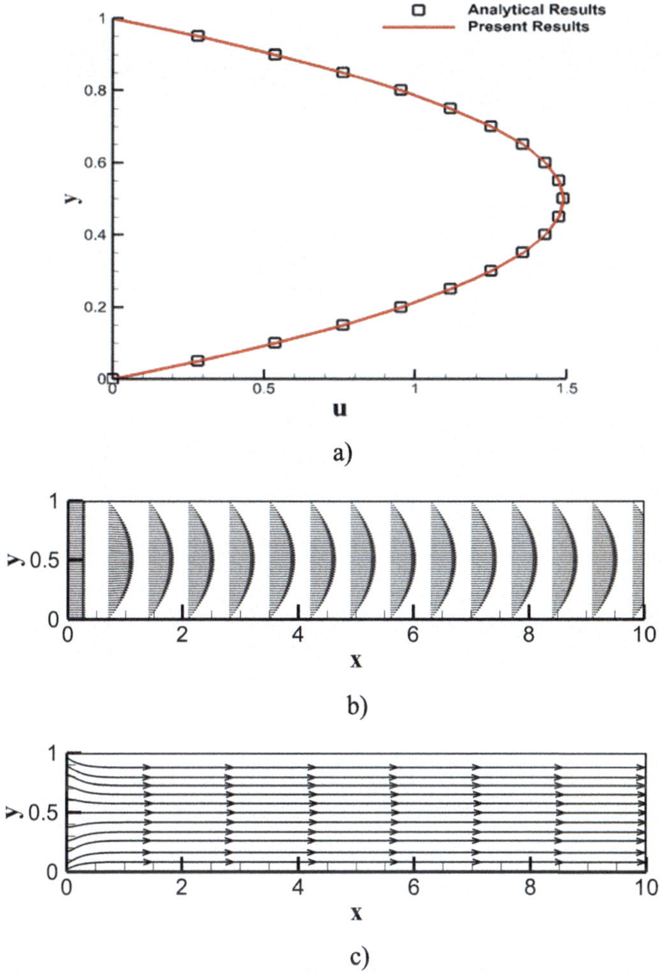

Fig. 1 **a** Validation plot for centerline velocity; **b** vector plot; **c** streamline plot for Poiseuille flow

In the next set of simulation, a rigid horizontal stationary filament with a non-dimensional length of 0.5 is placed under biomagnetic fluid flow within a channel, starting from the location at coordinates (3.0, 0.2). The focus is on studying the resulting fluid dynamics for a fixed Reynolds number of 30.0 and various magnetic numbers (Mn) such as 100.0, 300.0, and 600.0

Streamline plots in Fig. 3 depict the flow patterns for different Mn values, where an increase in Mn leads to the development of multiple recirculation zones within the channel. These recirculation zones originate from the location of the magnetic source and progressively multiply towards the right side of the source. Upon examining Figs. 2 and 3 it becomes evident that the presence of the filament causes a reduction

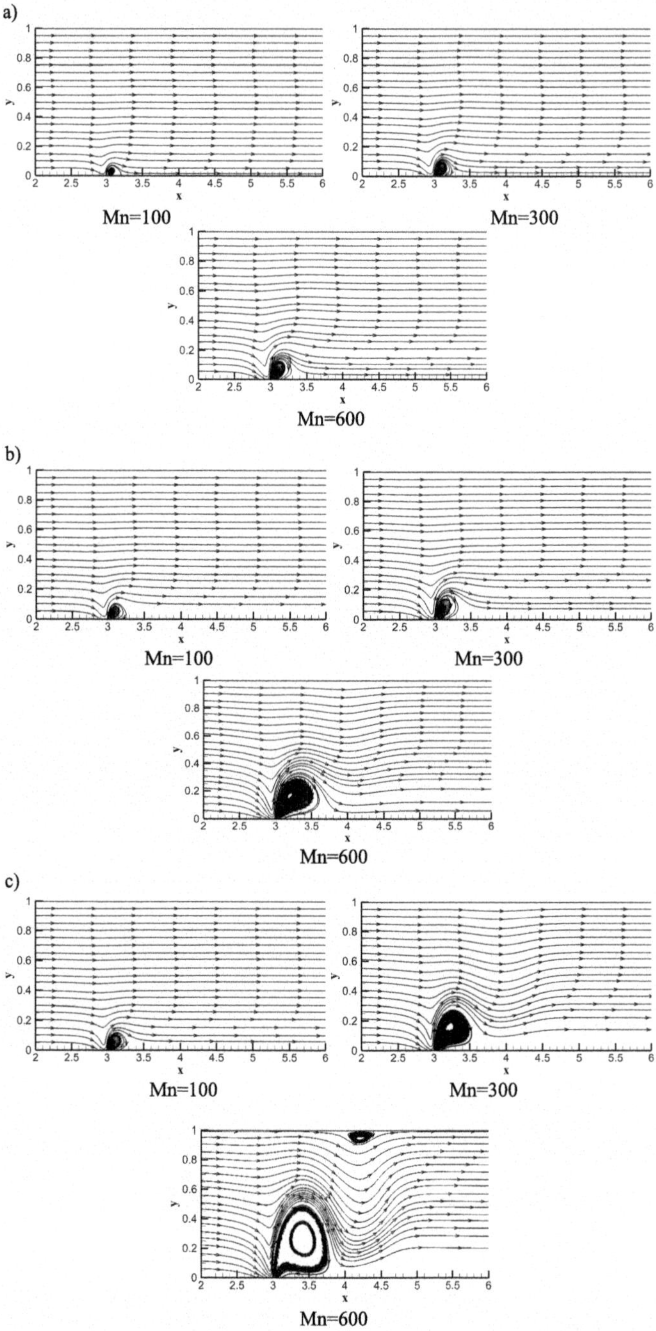

Fig. 2 **a** Reynolds number = 10 **b** Reynolds number = 30 **c** Reynolds number = 50

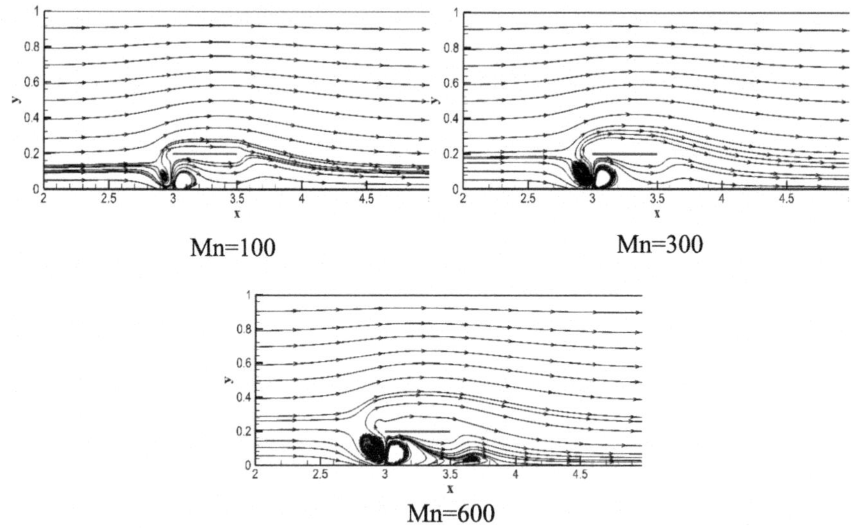

Fig. 3 Streamline plot for fixed Reynolds number = 30

in the size of the vortex. It is observed that the vortex formed rotates in opposite direction to each other.

This suggests that the effect of the magnetic field is limited to the region below the filament, as the magnetic force intensifies. Consequently, the formation of three vortices occurs below the filament due to the increased magnetic force (increase in magnetic number). Furthermore, Fig. 4 reveals a similar trend as the Reynolds number increases from Re = 10, 30, to 50. Keeping magnetic number constant to 100, the size of the vortex enlarges, indicating that lower viscous forces allow the magnetic force to exert a greater influence on the biomagnetic fluid flow.

In the fourth set of simulations, we examine the behaviour of an inclined filament placed at a 45° angle to horizontal axis of the channel, starting from the position (x = 3.0, y = 0.5). The filament has a non-dimensional length of 0.5. We analysed the results for fixed Re = 30 and varying magnetic number similar to case of horizontal filament. We observed intriguing phenomena related to the interaction between the magnetic field, the inclined filament, and the flow. Figure 5 illustrates the streamline plot for fixed Reynold number = 30.

As we increase the magnetic number from 100 to 600, we observed a significant deviation in the flow pattern below the inclined filament. This deviation occurs due to the combined effect of the magnetic force and the orientation of the filament (inclined). Notably, multiple recirculation regions form on both sides of the location where the magnetic source is placed, near the position of the inclined stationary filament, when we compare with biomagnetic flow without filament the size of recirculation is small in this case.

Furthermore, we notice that, as we increase the magnetic number to 300, there is formation of recirculation in the downstream of the flow near the filament, which grow

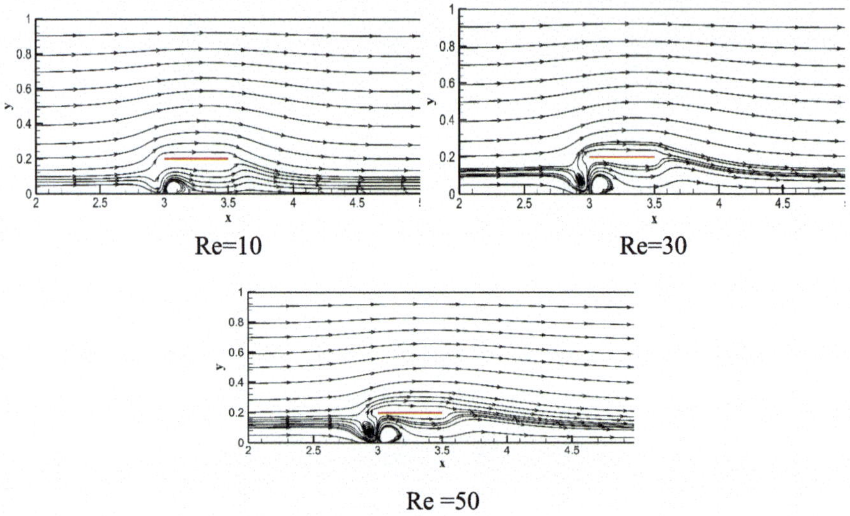

Fig. 4 Streamline plot for fixed magnetic number 100 and varying Reynolds number

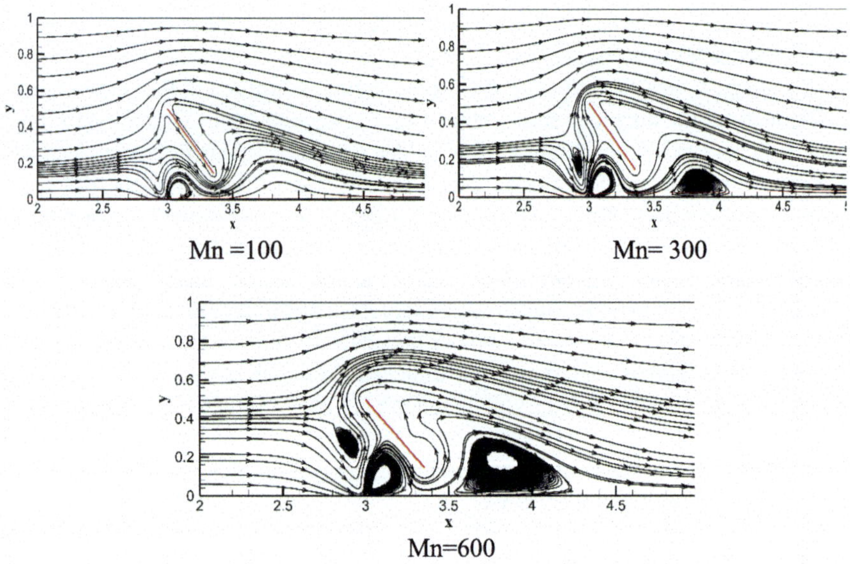

Fig. 5 Streamline plot for Reynolds number = 30

in size when we further increase the magnetic number to 600. This finding highlights the strong impact of the magnetic field and the presence of the inclined filament on the flow. Specifically, the fluid near the filament, resulting in the formation of even larger vortices compared to those formed at the location of the magnetic source.

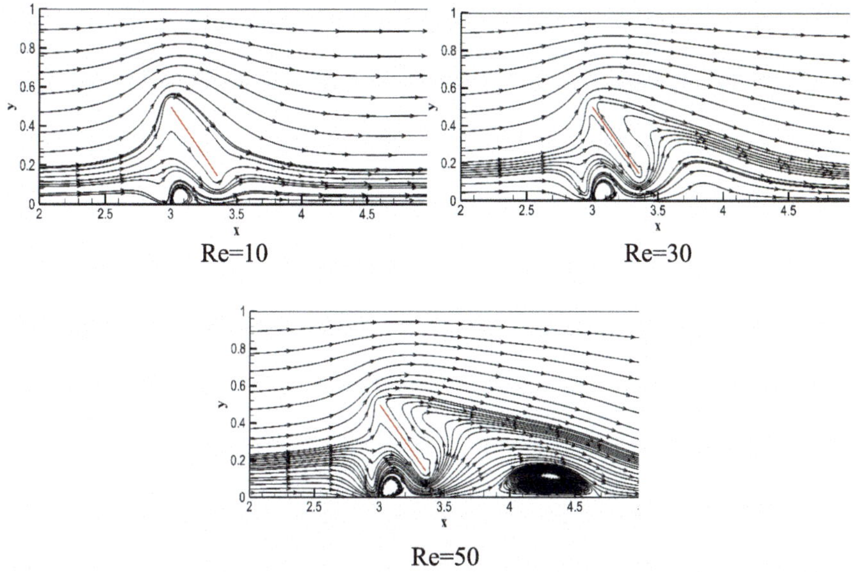

Fig. 6 Streamline plot for fixed magnetic number Mn = 100

From Fig. 6 we can observe stream line variation for the case varying Reynolds numbers while keeping the magnetic number fixed at 100. Specifically, we are investigating the effect of Reynolds numbers of 10, 30 and 50. We consistently observe an increase in the size of the vortex near the magnetic source as the Reynolds number increases. Interestingly, for a Reynolds number of 50, a larger vortex forms downstream in the flow. This occurrence can be attributed to the combined influence of the magnetic field and the presence of the inclined filament, leading to retardation in the flow. These findings highlight the intricate interactions between flow dynamics, magnetic field and immersed structure.

4 Conclusion

In this work, immersed boundary method based numerical model is constructed to study the biomagnetic flow in a channel under the presence of a rigid filament. The study compares the effect of presence of a filament in the biomagnetic fluid flow in terms of the formation and the size of the recirculation zones. It is found that under no filament case, increasing the magnetic number leads to the formation of larger recirculation regions near the magnetic source. The presence of a horizontal filament in the biomagnetic fluid flow reduces the size of the vortex and causes the vortices to rotate in opposite directions. When the filament is inclined, recirculation regions form on both sides of the magnetic source, altering the flow pattern. Additionally, increasing the Reynolds number results in larger vortices both near the magnetic source and downstream in the flow. These findings emphasize the interplay between flow dynamics, magnetic field and immersed object, highlighting the complexities of biomagnetic fluid behaviour within flow channels for the application of mixing and transporting phenomenon.

References

J. Basil, M. Kanchan and R. Maniyeri: Modeling rigid filament interaction under oscillatory flow using immersed boundary method. Materials Today: Proceedings, 56, (2022).

Dinesh Kumar Ravada and R. Maniyeri :Numerical simulation of two-dimensional biomagnetic shear flow. Materials Today: Proceedings https://doi.org/10.1016/j.matpr.2023.01.127. (2023).

M.R Habibi, M. Ghassemi,and A.Shahidian: Investigation of biomagnetic fluid flow under nonuniform magnetic fields. Nanoscale Microscale Thermophysical Engineering, 16(1), 64–77, (2012).

M. Kanchan and R. Maniyeri: Numerical analysis of the buckling and recuperation dynamics of flexible filament using an immersed boundary framework. International Journal of Heat and Fluid Flow, 77, 256–277 (2019).

M.Kanchan and R.Maniyeri: Numerical simulation of buckling and asymmetric behavior of flexible filament using temporal second-order immersed boundary method. International Journal of Numerical Methods for Heat & Fluid Flow, 30(3), 1047–1095, (2020).

S. Kang, and R.Maniyeri: Numerical study on bacterial flagellar bundling and tumbling in a viscous fluid using an immersed boundary method . Applied Mathematical Modelling, 38(14), 3567–3590, (2014).

S.Kenjeres:Numerical analysis of blood flow in realistic arteries subjected to strong non-uniform magnetic fields. International Journal of Heat and Fluid Flow, 29(3),752–764, (2008).

D.S. Kershaw: The incomplete Cholesky—conjugate gradient method for the iterative solution of systems of linear equations. Journal of Computational Physics, 26, 1,(1978).

M. C. Lai and C. S Peskin: An immersed boundary method with formal second-order accuracy and reduced numerical viscosity. Journal of Computational Physics, 160(2), 705–719, (2000).

V.C. Loukopoulos, E.E. Tzirtzilakis: Biomagnetic channel flow in spatially varying magnetic field. International Journal of Engineering Science, 42, 5–6, (2004).

R. Maniyeri and S Kang: S.merical study on the rotation of an elastic rod in a viscous fluid using an immersed boundary method. Journal of Mechanical Science and Technology, 26(5), 1515–1522,(2012).

S. M. Mousavi, M .Farhadi, and K .Sedighi: Effect of non-uniform magnetic field on biomagnetic fluid flow in a 3D channel. Applied Mathematical Modelling, 40, 15–16, (2016).

C. S. Peskin: Numerical analysis of blood flow in the heart. Journal of Computational Physics., 25(3), (1977).

E. E. Tzirtzilakis: Biomagnetic Fluid Flow Channel with Stenosis, Physic D Nonlinear Phenom., volume 237,1, (2008).

Experimental Investigations on Shock Train Oscillations Under the Influence of Downstream Perturbations

K. Abhishek, K. Ramachandra, S. L. N. Desikan, and R. Sriram

Abstract The dynamics of an isolator shock train with forced downstream pressure perturbations at different frequencies is experimentally investigated. Experiments are performed in an isolator test rig, in which an oblique shock train is formed at an incoming Mach number of 2.2. The oblique shock train's behavior is manipulated by an elliptical shaft positioned in the downstream diffuser section. Experimental investigations are conducted using high-speed flow visualization techniques and time-resolved pressure measurements. The results showed that, under the condition without the effect of downstream perturbations ($f_b \approx 0\,\text{Hz}$), the shock train exhibits large scale periodic oscillations at a dominant frequency of 76 Hz. It was observed that the amplitude of shock train oscillation decreased upon the introduction of the back pressure fluctuations in the isolator; the frequencies of oscillations were observed to be the forcing frequencies and/or their harmonics. The shock motion is found to be well correlated with back pressure fluctuations.

Keywords Shock train · Back pressure forcing · Shock oscillation

1 Introduction

In the internal flow field interaction between shock and the boundary layer intensifies the local separation, and this leads to the formation of shock train (Sullins and McLafferty 1992; Meier et al. 1990). The research shows that even without a change in back pressure, the shock position will move forward and backward, thus exhibiting "self-excitation oscillation" (Matsuo et al. 1999), which may change the flow field parameters such as field pressure and flow rate, cause the flutter of inlet pressure

K. Abhishek (✉) · K. Ramachandra · R. Sriram
Indian Institute of Technology Madras, Chennai, India
e-mail: ae21s019@smail.iitm.c.in

S. L. N. Desikan
Vikram Sarabhai Space Centre, Indian Space Research Organization, Thiruvananthapuram, India

G. Rajesh et al. (eds.), *Proceedings of the 15th International Symposium on Experimental and Computational Aerothermodynamics of Internal Flows*,
https://doi.org/10.1007/978-981-96-4082-9_18

of combustion chamber, which will affect the combustion stability (Huang et al. 2018; Tan et al. 2012). In scramjet and ramjet, precompression occurs through the isolator; The isolator is a rectangular conduit that connects the inlet to the combustion chamber, and its purpose is to contain the shock train motion caused by the increase of downstream pressure due to back pressure fluctuations from combustion side. However, in the combustion region, pressure often exhibits an unsteady nature due to factors such as improper fuel regulation, ignition delay, inefficient flame holding, and combustion instability. These factors can result in a blockage of fluid in the combustion chamber, leading to back pressure fluctuations which could act as a forcing on the shock train, and can push the shock train into the inlet resulting in the formation of bow shock upstream of inlet, ultimately ending up in the phenomenon known as "unstart" (Heiser et al. 1994; Curran et al. 1996). The unstart occurs when the isolator compression shock moves out of the inlet and eventually forms the bow shock. This bow shock obstructs the flow and does not allow the required mass flow which finally may lead to a loss of engine thrust. To ensure stable engine operation and prevent intake unstart, it is essential to have a clear understanding of how the shock train oscillates when subjected to downstream disturbances.

In an earlier work, oscillations of a single normal shock at transonic conditions was analyzed first by Culick and Rogers (1983). Bruce and Baninsky (2008) had done some analytical work to predict the amplitude of normal shock in the isolator under the back pressure fluctuations. The results showed that their theoretical model was able to predict the experimentally observed trend (in the same work) on the shock oscillations with varying excitation. At lower frequencies the shock exhibited large amplitude oscillations, but the amplitude was found to decrease with frequency. Similar trends were observe in the experiments with shock train with back pressure fluctuations at very low frequencies (0.59–3.75 Hz) by Klomparens et al. (2016) able to predict the trend between the excited amplitude of the shock wave well. During the cyclic movement of the shock train, a hysteresis effect was noted in the isolator response to fluctuations in back pressure. Few other researchers have carried out similar experiments (Cheng et al. 2019; Saravanan and Desikan 2020), related to back pressure fluctuations and observed that the introduction of such fluctuations leads to considerable changes in amplitude and frequencies of shock train oscillation.

The above studies on shock concern the effect of low back pressure frequencies, although in a practical scenario the forcing from combustion chamber can have a broad range of frequencies spanning various orders of magnitude. However, at forcing frequencies that are comparable with or higher than the inherent shock train oscillation frequencies, there are few reported data and there is no comprehensive understanding. Implementation of advanced data processing techniques such as modal analysis and time series analysis can help resolve many aspects. The present study was initiated with the objective of resolving the effect of back pressure forcing at various frequencies, particularly those comparable with shock oscillation frequencies. Some results from these investigations shall be presented in this paper (Fig. 1).

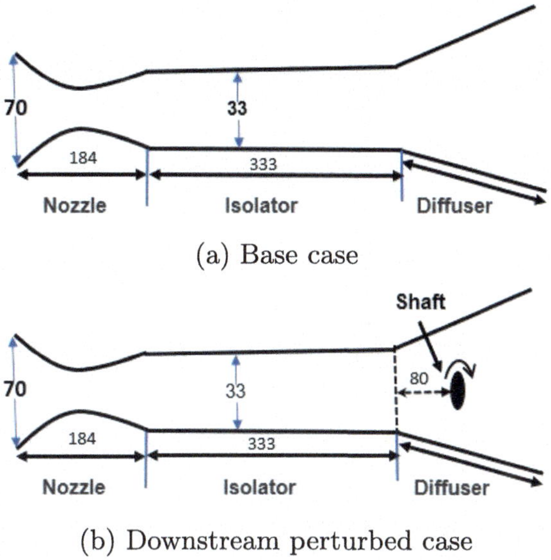

Fig. 1 Schematic diagram of the experimental setup

2 Experimental Set-Up

The experiments were conducted at the Gas Dynamics Laboratory, IIT Madras, using an isolator jet facility. In order to control and fix the mean position of the shock train within the isolator section, the total pressure was regulated in the settling chamber using a pressure-regulating valve. To simulate the desired Mach numbers, a contoured nozzle with a rectangular cross-section was connected to the settling chamber via a flange. Figure 3 depicts the schematic diagram of the base case. The isolator has a rectangular cross-section measuring 33 mm × 100 mm and has a length of 333 mm. To facilitate optical access, glass windows measuring 234 mm × 33 mm were affixed to the isolator's side walls. For the the base case, the Mach number in the isolator was 2.3, and the boundary layer thickness was determined to be 4 mm (using a pitot survey, at a pressure which just pushes the shock train entirely out of the isolator section) at a distance of 72 mm from the inlet of the isolator. To simulate the downstream forcing, the back pressure fluctuations are generated by means of an elliptic shaft of 15 mm major axis and 10 mm minor axis, placed at a distance of 80 mm from the isolator exit (in the diffuser section) as shown in Fig. 3. The shaft ended in circular section which was connected to bearings on the side walls, and was rotated from outside using a servo motor. The elliptic shaft is nominally set to rotate at frequencies between 120 rpm and 2400 rpm to produce an oscillatory back pressure in the isolator. The downstream pressure fluctuates at two cycles per rotation of the motor's shaft; hence, in every rotation of the elliptic shaft, the flow gets partially blocked twice.

To obtain time-resolved surface pressure measurements, three high-resolution Endevco pressure transducers with a rated pressure of 1 bar (absolute) were installed at the spanwise midplane on the upper wall of the isolator section. All flow signals were sampled at a rate of 250 kHz, for 4 s resulting in the recording of 1000k samples.

To capture the dynamic nature of the shock train, time-resolved shadowgraph visualization was employed. A high-speed camera, Photron FASTCAM SA4 Model 500K-M1, was utilized to capture time-resolved snapshots of the flow field. The camera operated at a frame rate of 20,000 frames per second (fps) with a shutter speed of 1/333,000 s. The spatial resolution of the images was 1024 × 128 pixels.

3 Results and Discussion

3.1 *Shadowgraph and Static Pressure Measurements*

The purpose of conducting a shadowgraph was to determine the pressure range in which the shock train remains within the isolator test section. Figure 2 shows the shadowgraph images of the flow field at three different time instants for the base case (without back pressure perturbation) at an inlet total pressure of 1.210 bar (gauge). The shock train exhibits self-excited oscillations. The flow configuration could be a potential factor contributing to the self-excited oscillations observed in the shock train. An asymmetric flow structure has been noted with respect to the horizontal centerline, even though symmetric geometries and inlet conditions were employed. The asymmetry manifests in the form of vertically skewed asymmetric shock structures, and in the offset of the shock interaction point away from center, both of which are observed to oscillate above and below the centerline. This oscillation results in the flapping of the separated flow regions in the top and bottom walls which in turn could be the source of the back-and-forth oscillations of the shock train. In the presence of downstream perturbation, it is observed that the downstream pressure perturbations propagate in an upstream direction, leading to oscillations of the oblique shock train.

Mean static pressure measurements were conducted at four different settling chamber pressures. As depicted in Fig. 3, it is evident that the time-averaged shock

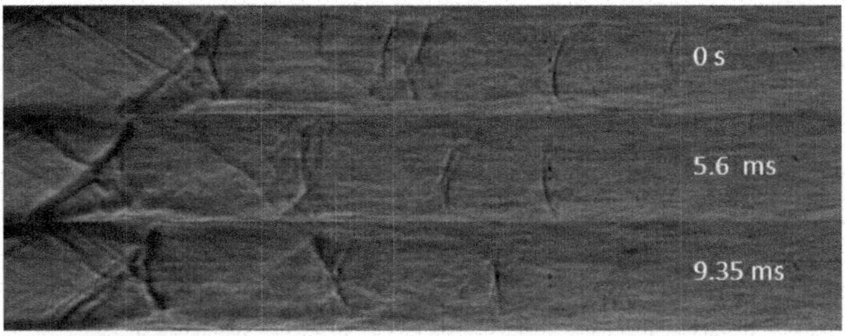

Fig. 2 Instantaneous shadowgraph images of shock train oscillations at different time instants (flow is from left to right)

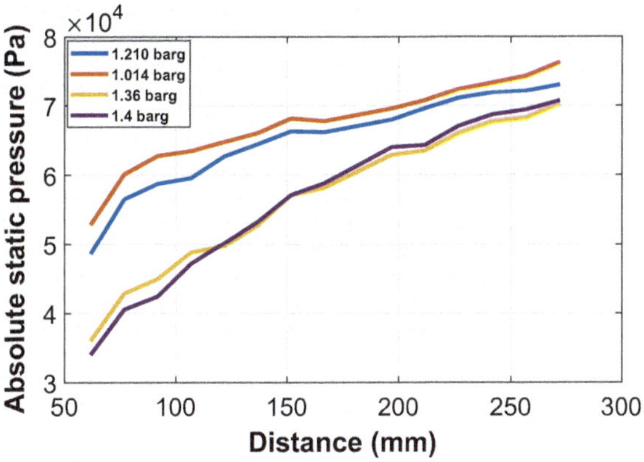

Fig. 3 Time-averaged surface pressure variation along the length of the isolator

train structures progressively move downstream towards the diffuser as the stagnation pressure increases. The time averaged position of the front most shock and the time averaged static pressure distribution were used to fix the operational pressure in the settling chamber such that the mean leading shock position is close to (but slightly upstream of) the middle of the isolator section, so that its dynamics can be entirely visualized within the shadowgraph window.

3.2 *Unsteady Pressure Measurements*

The total pressure at which the pressure rise initiates after the visible part of the isolator within the shadowgraph window was selected to investigate the dynamics using fast response diagnostics.

In the analysis of shock train dynamics within the isolator using Endevco pressure transducer, the following nomenclature was used: S1, S2, and S3 corresponded to sensor port distances of 55 mm, 142 mm, and 275 mm from the isolator inlet, respectively.

Base Case with and Without Stationary Shaft

Figure 4 shows the typical plot of the unsteady pressure variation versus time for base case and case with stationary shaft (placed with its major axis in the flow direction), respectively. Under the condition of no downstream pressure perturbations (Base case) and a consistent incoming flow, the shock train shows a quasi-periodic behavior and self-induced oscillations at a large scale as shown in Fig. 4a. Intermittency in the upstream sensor S1 is very much comparable to that in S2, which is indicative of the large scale shock motion.

The most interesting outcome of the experiments is that when the base case was repeated by having the elliptic shaft placed idle, i.e., without rotating ($f_b \approx 0\,\text{Hz}$), with its major axis placed horizontally and rigidly fixed, the amplitude of shock

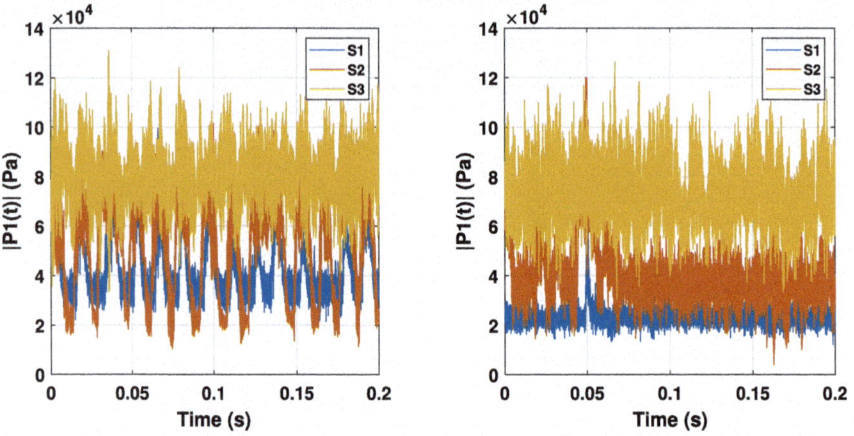

(a) Base case (b) Case with stationary shaft

Fig. 4 Unsteady pressure signal versus time at a different locations in the shock train region for the base case (left) and the case with stationary shaft (right)

oscillations was considerably reduced as shown in Fig. 4b compared to the base case discussed before. In this case S1 shows spikes occasionally but for most part, it reads the freestream pressure. Figures 5 and 6 show the results of the Fast Fourier Transform (FFT) for the base case and the case with a stationary shaft, respectively. These figures facilitate the examination of the relationship between variations in the positions of the oblique shock under these two different conditions. To determine the strength of the signal distributed across different frequencies, the Power Spectral Density (PSD) has also been calculated. The normalized power spectrum at different locations for these both is shown in Fig. 7a, b.

From the normalized power spectrum Fig. 7b, it is evident that in the case with a stationary shaft, the amplitude is smaller than that in the base case, and the signals are also relatively broadband. The mere presence of the shaft in the plane of symmetry, with its major axis aligned with the flow, shows a stabilizing effect on the shock train.

Under the Influence of Rotating Shaft

When subjected to periodic pressure perturbations in the downstream, the shock train was forced to oscillate in the channel. Variation in the downstream pressure perturbation exhibits sinusoidal periodic characteristics that are produced by rotating elliptical shafts. Two different forcing frequencies have been discussed: one at a lower forcing frequency (20 Hz) and the other at a higher forcing frequency (80 Hz).

Figure 8 shows the typical plot of the unsteady pressure variation versus time under the presence of forced downstream pressure perturbations.

At lower forced frequency (20 Hz) Fig. 8a highlights that at the farthest upstream location S1, the pressure exhibits periodic spikes for a short duration, after which it returns to the low pressures characteristic of the supersonic flow upstream. At an S2

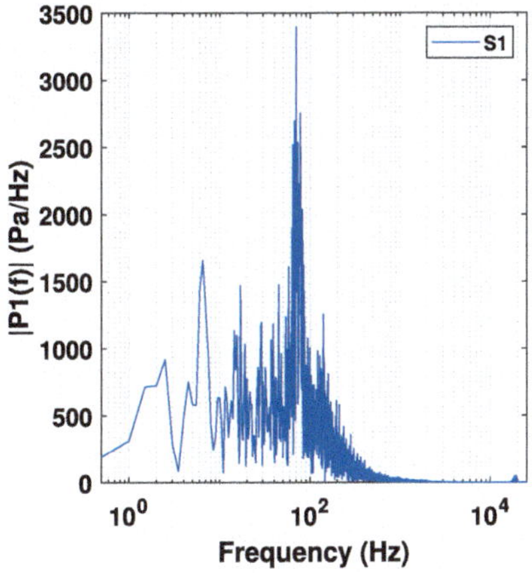

(a) S1

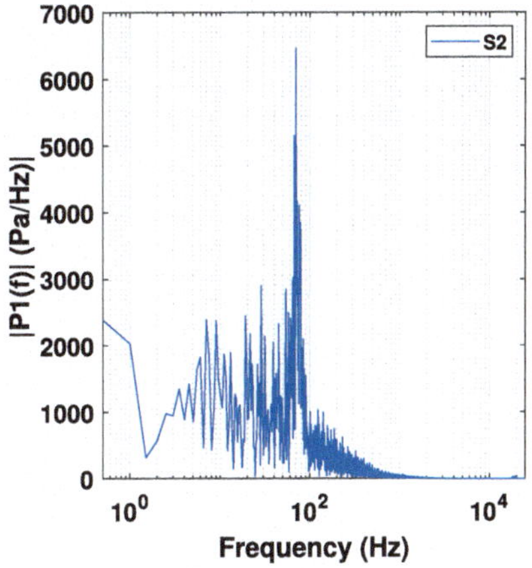

(b) S2

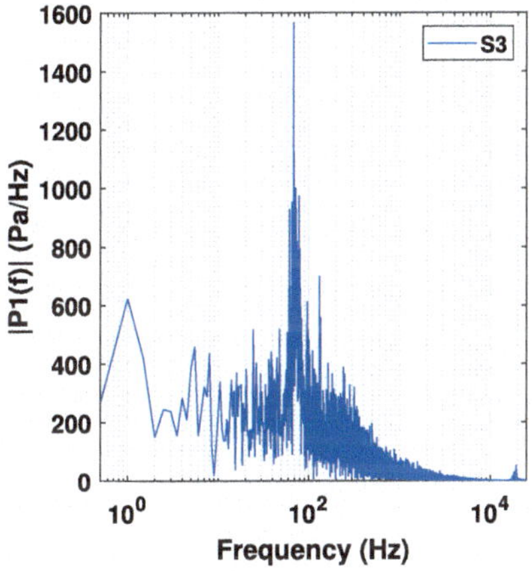

(c) S3

Fig. 5 FFT of unsteady pressure at different locations for base case

location, the pressure signal also exhibits intermittency, but in this case, the pressure remains mainly on the higher side and does not reach the low values observed in the upstream flow. Before the pressure starts to decrease, the shock rapidly moves upstream, leading to a significant increase in pressure at this station. The downstream sections of the isolator are located well beyond the leading shock and are situated within the pressurized flow that follows multiple shocks in the shock train, as a result, the S3 signal does not exhibit a prominent intermittent behavior. When encountering a high forced frequency (80 Hz), the time signal exhibits a similar trend as observed with a lower forced frequency, as shown in Fig. 8b. The quality disparity between the

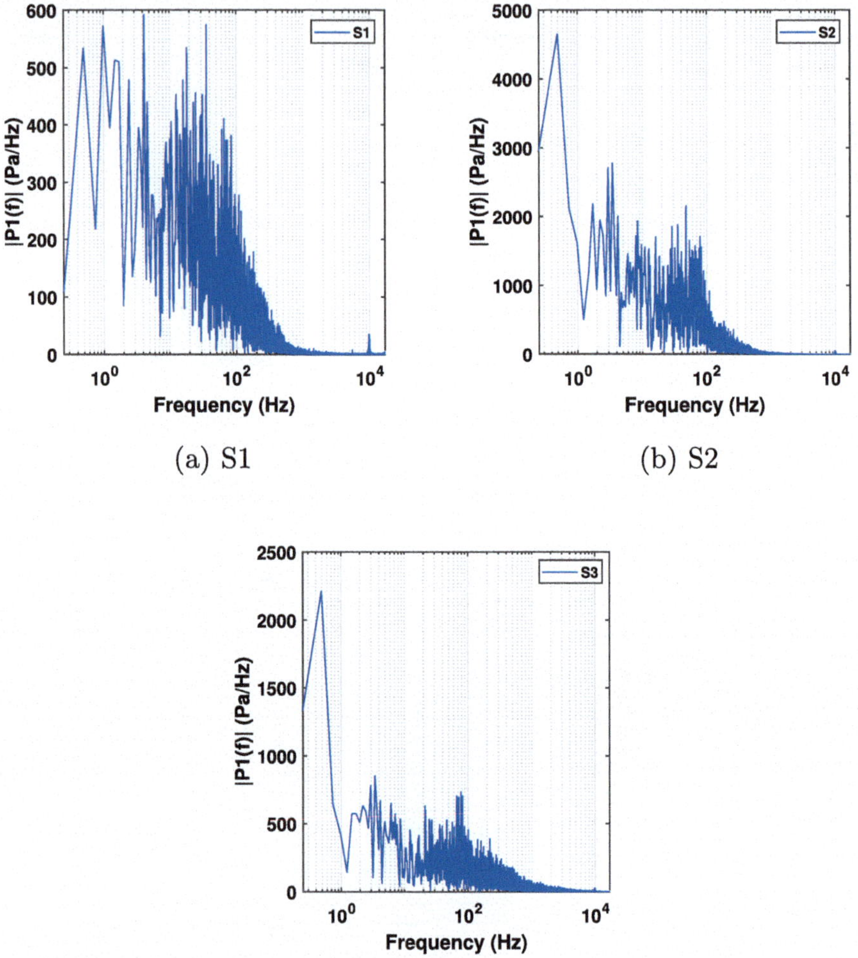

(a) S1

(b) S2

(c) S3

Fig. 6 FFT of unsteady pressure at different locations for the case with stationary shaft

temporal profiles of the pressure measurement signal in Fig. 8 for the right and left time histories is relatively inconspicuous. Upon closer observation, it becomes evident that the shape of the time signal undergoes rapid alterations when the rotational of the shaft increases, as compared to the shaft operating at a lower rotation.

Figures 9 and 10 show the results of the Fast Fourier Transform (FFT) for the back pressure frequencies of 20 Hz and 80 Hz, respectively, enabling the examination of the connection between variations in the positions of the oblique shock and downstream pressure perturbations. It is important to note that the high-frequency component of small pressure perturbations, which arises from the inherent unsteadiness of the

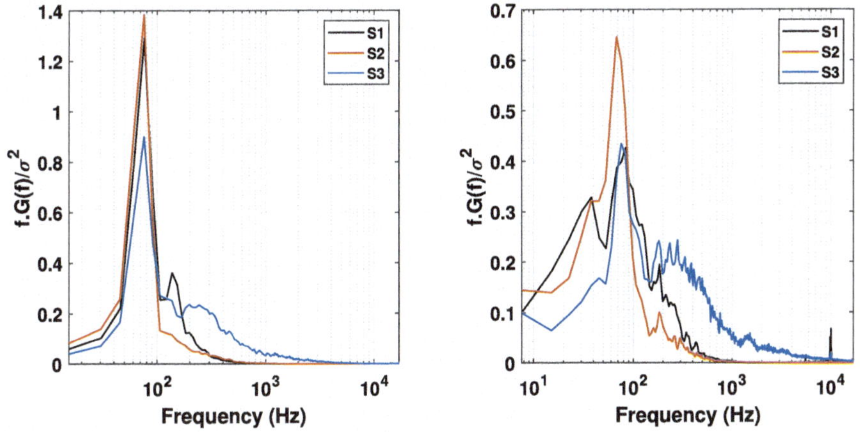

(a) Base case (b) Case with stationary shaft

Fig. 7 Normalized power spectral density of unsteady pressure versus frequency (Hz) for the base case (left) and case with stationary shaft (right)

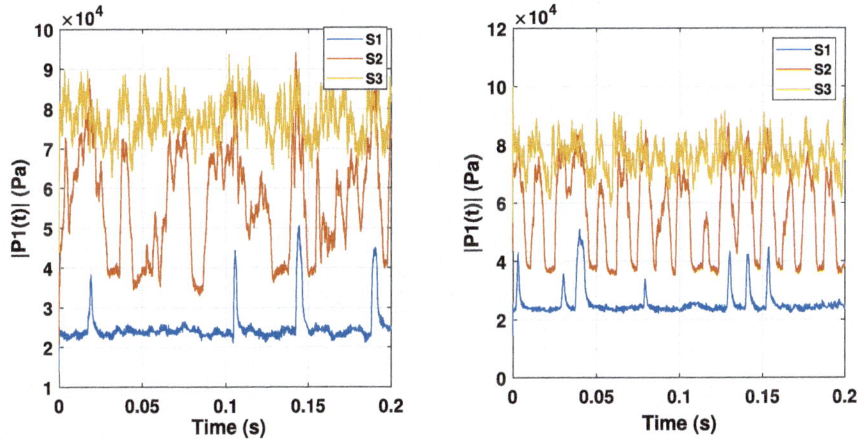

(a) 20 Hz (b) 80 Hz

Fig. 8 Unsteady pressure signal versus time at different locations in shock train region at different forcing frequency of the shaft—left (20 Hz); right (80 Hz)

flow system, is superimposed on the downstream sinusoidal pressure perturbations. An interesting observation is the intermittent nature observed in relatively upstream positions.

The normalized power spectrum at different locations for the lower forcing frequency is shown in Fig. 11a. At lower forcing frequency, the observations show that the dominant frequencies of the amplitudes at the middle and downstream positions are comparable but occur at distinct frequencies under 100 Hz. However, at

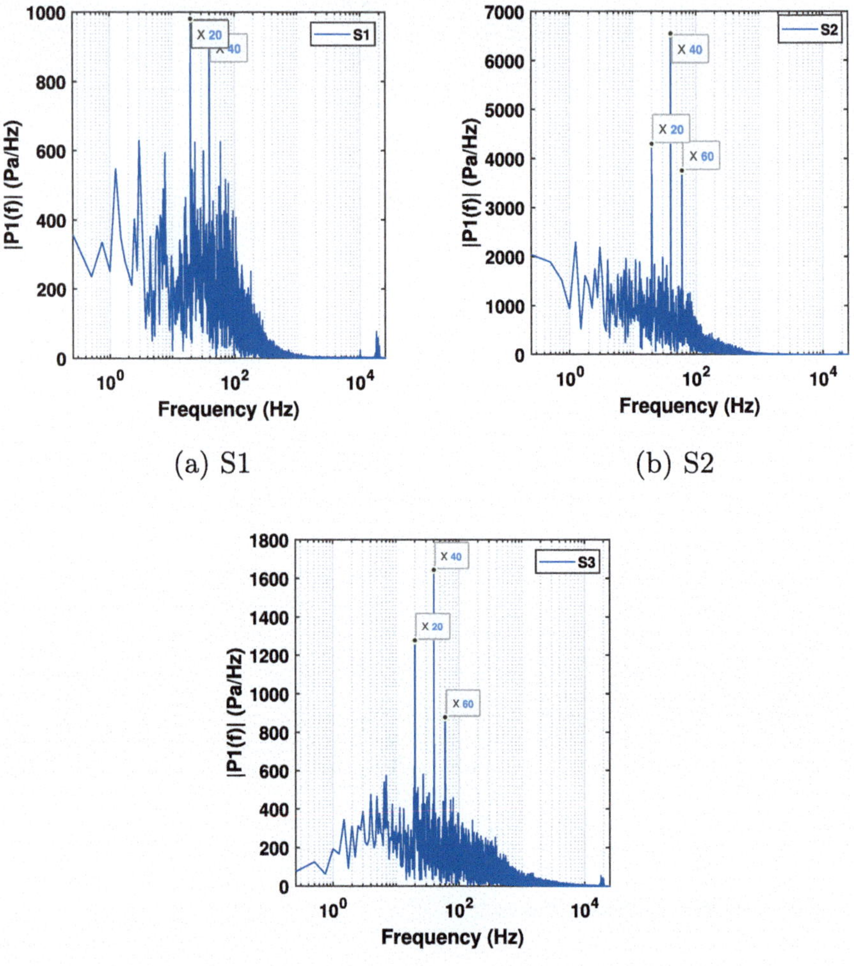

(a) S1 (b) S2

(c) S3

Fig. 9 FFT of unsteady pressure at different locations at 20 Hz

frequencies below 1000 Hz, the upstream position exhibits significant amplitude across various frequencies, showing a relatively broadband nature. At the higher speed, It can be seen in Fig. 11b that the upstream and downstream locations have much smaller amplitude than the middle one at the dominant frequency less than 100 Hz. At lower forcing frequency, the signals are also relatively more broadband. The cross-correlation coefficients have been computed for both scenarios, with the correlations depicted in a manner that indicates if a peak correlation is observed between locations 1 and 2 at a positive time lag, it means that the event at location 2

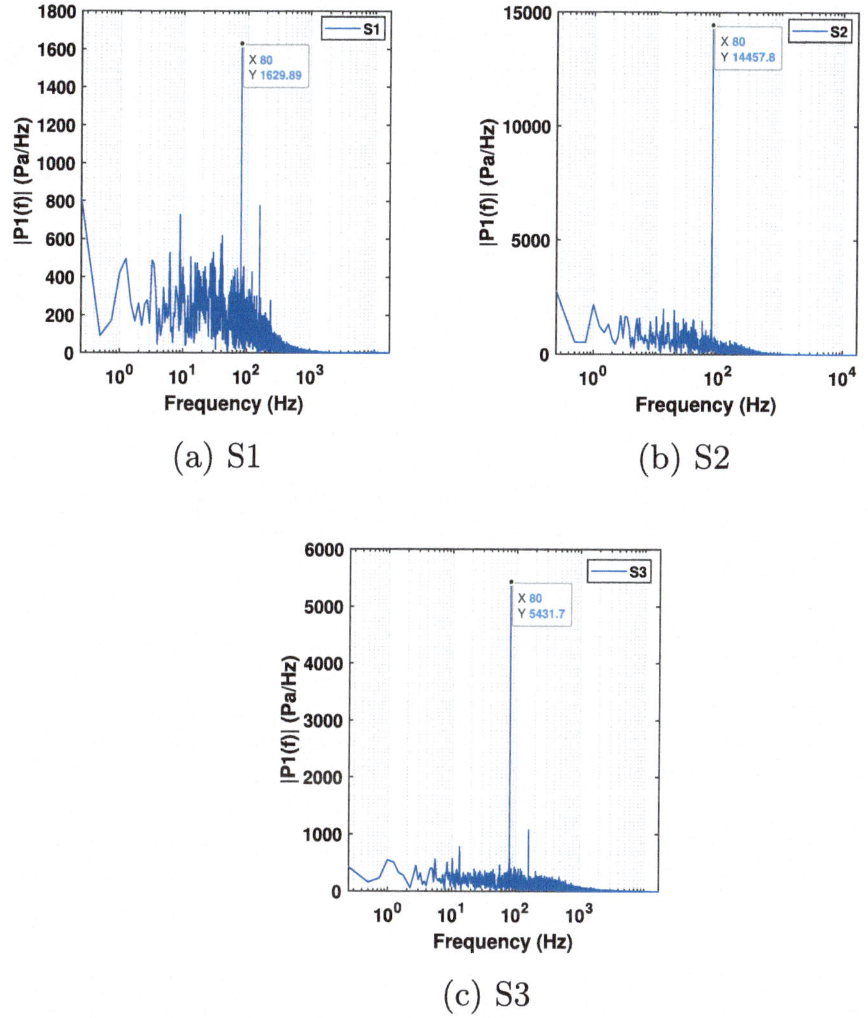

Fig. 10 FFT of unsteady pressure at different locations at a 80 Hz

follows the event at location 1 and vice versa. At lower forcing frequency both correlations exhibit a positive peak value, indicating synchronized movement of the shock train system, albeit at distinct magnitudes. This implies that a pressure increase or decrease at one location corresponds to a similar event at the other location. Whereas the higher forcing frequency disturbances show the quasi-periodicity of the shock train. To determine the frequency range where a significant linear relationship exists between the positions, magnitude-squared coherence versus frequency was plotted. For the lower forcing frequency, the coherence between S2 and S1 is found to be very high (0.58), around the range of 60 to 76 Hz. After 100 Hz, it starts to decrease before

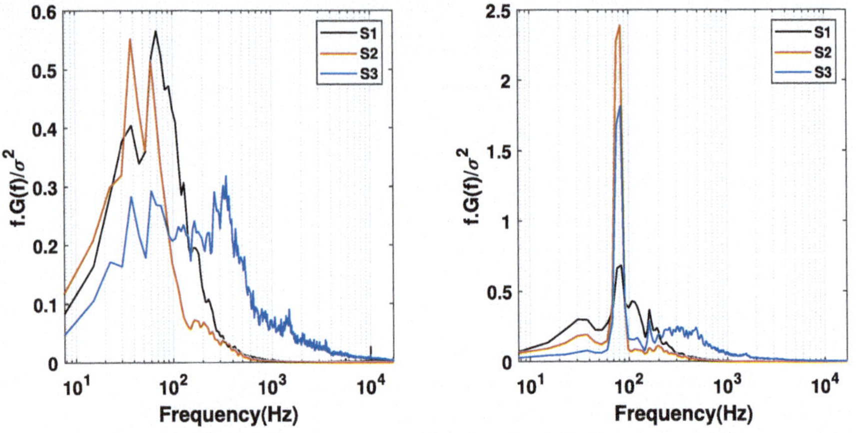

(a) 20 Hz (b) 80 Hz

Fig. 11 Normalized power spectral density of unsteady pressure versus frequency (Hz) at forcing frequency of the shaft—left (20 Hz); right (80 Hz)

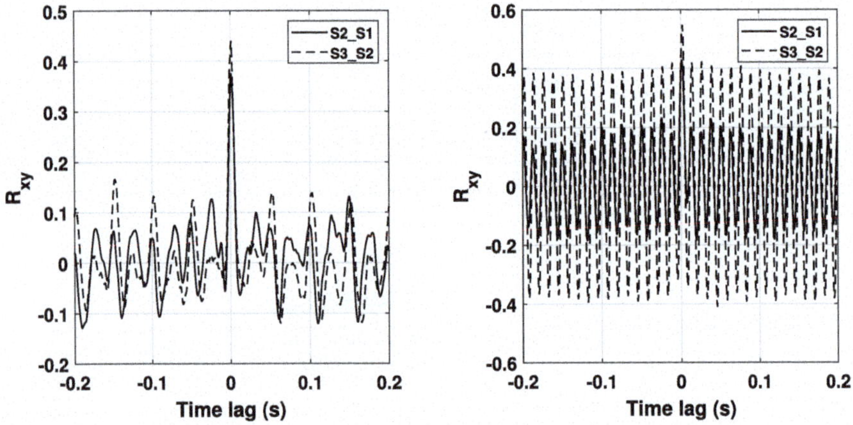

(a) 20 Hz (b) 80 Hz

Fig. 12 Cross-correlation coefficient versus time lag (in s) between different streamwise locations at different forcing frequencies of the shaft—left (20 Hz); right (80 Hz)

increasing again. The trend of the coherence also closely follows that of the power spectrum (refer to Fig. 11). At higher forcing frequency shows a high coherence of about 0.85 between S2 and S1 at lower frequencies (less than 100 Hz). The other combination (S3_S2) exhibits a similar trend, although it is less coherent compared to S2_S1, with S2_S1 showing a higher peak value. A similar pattern is observed in the cross-correlation analysis for both cases (Figs. 12 and 13).

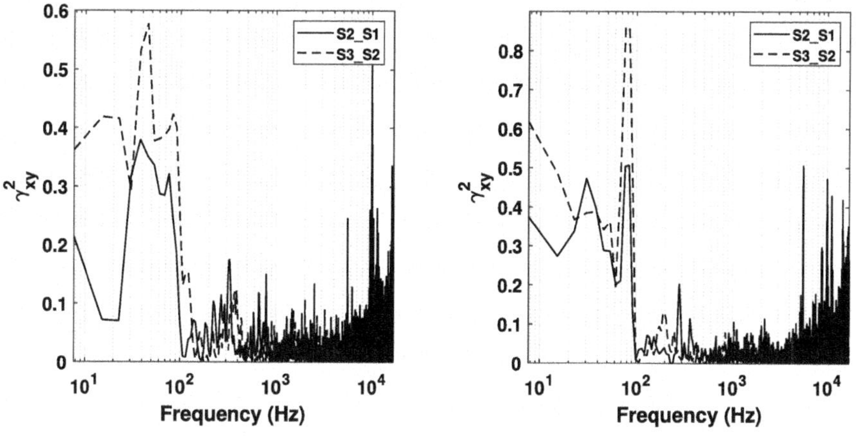

(a) 20 Hz (b) 80 Hz

Fig. 13 Magnitude-squared coherence versus frequency (Hz) between different streamwise locations at different forcing frequencies of the shaft—left (20 Hz); right (80 Hz)

4 Conclusions

Experimental investigations on shock train under back pressure forcing are conducted in the isolator jet facility using high-speed flow visualization techniques and time-resolved pressure measurements. The introduction of back pressure perturbations, which are produced by rotating the elliptical shaft into the isolator was noted to result in a reduction of the oscillation amplitude of the shock train. At lower shaft speeds, the power spectrum exhibits a broader band compared to higher shaft speeds. The correlation analyses indicate that the downstream events drive the upstream shock train in the presence of downstream perturbations.

Acknowledgements The authors would like to thank G. Rajesh and his group for letting us use some of their equipment. We are also thankful to our colleagues Mr. Sivaprasad, Mr. Vayala Siva, Mr. Harshit Bansal and Mr. Vinoth Kumar for their help in setting up and performing the experiments. The research work is supported by Indian Space Research Organisation through ISRO-IIT Madras Space Technology Cell, Project no. SP21221196AEISRO008894, Government of India.

References

Bruce, P., Babinsky, H.: Unsteady normal shock wave boundary layer interactions with control, 2008, p. 721. https://doi.org/10.2514/6.2008-721

Cheng, C., Wang, C., Cheng, K.: Response of an oblique shock train to downstream periodic pressure perturbations, 233 (1) (2019) 57-70. https://doi.org/10.1177/0954410017727028

Culick, F., Rogers, T.: The response of normal shocks in diffusers. AIAA journal 21 (10), 1382–1390 (1983). https://doi.org/10.2514/3.8257

Curran, E., Heiser, W., Prat, D.: Annual Review of Fluid Mechanics, 28(1), 323 (1996). https://doi.org/10.1146/annurev.fl.28.082106.100001

Heiser, W. H., Pratt, D. T., Daley, D. H.: Supersonic Combustion Flow Visualization at Hypersonic Flow. 1041–1047 (1994).

Huang, H., Tan, H., Zhuang, Y., Sheng, F., Sun, S.: Progress in internal flow characteristics of hypersonic inlet/isolator. 39 (10), 2252–2273 (2018). https://doi.org/10.13675/j.cnki.tjjs.2018.10.010

Klomparens, R., Driscoll, J. F., Gamba, M.: Response of a shock train to downstream back pressure forcing, 2016, p. 0078. https://doi.org/10.2514/6.2016-0078

Matsuo, K., Miyazato, Y., Kim, H.-D.: Shock train Progress in aerospace sciences 35 (1), 33–100 (1999). https://doi.org/10.1016/S0376-0421(98)00011-6

Meier, G.,Szumowski, A. Selerowicz, W. : Self-excited oscillations in internal transonic flows. 28(1), 145–200 (1990). https://doi.org/10.1016/0376-0421(90)90004-4

Saravanan, R., Desikan, S. L. N, Muruganandam, T.M.: Isolator characteristics under steady and oscillatory back pressures. 32 (9), 096104 (2020). https://doi.org/10.1063/5.0016360

Sullins, G., McLafferty, G.: Experimental results of shock trains in rectangular ducts. 1992. https://doi.org/10.2514/6.1992-5103

Tan, H., Sun, S., Huang, H.: Behavior of shock trains in a hypersonic inlet/isolator model with complex background waves. 53 (6), 1647–1661 (2012). https://doi.org/10.1007/s00348-012-1386-1

Numerical Investigation of Effects of Location of Cavity Behind Pylon on Supersonic Combustion of Ethylene

Anbarasan Sekar and Aravind Vaidyanathan

Abstract The effects of the location of the cavity behind the pylon on the supersonic combustion of the ethylene jet were studied numerically. Free-stream flight condition of Mach number 5 at an altitude of 25 km was considered, and the isolator inlet conditions were obtained using two-dimensional inlet simulation. The equivalence ratio of 0.15 was considered for this study. The position of the cavity behind the pylon was varied from 0 to 60 mm with a spacing of 20 mm. A steady Reynolds-Averaged Navier-Stokes equation set with species transport was used in this study. The SST K-ω turbulence model was taken as the closure model. The validation of the equation set was performed using experimental data from the literature. The simulation results indicate that the position of the cavity behind the pylon plays a vital role in combustion characteristics. The pylon and cavity with 0 mm separation and floor injection produced a normal shock train due to thermal choking, and the direct interaction of the cavity with the incoming boundary layer. The cavity with 20 mm separation showed better penetration and less spreading in the lateral direction compared to the 40 and 60 mm separation cases. The combustion efficiency and total temperature were higher for the 0 mm separation case, but the total pressure loss was also higher for the 0 mm separation case than the other cases. The total temperature rise of 300 K was achieved for the 0 mm case, whereas 150 K was achieved for the other cases towards the exit of the domain. A total pressure loss of almost 35% was observed for the 0 mm separation case. The 20 mm separation case showed the lowest total pressure loss among all the cases.

Keywords Scramjet · Supersonic ethylene combustion · Pylon-cavity flameholder

---

A. Sekar (✉) · A. Vaidyanathan
Department of Aerospace Engineering, Indian Institute of Space Science and Technology, Thiruvananthapuram, Kerala, India
e-mail: anbumsr@gmail.com

G. Rajesh et al. (eds.), *Proceedings of the 15th International Symposium on Experimental and Computational Aerothermodynamics of Internal Flows*,
https://doi.org/10.1007/978-981-96-4082-9_19

1 Introduction

The realization of hypersonic vehicles with an air-breathing engine depends on the heat addition process, which is ineffective when performed at subsonic speeds due to the non-equilibrium effects and enormous total pressure loss. Supersonic combustion is challenging due to the low residence time, flame-holding, and complete combustion within a limited combustor length. Hydrocarbon fuels must be used for long-range hypersonic vehicles due to high energy density per unit volume. Ethylene would be a suitable option since it has the lowest ignition delay time among hydrocarbons (Colket III and Spadaccini 2001). The chemical time scale is smaller than the flow time scale at low hypersonic Mach numbers, so the supersonic combustion process is mixing-controlled (Ferri 1973). Many injection techniques, including transverse/angled injection, cavity-based, strut-based, physical/aero ramp, pylon-aided, and a combination of these, were studied by researchers (Schetz and Billig 1966; Tan et al. 2018; Sekar and Vaidyanathan 2023; Shi et al. 2021). Among these, pylon-aided injection is a better option because it provides better penetration with a low momentum-flux ratio and better near-field mixing than transverse injection. It also produces less stagnation pressure loss due to the sharp leading edge and low blockage compared to a strut injector. The fuel has to be injected into the core of the main flow to get effective combustion, which was the reason for the employment of strut-based injection. This could also be achieved using pylon-aided injection. When hydrocarbon fuels are used in scramjet applications, the cavity plays a vital role in flame holding. Hydrocarbon fuels usually have high ignition time delays and low laminar flame speeds. Thus cavities could be used to improve the residence time available for the fuel. However, using a cavity comes with an issue of pressure oscillations which would affect the combustion dynamics (Lin et al. 2010; Wang et al. 2013). A recent study that employed a pylon-cavity configuration showed the absence of high-frequency oscillations. Adding a pylon to the cavity also improves mass exchange between the main flow and the cavity (Oamjee and Sadanandan 2020). The current study aims to address the effects of the location of the cavity behind the pylon on the combustion characteristics of ethylene.

2 Numerical Methodology

2.1 Geometric Parameters

The numerical simulation of the supersonic combustion of ethylene was carried out for the pylon with cavity configuration. The distance between the pylon and the cavity was varied to study the effects on supersonic combustion. A typical isolator dimension of 25.4×25.4 mm was considered with a standard pylon and a cavity. The pylon with a 10 mm length, 5 mm width and 45° leading edge angle was considered in this study. A cavity with an L/D of 3, a depth of 10 mm and an aft wall angle of

45° was selected. The distance between the pylon and the cavity was varied as 0, 20, 40, and 60 mm, and these are hereafter denoted as 0-X/D, 10-X/D, 20-X/D, and 30-X/D, respectively. The injection diameter was 2 mm, and placed 3 mm from the trailing edge of the pylon. For the 0-X/D case, the fuel was injected from the floor of the cavity. The bottom wall after the cavity was given a divergence angle of 3°.

2.2 Fluid Domain and Grid Details

Figure 1 shows the fluid domain and the structured grid for a pylon-cavity configuration. Although the geometric symmetry does not confirm the flow symmetry, the mean flow field obtained using high-fidelity simulations and steady RANS simulations are comparable. Recent studies also showed reliable results using symmetric conditions. A fluid domain with a total length of 235 mm was considered, and the pylon was placed 70 mm from the main flow inlet. The constant area upstream of the pylon acts as an isolator. The injector was extended outside to avoid the effect of back pressure on the ethylene mass flow rate. The structured grid was generated using ICEM CFD with approximately 4.5 million nodes. The Y-grid and O-grid options were utilized for the pylon region and the injector region to improve the orthogonal quality of the grid, respectively. Since the equations were solved to the wall, the first layer thickness from the wall was kept at 2×10^{-6} m.

2.3 Solver Setup and Boundary Conditions

A commercial solver, CFD++ 18.1, was used in this study. A steady-state multi-species density-based solver with a species transport equation set was used for this

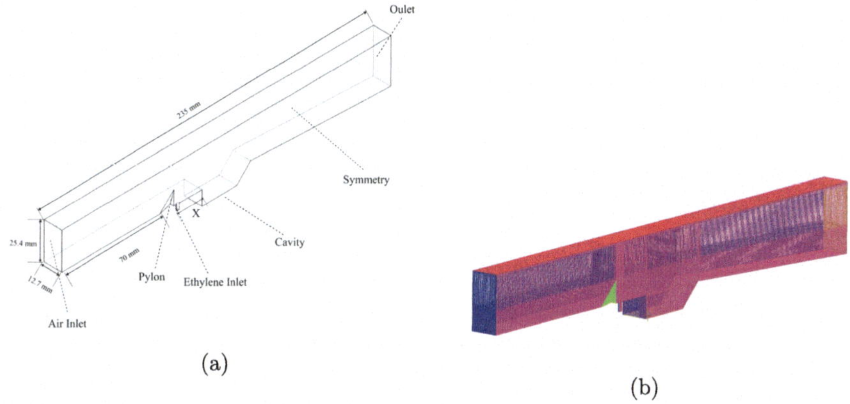

Fig. 1 Details of **a** fluid domain and **b** structured grid for a pylon-cavity configuration

Table 1 6-species 3-step ethylene-air reaction mechanism

Reactions	A[a]	b	T_a (K)
$C_2H_4 + O_2 \rightleftharpoons 2\,CO + 2\,H_2$	2.1×10^{14}	0	18015.3
$2\,CO + O_2 \rightleftharpoons 2\,CO_2$	3.4×10^{11}	2	10134.9
$2\,H2 + O_2 \rightleftharpoons 2\,H_2O$	3×10^{20}	−1	0

[a] Units are a multiple of cm^3/mol/s

Table 2 Boundary conditions at main inlet and secondary inlet

Boundary	$p(bar)$	$T(K)$	$U(m/s)$	M	Y_{N_2}	Y_{O_2}	$Y_{C_2H_4}$
Air inlet	1.0838	745	1044	–	0.79	0.21	0
Ethylene inlet	3	265	–	1	0	0	1

study. The 6-species, 3-reaction combustion scheme of Mawid (Baurle and Eklund 2002; Kannaiyan 2020) was used for ethylene combustion in air, and the reaction mechanism is detailed in Table 1. The rate constant A was modified as suggested in the literature (Goldberg et al. 2010) due to the low enthalpy condition considered in this study.

Specific heat capacity at constant pressure was taken as a function of temperature, and thermal conductivity and viscosity were found using Sutherland's formula for each species. This study used two equation SST K-ω turbulence model with the turbulent Prandtl number and Schmidt number of 0.9 and 0.7, respectively. For the simulation of pylon cavity configurations free-stream condition of Mach number 5 at an altitude of 25 km was considered. The isolator inlet conditions were calculated from a 2D simulation of a typical 3-step inlet designed for decelerating free-stream Mach number of 5 to 2. The ethylene equivalence ratio of 0.15 was taken with a stagnation temperature of 300 K. Sonic condition was applied to the ethylene inlet. The outlet was considered a supersonic outlet where the solution is extrapolated from the interior cells using a backward difference scheme. Instead of assuming an adiabatic wall condition, the walls are considered isothermal with a temperature of 600 K to have slightly realistic conditions. The boundary conditions at both inlets are given in Table 2.

2.4 Validation

The numerical scheme was validated using the experiment carried out using an axisymmetric model from the literature (Vanyai et al. 2019). Low enthalpy condition with an equivalence ratio of 0.14 and with shock tunnel exit pressure, temperature, and Mach number of 33.7 kPa, 668 K, and 4, respectively, was considered for the validation. The ethylene fuel was injected with a stagnation temperature of 300 K. One-fourth of the full-scale model was considered for the simulation with two axes symmetry. The injectors were extended 4 mm outside to avoid the back pressure

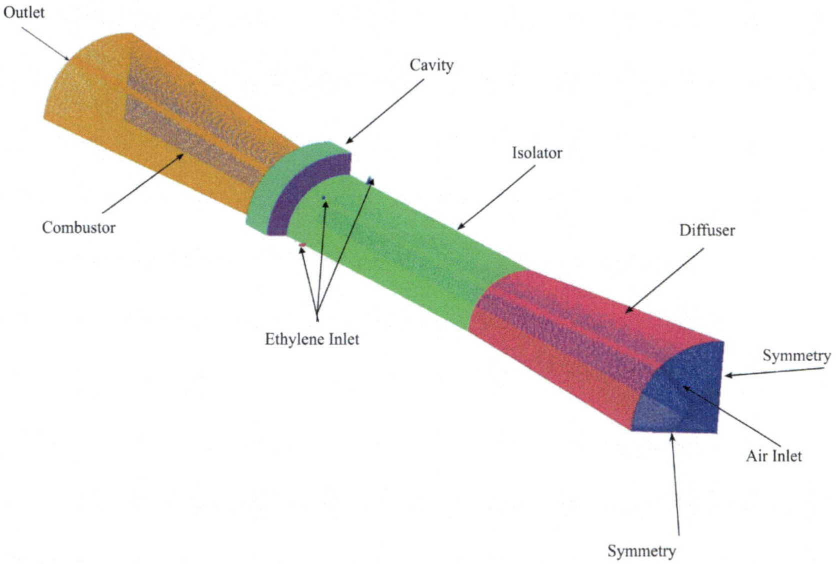

Fig. 2 Structured grid of one fourth of full scale experimental model (Vanyai et al. 2019)

Fig. 3 Comparison of wall pressure from experiment (Vanyai et al. 2019) and simulation

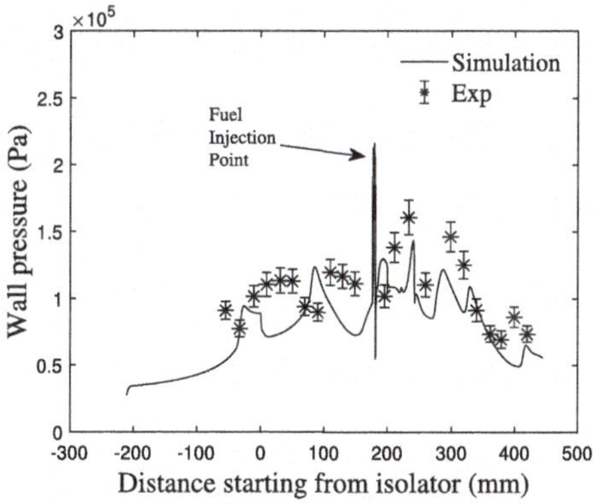

effects on the fuel mass flow rate. Figures 2 and 3 show the structured mesh that was created using ICEM CFD with approximately 5.2 million nodes and the comparison of wall pressure from experiment and simulation, respectively.

3 Results and Discussion

The steady RANS reactive flow simulations were carried out for an equivalence ratio of 0.15 for all the cases. The results obtained from the simulations are presented in this section. Figure 4 shows the Mach number contour on the symmetry plane for

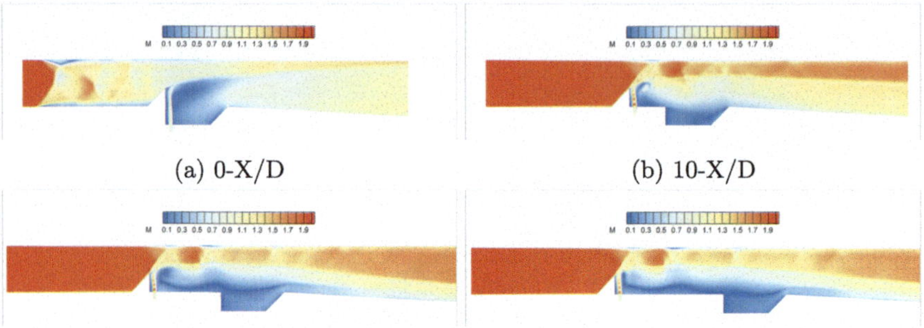
(a) 0-X/D (b) 10-X/D

(c) 20-X/D (d) 30-X/D

Fig. 4 Mach number contours on the symmetry plane for the different cavity placement from pylon

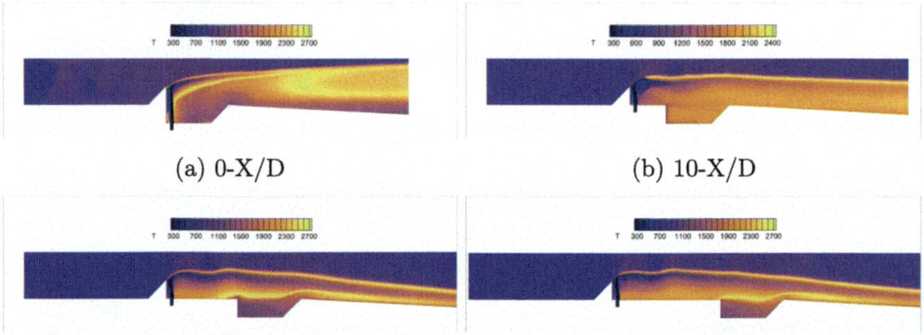
(a) 0-X/D (b) 10-X/D

(c) 20-X/D (d) 30-X/D

Fig. 5 Temperature contours on the symmetry plane for the different cavity placement from pylon

different cavity positions behind the pylon. All cases except the 0-X/D case showed bow shock attached to the pylon. The normal shock train in front of the pylon was observed for the 0-X/D case. These contours also show that the subsonic region in the wake was different for different cases. The normal shock train for the 0-X/D case could be due to the separation of the incoming boundary layer because of the sudden rise in pressure inside the cavity due to combustion or the thermal choking because of the enhanced combustion due to the effective fuel mixing inside the cavity. The coupling between these two effects could also be possible. For the other three cases, the fuel was injected behind the pylon, and the cavity was separated and hence the pressure rise inside the cavity did not directly affect the incoming boundary layer.

Figure 5 shows the temperature contour for different cases on the symmetry plane. The 0-X/D case showed better combustion due to the subsonic flow produced due to the normal shock train. Among the other three cases, 10-X/D showed a better distribution of temperature and the high-temperature region also stayed parallel till the exit of the domain. However, the 20-X/D and 30-X/D cases showed a declining trend in the high-temperature region. A thin high-temperature zone above the cavity was also observed for these cases, indicating shear layer oscillations, which should

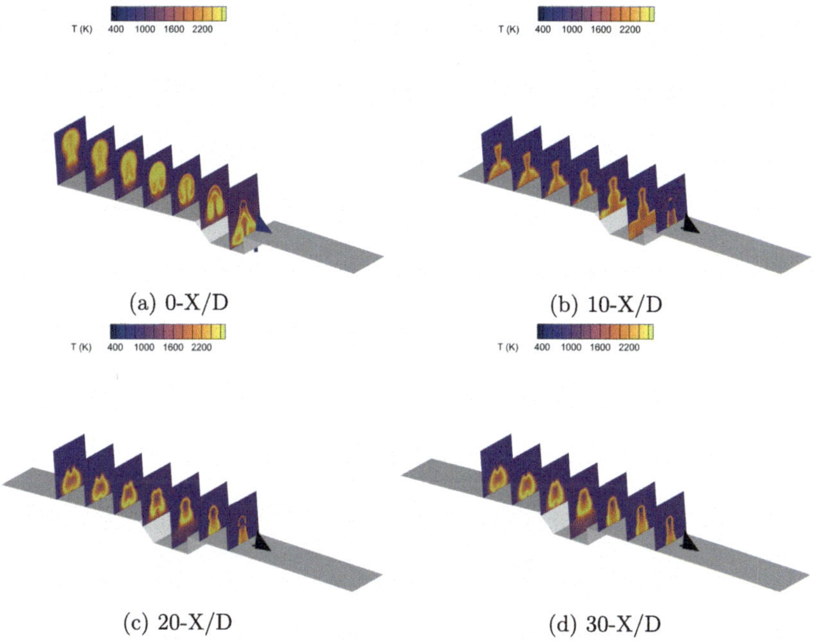

Fig. 6 Temperature contours on the symmetry plane for the different cavity placement from pylon

be analyzed using unsteady simulation. This could be due to the weak interactions of reacting flow inside the cavity and the upstream flow field. The temperature inside the cavity also was lower for the last two cases as compared to the other cases, indicating that the mixing and combustion inside the cavity were inefficient.

Figure 6 shows the temperature contour on the cross-sectional planes starting from $x/d_j = 5$ for all the cases. The 0-X/D case clearly shows the better penetration and spreading of the combustion region throughout the cross-sectional planes towards the exit of the domain when compared to the other cases. Among the other three cases, the 20-X/D and 30-X/D cases show similar trends and show better spreading of high-temperature zone than the 10-X/D case. The penetration of the high-temperature region was better for the 10-X/D case than the other two cases, as seen from the contours on the symmetry plane. However, these cross-sectional planes show that the penetration was limited to a small region in the lateral direction. The planes across the cavity also show better combustion inside the cavity for 10-X/D than for the 20-X/D and 30-X/D cases. This was caused by the jet wake combustion for the 20-X/D and 30-X/D cases which reduces the amount of ethylene entering the cavity, thereby reducing the effectiveness of the cavity.

Figure 7 shows the comparison of mass-averaged total temperature 'T_t' and combustion efficiency 'η_c' along the flow direction. The combustion efficiency was calculated using the available fuel mass-flow rate at a particular cross section and the supplied fuel mass-flow rate. The mass-averaged total temperature value of 1200 K

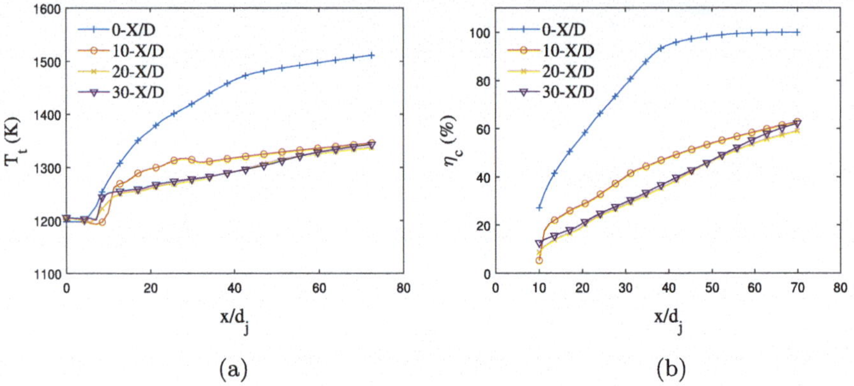

Fig. 7 Comparison of **a** total temperature and **b** combustion efficiency of different cases

began to increase after the pylon where the reaction had started. The 0-X/D case showed the highest increase at all the locations, and towards the exit of the domain, the value even reached about 1500 K. The 10-X/D case showed better higher near-field temperature than the other three cases. All three cases reached a similar total temperature value towards the exit of the domain. This trend has also been seen in the combustion efficiency plot. The 0-X/D case almost reached complete combustion because of the subsonic velocities created by the normal shock train. The better combustion efficiency of the 10-X/D case than the other two cases near-field was due to the effective combustion inside the cavity. The slope of the combustion efficiency drops after x/$d_j = 30$ for the 10-X/D case, which could be due to less lateral spread of the combustion region. The better spreading of the combustion region of the 20-X/D and 30-X/D cases helped them reach the same combustion efficiency towards the exit of the domain. Even though the combustion efficiency was better for some cases, the full effectiveness of combustion should also be derived based on stagnation pressure loss 'Λ'. Figure 8 shows the total pressure loss among all the cases. The normal shock train close to the isolator entrance caused a huge total pressure loss within x/$d_j = 10$ from the entrance for the 0-X/D case. It continued to increase along the flow direction and reached almost 35% towards the exit. All the other three cases showed a similar trend till the pylon and then a slight variation towards the exit. The 30-X/D case showed the highest total pressure loss among the three cases.

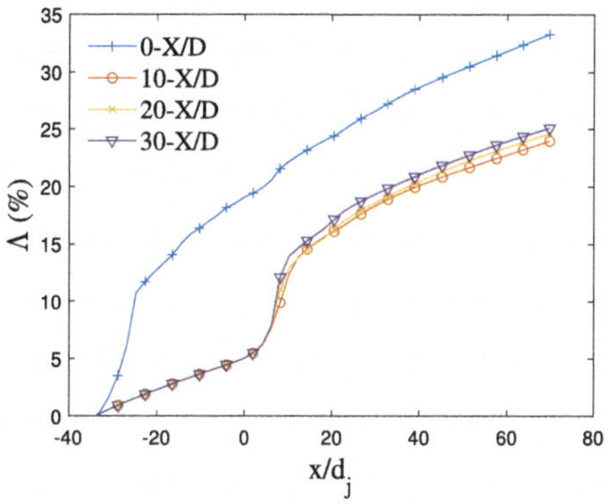

Fig. 8 Comparison of total pressure loss for all the cases

4 Conclusions

Steady RANS reactive flow simulations were carried out for different cavity positions behind the pylon with an equivalence ratio of 0.15 using ethylene. The results are discussed briefly in the previous section. The following conclusions are also drawn from the results.

- The pylon-cavity configuration with floor injection was found to interact with the incoming boundary layer and make it separate. The improved combustion causes thermal choking and forms a normal shock train.
- Among the configurations with a cavity separated from the pylon, the 10-X/D configuration performs better with better near-field combustion and lower total pressure loss than the other two cases.
- The effectiveness of the cavity for the 20-X/D and 30-X/D cases was less when compared to the 10-X/D case due to the less entrainment of ethylene into the cavity.

References

Baurle, R. A., and Eklund, D. R.: Analysis of dual-mode hydrocarbon scramjet operation at Mach 4-6.5. Journal of Propulsion and Power, 18(5), 990–1002 (2002).

Colket III, M. B., Spadaccini, L. J.: Scramjet fuels autoignition study. Journal of propulsion and power 17 (2), 315–323 (2001).

Ferri, A.: Mixing-controlled supersonic combustion. Annual Review of Fluid Mechanics 5 (1), 301–338 (1973).

Goldberg, U. C., Palaniswamy, S., Batten, P., and Gupta, V.: Variable turbulent Schmidt and Prandtl number modeling. Engineering Applications of Computational Fluid Mechanics, 4(4), 511–520 (2010).

Kannaiyan, K.: Numerical investigation of the local and global supersonic combustion characteristics of ethylene fuel. Aerospace Science and Technology 106, 106178 (2020).

Lin, K.C., Jackson, K., Behdadnia, R., Jackson, T.A., Ma, F. and Yang, V.: Acoustic characterization of an ethylene-fueled scramjet combustor with a cavity flameholder. Journal of Propulsion and Power, 26(6), pp.1161–1170 (2010).

Oamjee, A., and Sadanandan, R.: Effects of pylon geometry on mixing enhancement in a scramjet pylon-cavity flameholder. The Aeronautical Journal, 124(1278), pp.1262–1280 (2020).

Schetz, J.A. and Billig, F.S.: Penetration of gaseous jets injected into a supersonic stream. Journal of Spacecraft and Rockets, 3(11), pp.1658–1665 (1966).

Sekar, A. and Vaidyanathan, A.: Mixing enhancement of ethylene secondary jet injected into supersonic cross-flow using curved pylon. Acta Astronautica, 210, pp.253–267 (2023).

Shi, W., Tian, Y., Le, J. and Zhong, F.: Effect of pilot hydrogen on the formation of dynamic flame in an ethylene-fueled scramjet with a cavity. Physics of Fluids, 33(5), p.055130 (2021).

Tan, J., Zhang, D. and Lv, L.: A review on enhanced mixing methods in supersonic mixing layer flows. Acta Astronautica, 152, pp.310–324 (2018).

Vanyai, T., Grieve, S., Street, O., Denman, Z., McIntyre, T., Veeraragavan, A., Wheatley, V., and Smart, M.: Fundamental scramjet combustion experiments using hydrocarbon fuel. Journal of Propulsion and Power 35 (5), 953–963 (2019).

Wang, H., Wang, Z. and Sun, M.: Experimental study of oscillations in a scramjet combustor with cavity flameholders. Experimental thermal and fluid science, 45, pp.259–263 (2013).

Study of Mach Reflection in the Presence of an Upstream Expansion Fan Which Interacts with the Incident Shock and the Mach Stem

S. R. Siva Prasad Kochi and M. Ramakrishna

Abstract In this paper, the off-design condition of a supersonic intake where the lip generated expansion fan interacts with the incident shock and the Mach stem (called Type 2 Mach reflection) is investigated. Because of the presence of the expansion fan, the incident shock becomes curved and the curvature of the Mach stem changes in a significant fashion. The flow in the region downstream of the incident and reflected shocks is also affected by the presence of the expansion waves transmitted from the incident shock. An analytical solution based on the procedures of Gao and Wu (2010) and Bai and Wu (2017) taking into account all these effects has been obtained which gives the shapes of the incident shock, reflected shock, slipline and the Mach stem. The results are compared with the numerical solution obtained using a higher order Runge-Kutta discontinuous Galerkin method.

Keywords Mach reflection · Expansion fan · Incident shock · Mach stem · discontinuous Galerkin method

1 Introduction

There are two types of wave configurations that can occur when a straight oblique shock reflects over a wall—regular reflection (RR) and Mach reflection (MR). The straight oblique shock can become curved due to interference with an upstream expansion fan and this changes the reflection configuration in a significant fashion especially for Mach reflection. When this occurs, we get three types of Mach reflection structures—intersection of the expansion fan with only the incident shock wave (called Type 1), intersection of the expansion fan with both the incident shock and the Mach stem (called Type 2), intersection of the expansion fan with only the Mach stem (called Type 3). These three configurations are shown in Fig. 1.

S. R. S. P. Kochi (✉) · M. Ramakrishna
Department of Aerospace Engineering, IIT Madras, Chennai, India
e-mail: siva.ksr@gmail.com

G. Rajesh et al. (eds.), *Proceedings of the 15th International Symposium on Experimental and Computational Aerothermodynamics of Internal Flows*,
https://doi.org/10.1007/978-981-96-4082-9_20

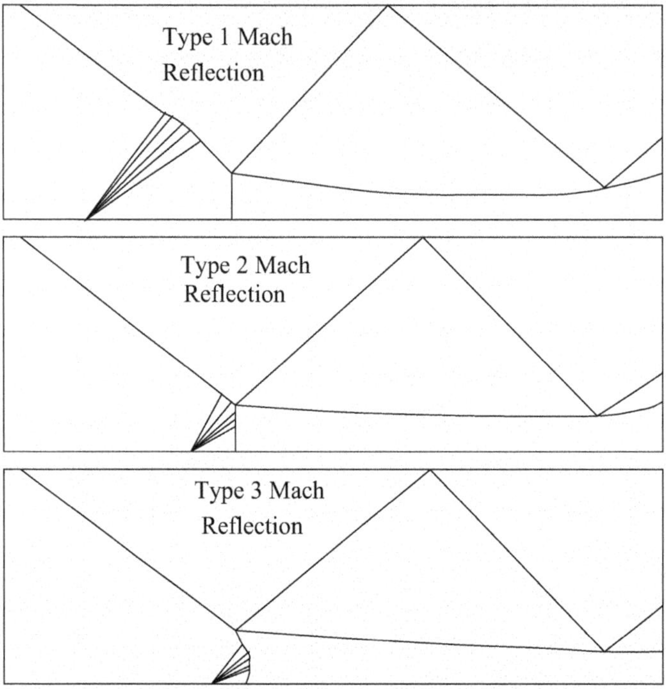

Fig. 1 The three types of Mach reflections (Type 1—expansion fan intersects with only the incident shock, Type 2—expansion fan intersects with both the incident shock and the Mach stem, Type 3—expansion fan intersects with only the Mach stem) that can occur in the presence of an upstream expansion fan. Here, we show only the essential features of the configurations

These configurations were studied in detail using CFD and method of characteristics by Hillier (2007). Yao et al. (2013) studied these shock reflection problems as applicable to the off-design condition of a supersonic air intake. They have also developed a theoretical model using Gao and Wu's (2010) method for Type 1 Mach reflection. In this paper, we develop a theoretical model using a combination of the methods of Gao and Wu (2010) and Bai and Wu (2017) for Type 2 Mach reflection. We also compare our results with CFD simulations obtained using a higher order Runge-Kutta discontinuous Galerkin method.

We look at the off-design condition of a supersonic intake as considered by Yao et al. (2013). An example intake with all the important parameters is shown in Fig. 2. Under the off-design condition, there is a lip generated expansion fan which intersects with the shock coming from a wedge of angle θ_w due to an incident Mach number M_0. The height of the intake is H, length of the wedge is w, the distance of the lip from the leading edge of the wedge is x_l, and the length of the cowl is L as shown in Fig. 2. For the Type 2 Mach reflection shown in the figure, we want to estimate the shape of the incident shock, the reflected shock, the slipline and the height of the

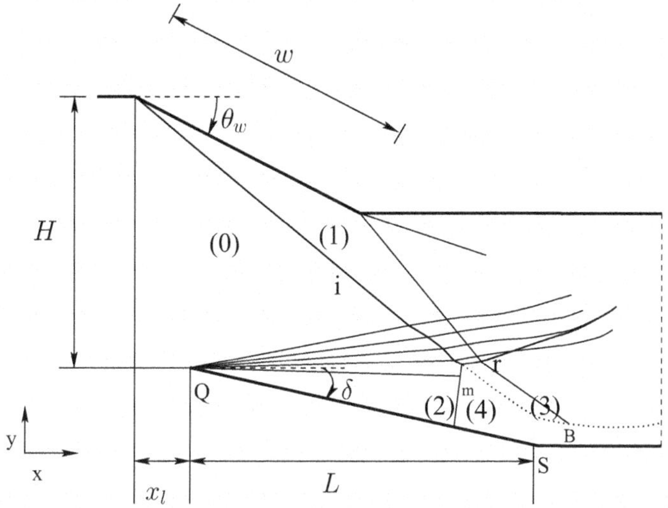

Fig. 2 Configuration for Mach reflection with wedge angle θ_w and the turning angle of the cowl of the intake given by δ. Here i is the incident shock wave from wedge leading edge, r is the reflected shock, m is the Mach stem and s is the slip line. The important regions of interest (0), (1), (2), (3), and (4) are also shown. B is the point where the leading characteristic from the wedge intersects the slipline. This configuration is similar to that used in Yao and Wu (2013)

Mach stem. We use a combination of the procedures given by Gao and Wu (2010) and Bai and Wu (2017) for this purpose.

2 Solution Approach

Before we describe our algorithm for the determination of the height of the Mach stem, we first describe the main ingredients of our procedure.

2.1 Three-Shock Theory

For oblique shock waves, the following classical shock relations hold:

$$M_1^2 = f(M_0, \beta_1) \tag{1}$$

$$p_1/p_0 = g(M_0, \beta_1) \tag{2}$$

$$\rho_1/\rho_0 = h(M_0, \beta_1) \tag{3}$$

$$a_1/a_0 = A(M_0, \beta_1) \tag{4}$$

$$S(M_0, \beta_1, \theta_1) = 0 \tag{5}$$

where the upstream flow has subscript 0 and the downstream flow has subscript 1. We have

$$f(M,\beta) \equiv \frac{M^2 + \frac{2}{\gamma-1}}{\frac{2\gamma}{\gamma-1}M^2\sin^2\beta - 1} + \frac{M^2\cos^2\beta}{\frac{\gamma-1}{2}M^2\sin^2\beta + 1} \tag{6}$$

$$g(M,\beta) \equiv \frac{2\gamma}{\gamma+1}M^2\sin^2\beta + \frac{\gamma-1}{\gamma+1} \tag{7}$$

$$h(M,\beta) \equiv \frac{(\gamma+1)M^2\sin^2\beta}{(\gamma-1)M^2\sin^2\beta + 2} \tag{8}$$

$$A(M,\beta) \equiv \frac{\left[(\gamma-1)M^2\sin^2\beta + 2\right]^{1/2}\left[2\gamma M^2\sin^2\beta - (\gamma-1)\right]^{1/2}}{(\gamma+1)M\sin\beta} \tag{9}$$

$$S(M,\beta,\theta) \equiv 2\cot\beta\frac{M^2\sin^2\beta - 1}{M^2(\gamma+\cos 2\beta) + 2} - \tan\theta \tag{10}$$

For the three-shock theory, the shock relations given above are applied at the triple point in three locations—across the incident shock for a weak shock solution, across the reflected shock for a weak shock solution normally (if Mach number across the reflected shock is less than 1 for a strong shock solution) and across the Mach stem for a strong shock solution. The pressure across the slipline is continuous and the flow in the regions after the reflected shock and after the Mach stem are parallel to each other. The equations can be solved by an iterative process.

2.2 *Interaction Between the Incident Shock Wave and the Expansion Wave*

The expansion wave (R_l) transmitted through the incident shock wave (i) (see Fig. 3) produces a transmitted expansion wave (R_i), a slip line (s) and a transmitted shock wave (i'). Before they interact, the flow parameters immediately behind i satisfy the following equations.

$$M_1^2 = f(M_0, \beta_1), \quad p_1/p_0 = g(M_0, \beta_1), \quad S(M_0, \beta_1, \theta_1) = 0 \tag{11}$$

The flow parameters immediately behind R_l satisfy the equations

$$\nu(M_2) - \nu(M_0) = \theta_2, \quad p_2/p_0 = \chi(M_0, M_2) \tag{12}$$

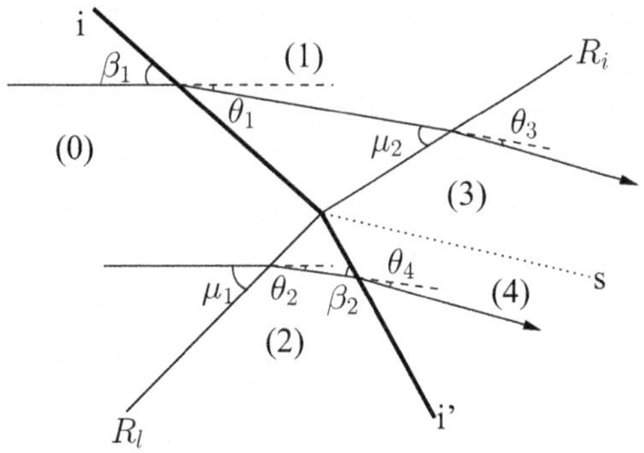

Fig. 3 Interaction of incident shock wave and the expansion fan

where

$$\nu(M) = \sqrt{\left(\frac{\gamma+1}{\gamma-1}\right)} \tan^{-1}\left(\sqrt{\left[\frac{\gamma-1}{\gamma+1}(M^2-1)\right]}\right) - \tan^{-1}(\sqrt{(M^2-1)}) \quad (13)$$

and

$$\chi(M_0, M_1) = \left(\frac{2+(\gamma-1)M_0^2}{2+(\gamma-1)M_1^2}\right)^{\gamma/(\gamma-1)} \quad (14)$$

We can obtain the flow parameters in regions (1) and (2) using the above equations. Across the transmitted expansion wave and the transmitted shock, we have

$$\nu(M_3) - \nu(M_1) = \theta_3, \quad p_3/p_1 = \chi(M_1, M_3) \quad (15)$$

and

$$M_4^2 = f(M_2, \beta_2), \quad p_4/p_2 = g(M_2, \beta_2), \quad S(M_2, \beta_2, \theta_2) = 0 \quad (16)$$

The five relations given above are for the seven parameters $M_3, p_3, \beta_3, \theta_3, M_4, p_4,$ and θ_4. The additional two required equations are given by

$$\theta_1 + \theta_3 = \theta_2 + \theta_4, \quad p_3 = p_4 \quad (17)$$

From these equations, we can find all the flow parameters when a shock interacts with an expansion wave.

2.3 *Interaction Between Two Transmitted Expansion Waves*

We consider the interaction between the expansion waves transmitted from the shock and the expansion waves originating from the wedge as shown in Fig. 4. The interaction between the expansion wave R_i and the expansion wave R_r produces two more expansion waves R_i' and R_c. The intensity and direction of the incident waves are

Fig. 4 Interaction between the expansion waves from the wedge and those from the incident shock wave

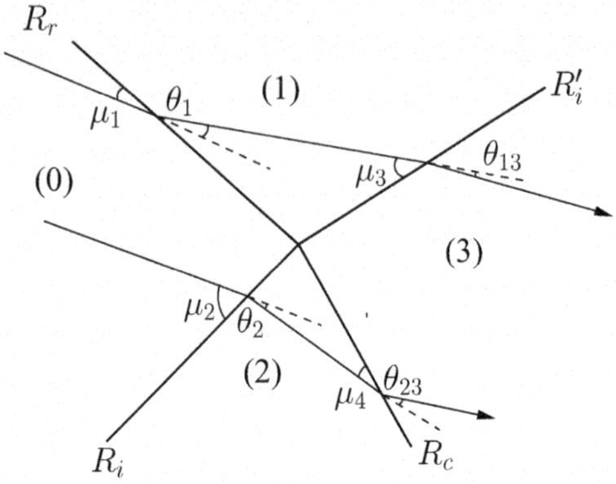

known. The flow parameters in regions (1) and (2) satisfy

$$\nu(M_1) - \nu(M_0) = \theta_1, \quad p_1/p_0 = \chi(M_0, M_1) \tag{18}$$

and

$$\nu(M_2) - \nu(M_0) = \theta_2, \quad p_2/p_0 = \chi(M_0, M_2) \tag{19}$$

Similarly, across the transmitted expansion waves, we have

$$\nu(M_3) - \nu(M_1) = \theta_{13}, \quad p_3/p_1 = \chi(M_1, M_3) \tag{20}$$

and

$$\nu(M_3) - \nu(M_2) = \theta_{23}, \quad p_3/p_2 = \chi(M_2, M_3) \tag{21}$$

Only three equations of the above four are independent as the flow is isentropic. We need one more relation given by

$$\theta_1 - \theta_{13} = \theta_{23} - \theta_2 \tag{22}$$

These equations give us all the flow parameters when two expansion waves interact with each other.

2.4 Overall Procedure

Now we describe the global algorithm used to obtain the shapes of the incident, reflected shocks, Mach stem and the slipline. We mostly follow the approach of Gao and Wu (2010) and initially assume a Mach stem height H_m, from which the coordinates of the triple point can be found. We should be careful to include the angle

δ so that the correct coordinates are used. We now find the flow properties near the triple point using the three shock theory (as described briefly in Sect. 2.1) assuming the absence of the expansion fan. Now, referring back to Fig. 2, the flow conditions in region (2) are found using Prandtl-Meyer equations (John and Keith 2006). By knowing the flow conditions in region (2), we calculate the flow properties near the triple point and the foot of the Mach stem in region (4). We use the relations given in Sect. 2.2 for the interaction of an expansion wave and a shock to find the deflection of the Mach stem because of the expansion fan. Like in Gao and Wu (2010), all the expansion fans are represented by a finite number of Mach waves.

Now, using quasi-one-dimensional-flow theory, we find the height of the sonic throat H_s^* using

$$\frac{H_m}{H_s^*} = \frac{1}{\overline{M}_m}\left(1 + \frac{\gamma - 1}{2}\overline{M}_m^2\right)^{(\gamma+1)/(2(\gamma-1))} \tag{23}$$

where $\overline{M}_m$ is the average Mach number immediately behind the Mach stem which is calculated from the aforementioned calculation in region (4). Now the approximate location of the point B (shown in Fig. 2) where the leading Mach wave from the wedge intersects the slipline is calculated using the procedure given in Bai and Wu (2017). Knowing this approximate location, we now can calculate the flow properties in region (4) till that location using the procedure in Gao and Wu (2010) by making sure that the increase in Mach number from the expansion fan in region (4) is accounted for.

We now find the flow properties in region (1). We do this in two steps. In the first step, we calculate the flow properties and the strength and direction of transmitted expansion waves by using the method given in Sect. 2.2. In step 2, we find the flow properties where the expansion waves from the wedge interact with the transmitted expansion waves using the method given in Sect. 2.3. Now the shape of the reflected shock and the flow properties in region (3) are found by the same procedure given in Gao and Wu (2010) by including the interactions of the expansion waves emanating from the slip line, the lip generated expansion waves and the wedge generated expansion waves.

Now, the flow properties in region (4) after the point B can be determined. This is done as given in Bai and Wu (2017). This gives us the position and shape of the slipline. The sonic flow must occur at a section of minimum area. We therefore examine whether the minimum distance between the slipline and the reflecting surface is equal to H_s^* calculated above. If they are not equal, we update the value of H_m and repeat the whole procedure until the correct value of H_m is obtained. Once the correct value of Mach stem height is obtained, we can be sure of the reasonable accuracy of the shapes of the incident shock, reflected shock, Mach stem and slipline calculated using the above procedure.

3 Description of Numerical Method

Consider the Euler equations in conservative form as given by

$$\frac{\partial \mathbf{Q}}{\partial t}+\frac{\partial \mathbf{F}(\mathbf{Q})}{\partial x}+\frac{\partial \mathbf{G}(\mathbf{Q})}{\partial y}=0 \quad \text{in the domain} \quad \Omega \tag{24}$$

where $\mathbf{Q}=(\rho,\rho u,\rho v,E)^T$, $\mathbf{F}(\mathbf{Q})=u\mathbf{Q}+(0,p,0,pu)^T$ and $\mathbf{G}(\mathbf{Q})=v\mathbf{Q}+(0,0,p,pv)^T$ with $p=(\gamma-1)(E-\frac{1}{2}\rho(u^2+v^2))$ and $\gamma=1.4$. Here, ρ is the density, (u,v) is the velocity, E is the total energy and p is the pressure. We approximate the domain Ω by K non overlapping elements given by Ω_k.

We look at solving the above equations using the discontinuous Galerkin method. We approximate the local solution in an element Ω_k, where k is the element number, as a polynomial of order N which is given by:

$$Q_h^k(r,s)=\sum_{i=0}^{N_p-1} Q_i^k \psi_i(r,s) \tag{25}$$

where $N_p=(N+1)(N+1)$ and r and s are the local coordinates. Here, the subscript i represents the particular degree of freedom, h represents the grid size, and the superscript k is the element number. The polynomial basis used $(\psi_i(r,s))$ is the tensor product orthonormalized Legendre polynomials of degree N. The number of degrees of freedom are given by $N_p=(N+1)(N+1)$. Now, using $\psi_j(r,s)$ as the test function, the weak form of the above equation is obtained as

$$\sum_{i=0}^{N_p-1}\frac{\partial Q_i^k}{\partial t}\int_{\Omega_k}\psi_i\psi_j d\Omega+\int_{\partial\Omega_k}\hat{F}\psi_j ds-\int_{\Omega_k}\mathbf{F}\cdot\nabla\psi_j d\Omega=0 \quad j=0,\ldots,N_p-1 \tag{26}$$

where $\partial\Omega_k$ is the boundary of Ω_k, $\mathbf{F}=(\mathbf{F}(\mathbf{Q}),\mathbf{G}(\mathbf{Q}))$ and $\hat{F}=\bar{F}^*\cdot\hat{n}$ where $\bar{F}^*$ is the monotone numerical flux at the interface which is calculated using an exact or approximate Riemann solver and $\hat{n}$ is the unit outward normal. We have used the Lax-Friedrichs numerical flux for all our calculations. This is termed to be $\mathbf{P}^N$ based discontinuous Galerkin method.

The above equation is integrated using an appropriate Gauss Legendre quadrature and is discretized in time by using the fifth order Runge-Kutta time discretization given in Butcher (2016) unless otherwise specified. To control spurious oscillations which occur near discontinuities, a limiter is used with a troubled cell indicator. We have used the KXRCF troubled cell indicator (Krivodonova et al. 2004) and the compact subcell WENO (CSWENO) limiter proposed in Siva Prasad Kochi and Ramakrishna (2021) for all our calculations. The computational domain used for the solution of the equations is shown in Fig. 5.

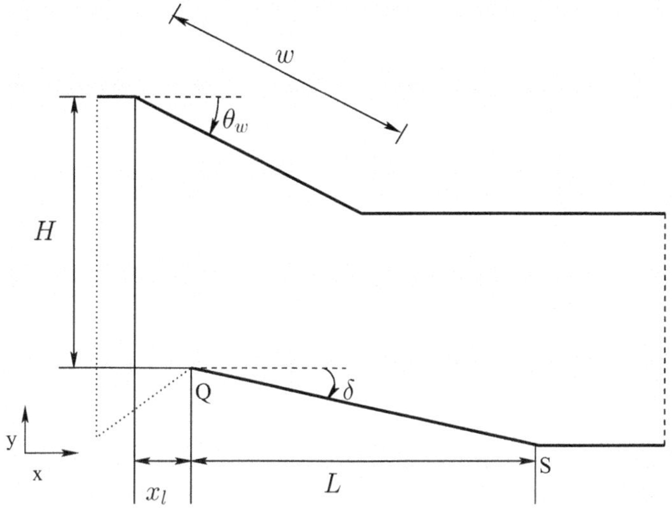

Fig. 5 Computational Domain for Mach reflection with wedge angle θ_w and the turning angle of the cowl of the intake given by δ. To the left, we apply the inflow boundary conditions (indicated by dotted lines), to the right we apply the outflow boundary conditions (indicated by dashed lines) and the remaining are wall boundary conditions (indicated by solid lines). This configuration is similar to that used in Yao and Wu (2013)

4 Results and Discussion

For the considered problem of determination of Mach stem height for flows where an expansion fan interacts with the incident shock and the Mach stem, we consider three Mach numbers—2.0, 4.0, and 5.0. Conditions for Mach 2.0 flow were analysed extensively by Hillier (2007) and we consider some cases studied by him. The conditions for Mach 4.0 and Mach 5.0 flow were given in Yao and Wu (2013). The values of Mach stem height obtained (numerically and analytically) for various Mach numbers, wedge angles, expansion fan turning angles are given in Table 1. We also show the shapes of incident shock, reflected shock, Mach stem and the slipline obtained analytically (shown as solid black line) across the numerical solution for the first case given in Table 1 (Mach 2.0 flow with $\theta_w = 20.44$, $\delta = 20.44$ and $x_l/H = 0.4$) in Fig. 6. This shows a good agreement between analytical and numerical solutions. From the results shown in Table 1, we can also see that the analytical method gives accurate results though it overpredicts the value of Mach stem height for all the cases.

Table 1 Mach stem heights obtained for different flow conditions

S. no	Mach number	x_l/H	w/H	L/H	θ_w	δ	H_m/H (numerical)	H_m/H (analytical)
1	2.0	0.4	1.1	1.4	20.44	20.44	0.139	0.142
2	2.0	0.5	1.1	1.4	20.44	20.44	0.124	0.126
3	2.0	0.6	1.1	1.4	20.44	20.44	0.119	0.125
4	4.0	0.5	1.3	1.6	26.0	12.0	0.166	0.171
5	4.0	0.5	1.3	1.6	26.0	15.0	0.182	0.195
6	5.0	0.6	1.1	1.8	24.0	12.0	0.225	0.238
7	5.0	0.6	1.1	1.8	24.0	15.0	0.232	0.255

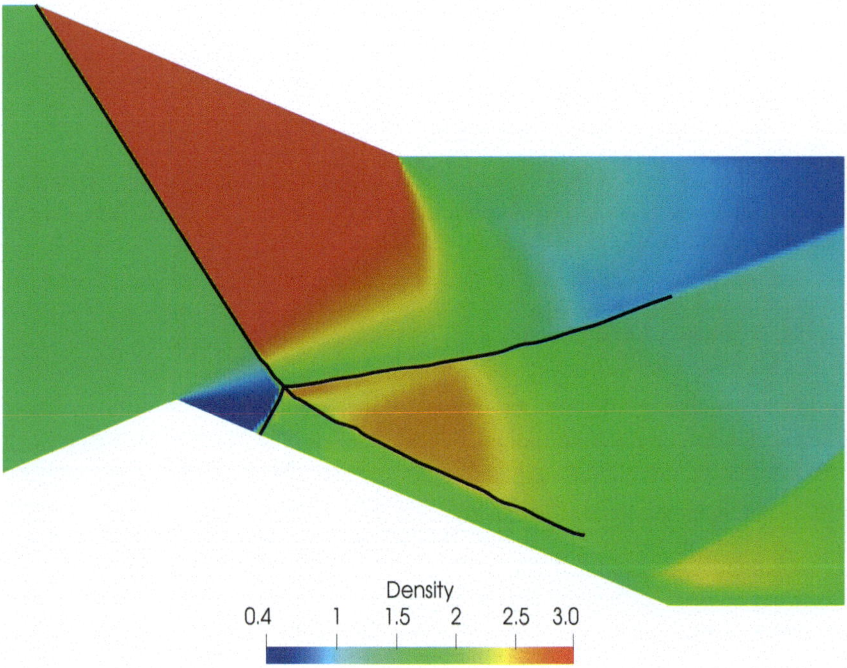

Fig. 6 Analytically calculated solution shown as a solid black line showing the locations of incident shock, reflected shock, Mach stem, and the slipline. Density solution for Mach 2.0 flow over a 20.44° wedge with an expansion fan at $x_l/H = 0.4$ and $\delta = 20.44$ with $\mathbf{P}^4$ based discontinuous Galerkin method is also shown for comparison

5 Conclusion

We have considered the off-design condition of a supersonic intake where the lip generated expansion fan interacts with the incident shock and the Mach stem (called Type 2 Mach reflection) and developed an analytical method to estimate the shape of the incident shock, the reflected shock, the slipline and the height of the Mach stem. We have calculated the height of the Mach stem for various Mach numbers, wedge angles and the expansion fan turning angles and compared it with a numerical solution obtained using a higher order discontinuous Galerkin method. By comparing the results from analytical and numerical methods, we found that the accuracy of the analytical method is quite good and it overpredicts the Mach stem height.

References

Bai C.Y., Wu Z.N.: Size and shape of shock waves and slipline for Mach reflection in steady flow. Journal of Fluid Mechanics 818, 116–140 (2017).

Butcher, J.C.: Numerical Methods for Ordinary Differential Equations. 3rd edn. John Wiley and Sons, UK (2016).

Gao B., Wu Z.N.: A study of the flow structure for Mach reflection in steady supersonic flow. Journal of Fluid Mechanics 656, 29–50 (2010).

Hillier R.: Shock-wave/expansion-wave interactions and the transition between regular and Mach reflection. Journal of Fluid Mechanics 575, 399–424 (2007).

John, J.E., Keith, T.G.: Gas Dynamics. 3rd edn. Pearson Prentice Hall, New Jersey (2006).

Krivodonova L., Xin J., Remacle J.-F, Chevaugen N., Flaherty J.: Shock detection and limiting with discontinuous Galerkin methods for hyperbolic conservation laws. Applied Numerical Mathematics 48, 322–338 (2004).

Siva Prasad Kochi S.R., Ramakrishna M.: A compact subcell WENO limiting strategy using immediate neighbours for Runge-Kutta discontinuous Galerkin methods. International Journal of Computer Mathematics 98, 608–626 (2021).

Yao Y., Li S.G., Wu Z.N.: Shock reflection in the presence of an upstream expansion wave and a downstream shock wave. Journal of Fluid Mechanics 735, 61–90 (2013).

Study of Shock Wave Propagation and Mitigation Effects at the Air–Water Interface

Jayabal Rajasekar and Heuy Dong Kim

Abstract This research undertakes a numerical analysis of the behavior of shock waves near the air–water interface, specifically examining phenomena such as transmission, reflection, attenuation, and propagation of the shock wave. The numerical simulation considers air, water, and TNT (tri nitro toluene) as the modeling materials, employing the ideal gas, Mie-Gruneisen (shock), and Jones-Wilkins-Lee (JWL) equations of state, respectively. Two distinct cases are investigated: one involving only air and the other with air and water separated by an interface. In the first case, the study explores the propagation and attenuation of shock waves in both transverse and longitudinal directions. In the second case, the focus shifts to the impact of the shock wave impinging on the water surface and the subsequent effects on shock wave propagation and attenuation. Both qualitative and quantitative analyses are employed to understand the characteristics of shock wave behavior. The numerical results reveal that the presence of a water surface significantly contributes to the mitigation of the shock wave. Thus, each case is visually represented through numerical simulations, and the results are compared for further insights.

Keywords Air–water interface · Unsteady · Multiphase flow

1 Introduction

High explosives (HEs) are known for their ability to rapidly release a significant amount of energy (Needham 2018). These explosives find widespread applications in various industries such as mining and explosive work, as well as in medical procedures like extracorporeal shock wave lithotripsy (Oshita et al. 2014). However, when an explosion occurs unexpectedly, it results in the rapid release of high-pressure and high-temperature gases (Cole 1948). Unfortunately, accidents involving explosions can happen during industrial operations, such as when natural gas or coal dust leaks

J. Rajasekar · H. D. Kim (✉)
Department of Mechanical Engineering, Andong National University, Andong, Korea
e-mail: kimhd@anu.ac.kr

G. Rajesh et al. (eds.), *Proceedings of the 15th International Symposium on Experimental and Computational Aerothermodynamics of Internal Flows*,
https://doi.org/10.1007/978-981-96-4082-9_21

due to mechanical errors during excavator operations or when high explosives are used for blasting in tunnels (Jeon 2017; Sugiyama et al. 2019). During the process of gas expansion following an explosion, a shock wave is generated. This shock wave poses a threat to individuals and has the potential to cause extensive damage to mining equipment within tunnels and surrounding property (Cheng et al. 2005; Keenan and Wager 1992; Malvar and Tancreto 1998).

Despite the discovery of shockwave mitigation techniques several years ago, ongoing research in this field remains active within the scientific community (Cheng et al. 2005; Rajasekar et al. 2022a, b; Wu et al. 2013; Chen et al. 2016; Sugiyama et al. 2018). Specifically, the interaction of shockwaves at the interface between air and water continues to be a focus of investigation. When a shockwave travels from a medium with lower acoustic impedance, like air, to a medium with comparable impedance, such as water, variations in acoustic impedance cause the shockwave to undergo reflection, refraction, and transmission as it transitions between the two media. This behavior is primarily influenced by the rigid surface-like properties of water's surface. Consequently, some of the energy carried by the shockwave is reflected and transmitted near the air–water interface, resulting in the attenuation of the shockwave. There is a limited body of literature available on this particular area of research, with most of the existing studies failing to address shockwaves at the air–water interface.

The present study aims to investigate the propagation of shockwaves at this interface and the methods employed to mitigate the effects of these shockwaves using water. The research involves a computational analysis of the characteristics associated with shockwave mitigation. It quantitatively and qualitatively examines phenomena such as reflection, transmission, and shockwave mitigation. Computational fluid dynamics simulations were conducted using a shock tube attached to the left side of the domain, as depicted in Fig. 1. The shock tube comprises a driver section and a driven section, with explosives utilized in the driver section to simulate real-world applications in the mining industry. The domain encompasses both air and water, with water possessing a higher acoustic impedance, resulting in significant shockwave attenuation. Consequently, the study numerically simulated and compared the mitigation of shockwaves in domains with and without water to determine the effectiveness of water in reducing shockwave effects. This section is followed by 2. Numerical methods, 3. Results and Discussion and 4. Conclusion.

2 Numerical Methods

2.1 Governing Equations

This study employed the Finite Volume Method (FVM) to solve the Euler equations. Within this methodology, the FVM solver calculated flow field variables at each grid point over different time intervals. By utilizing the FVM approach, the governing

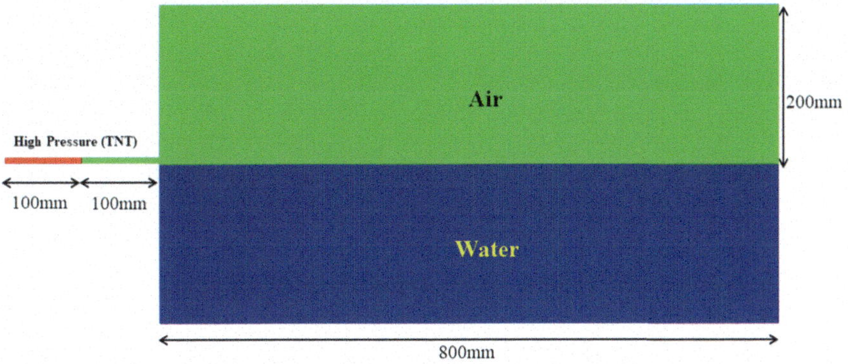

Fig. 1 Schematic diagram of computational domain

equations, including the conservation of mass, momentum, and energy equations, were effectively solved. The equations considered in this study are outlined below:

$$\frac{d}{dt}\int_{v} \rho dV + \int_{A} \rho(u.n)dA = 0 \quad (1)$$

$$\frac{d}{dt}\int_{v} \rho u_i dV + \int_{A} \rho u_i(u.n)dA = -\int_{A} pn_i dA \quad (2)$$

$$\frac{d}{dt}\int_{v} \rho E dV + \int_{A} \rho E(u.n)dA = -\int_{A} u.pndA \quad (3)$$

where pressure, velocity, area, volume, density and specific total energy is denoted as P, u, A, V, ρ and E, respectively. The normal vector of each element face is represented by n. For a inviscid 2D compressible flow, we have

$$\frac{\partial U}{\partial t} + \frac{\partial F(U)}{\partial x} + \frac{\partial G(U)}{\partial y} = 0 \quad (4)$$

For gas flow, the equation is given as follows,

$$U = \begin{bmatrix} \rho \\ \rho u \\ \rho v \\ E \end{bmatrix}, F = \begin{bmatrix} \rho u \\ \rho u^2 + p \\ \rho uv \\ (E+p)u \end{bmatrix}, G = \begin{bmatrix} \rho v \\ \rho uv \\ \rho v^2 + p \\ (E+p)v \end{bmatrix} \quad (5)$$

where u and v are flow velocities in the x and y directions, respectively.

Explosives were modeled using Jones, Wilkins, and Lee (JWL) equation of state. The equation is as follows:

$$P = C_1\left(1 - \frac{\omega}{R_1 \upsilon}\right)e^{-R_1\upsilon} + C_2\left(1 - \frac{\omega}{R_2 \upsilon}\right)e^{-R_2\upsilon} + \frac{\omega E}{\upsilon} \quad (6)$$

where E and v are specific internal energy and specific volume, respectively. C_1, C_2, R_1, R_2, and ω are material constants. In this study, the commercial hydrocode software is used and values of all other constants available in the hydrocode package (Jaiswal 2011). Water and Air were modeled using Mie-Gruneisen (shock) and ideal gas equation of state (Rajasekar et al. 2020) and the equations are as follows:

$$P = P_r(\varepsilon_v) + \frac{\Gamma(v)}{v}[e - e_r] \quad (7)$$

$$P = (\gamma - 1)\frac{\rho}{\rho_0}E \quad (8)$$

where Γ(v) is Gruneisen gamma which is the thermodynamic property and γ is the specific heat ratio ($\gamma = 1.4$).

2.2 *Computational Domain*

Figure 1 presented the computational diagram depicting the setup used in this study. The domain dimensions were 800 mm in length and 400 mm in height. The lower half of the domain was filled with water, while the upper half contained air, separated by an interface. Two distinct cases were examined in this research: the first case involved only air, while the second case involved both air and water. To generate a high-impact shockwave, a shock tube measuring 200 mm in length was attached to the domain and filled with TNT. The Euler sub-grid technique was employed to analyze the flow properties, with an equally spaced grid allowing the flow to pass through the material. Gauge points were strategically positioned near the interface to monitor the history of flow field variables. To model the numerical domain, a combination of wall and transmitting boundary conditions was utilized, effectively reducing stress wave reflection from the boundaries. The commercial software ANSYS Workbench was employed to conduct the simulations.

3 Results and Discussion

This research examined two different domains to investigate the propagation and mitigation effects of shockwaves on the water surface. The first domain consisted solely of air, while the second domain featured a separation between air and water at the interface. In this research, the computational domain is connected to a shock tube. The shock tube used has dimensions of 200 mm in length and 10 mm in height. Half of the tube is filled with TNT, which generates a planar shock wave directed towards the computational domain. The objective of this study is to investigate the propagation and transmission of the shock wave along the surface of water. To achieve this, a high-intensity planar shock wave is utilized in the computational setup. To assess the flow field parameters, monitoring points were strategically placed at 100 mm intervals in both scenarios. The obtained results were then compared using qualitative and quantitative analyses.

3.1 Shock Wave Propagation and Attenuation

In the first scenario (Case 1), the domain contains only air. When the explosion occurs within the shock tube, a planar shock wave is generated and propagates towards the domain. Upon reaching the domain, the shock wave travels as a planar wave along the bottom surface, while simultaneously attempting to expand in the transverse direction. Consequently, the shock wave gradually weakens and loses its intensity over time. At $t = 0.15$ ms, the shock front remains planar and moves at a higher speed along the longitudinal direction compared to the transverse direction, as depicted in Fig. 2. This behaviour is a result of the alignment between the shock tube and the computational domain, as illustrated in Fig. 1. By $t = 0.3$ ms, the shock front dissipates progressively in both the longitudinal and transverse directions, as shown in Fig. 3.

In the second scenario (Case 2), the shock wave encounters the computational domain consisting of both air and water, with an interface separating the two. When the shock wave strikes the water surface, it attempts to transmit through the water and reflect back into the air. At standard atmospheric pressure (1 atm) and a temperature of 15 °C, the acoustic impedance of air is 410 kg m^{-2} s^{-1}, while that of water is 1.5 × 106 kg m^{-2} s^{-1}. These values indicate that the acoustic impedance of water is approximately 3500 times higher than that of air. Consequently, the water surface initially behaves similar to a solid surface due to the significant impedance mismatch. At $t = 0.15$ ms, the planar shock wave exerts a substantial impact, leading to significant transmission of the shock wave inside the water, as illustrated in Fig. 2. The shock front propagates along the surface of the water, as depicted in the figure. Comparing Case 1 and Case 2 in the air region, at $t = 0.15$ ms, the shock wave in Case 1 reaches a distance of 0.39 m, while in Case 2, it reaches approximately 0.375 m. The disparity in acoustic impedance between air and water causes the shock wave to propagate faster in water than in air, as shown in Fig. 2.

In Fig. 3, at $t = 0.3$ ms, the propagation of the shock wave in Case 1 reaches a distance of 0.515 m, while in Case 2, it reaches approximately 0.495 m. The presence of water contributes to the attenuation of the shock wave. Conversely, the shock wave propagates in water at a speed 3 to 4 times faster than in air, reaching a distance of nearly 0.64 m. Thus, the presence of water has a significant impact on both the propagation and mitigation effects of the shock wave. Figure 4 illustrates the temperature distribution in both cases. At $t = 0.3$ ms in Case 1, the temperature

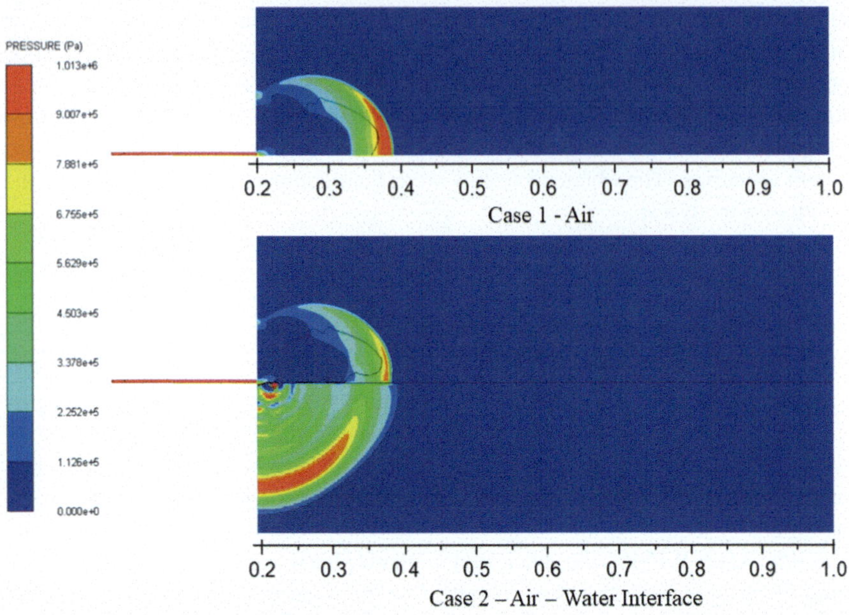

Fig. 2 Pressure Contours of shock wave in case (1) air and case (2) air–water interface at time t = 0.15 ms

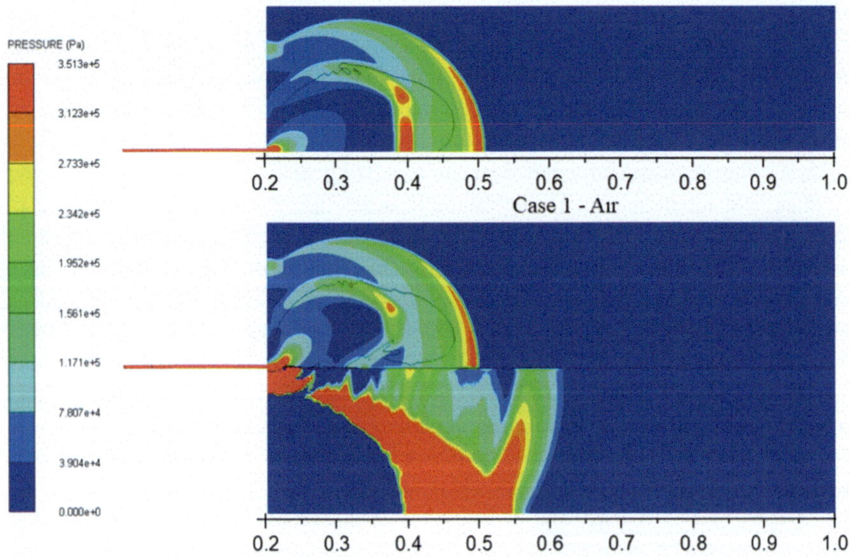

Fig. 3 Pressure contours of shock wave in case (1) air and case (2) air–water interface at time t = 0.3 ms

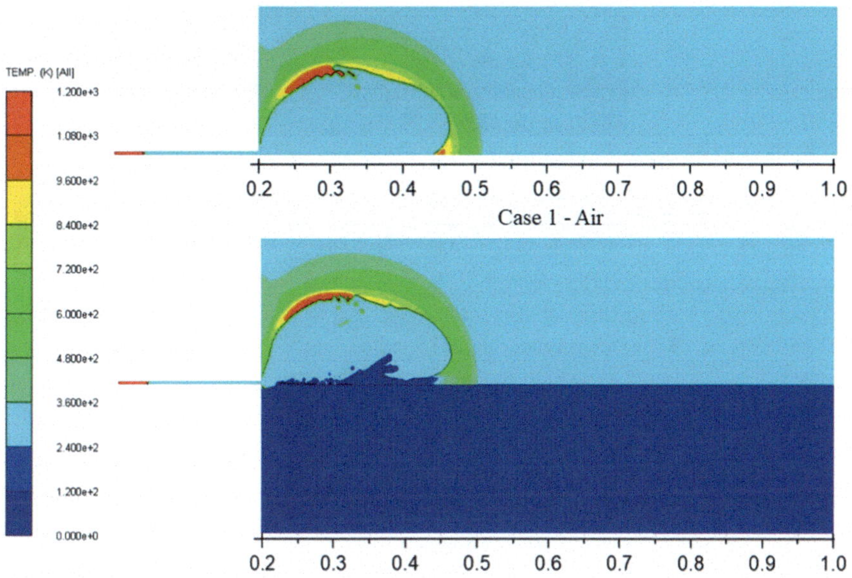

Fig. 4 Temperature contours of shock wave in case (1) air and case (2) air–water interface at time t = 0.3 ms

near the shock front measures around 1200 K, whereas in Case 2, it is approximately 480 K. When the shock wave encounters the interface, it deforms the water surface, and the water absorbs the temperature and internal energy of the shock front. As a result, the shock strength is considerably reduced, as depicted in Fig. 4.

3.2 Analysis of Flow Field Variables at Different Monitoring Points

Figure 5a illustrates the computational domain for Case 1, with monitoring points fixed at 100 mm intervals along the surface. Similarly, monitoring points were placed along the water surface for Case 2. Figure 5b compares the flow field variables, such as pressure, temperature, and internal energy, at different monitoring points for both cases. Upon the shock wave reaching the computational domain, the pressure recorded at monitoring point 1 in Case 1 is 9.897 MPa, while in Case 2, it measures 7.16 MPa. This discrepancy arises because, in Case 2, the shock wave impacts the water surface, leading to transmission and reflection in the air. Consequently, Case 2 exhibits higher pressure at point 1 compared to subsequent monitoring points. The pressure reduction at point 1 is approximately 27.6%, followed by 80.3%, 63.9%, and 41.1% at successive monitoring points.

Regarding temperature and internal energy, when the shock wave encounters the air–water interface, the water surface absorbs the temperature and energy along

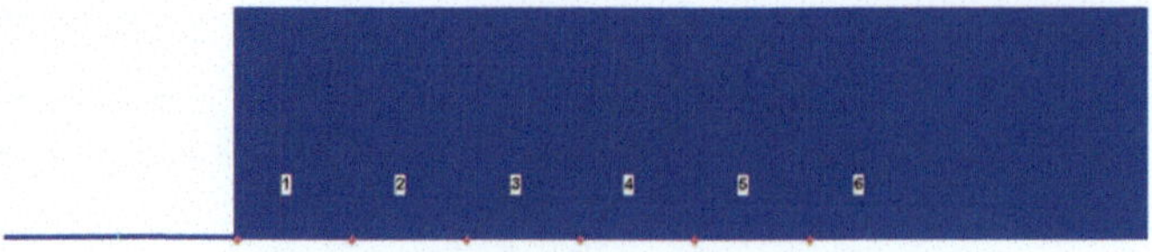

(a)

Monitoring Point (MP)	Air			Air-Water Interface			% Reduction of variables		
Flow Field Parameters	Pressure (MPa)	Temp (K)	Internal Energy (MJ/Kg)	Pressure (MPa)	Temp (K)	Internal Energy (MJ/Kg)	Pressure (MPa)	Temp (K)	Internal Energy (MJ/Kg)
MP 1	9.897	9286	5.214	7.16	511	1.117	27.6%	94.4%	78.5%
MP 2	1.987	2199	1.048	0.3912	218	0.3164	80.3%	90.08%	69.8%
MP 3	0.7732	1196	0.8057	0.2789	186	0.2024	63.9%	84.4%	74.8%
MP 4	0.3267	1039	0.7055	0.1925	2.8	0.0156	41.1%	99.7%	97.7%

(b)

Fig. 5 **a** Computational domain with monitoring points for case (1) air **b** comparison of flow field variables at different monitoring points

its surface, resulting in significant attenuation of the shock front. The quantitative results obtained at different points also reflect this behaviour, as depicted in Fig. 5b. The temperature exhibits a substantial drop of 94.4% at the first point, followed by reductions of 90.08, 84.4, and 99.7% at subsequent points. Similarly, the internal energy recorded at point 1 is 78.5%, followed by 69.8, 74.8, and 97.7% at successive monitoring points. These quantitative results highlight the significant impact of the presence of water in attenuating the shock wave.

3.3 *Comparison of Results*

Figure 6 provides a comparison of the percentage reduction of shock wave pressure, temperature, and internal energy at each monitoring point. The initial percentage of shock pressure at monitoring point 1 is quite low, possibly due to the acoustic impedance disparity between air and water. As a result, the shock wave impinges on the water surface and undergoes considerable reflection at point 1. In subsequent points, the reflection of the shock wave diminishes, leading to higher percentage reductions in shock pressure, as depicted in Fig. 7. Regarding temperature and internal energy, the water surface absorbs a significant amount of the shock front's temperature and internal energy at point 1, owing to the initial impact on the water surface. As the shock front propagates along the interface, the water's nature leads to minimal mitigation at points 2 and 3, followed by a substantial increase in the

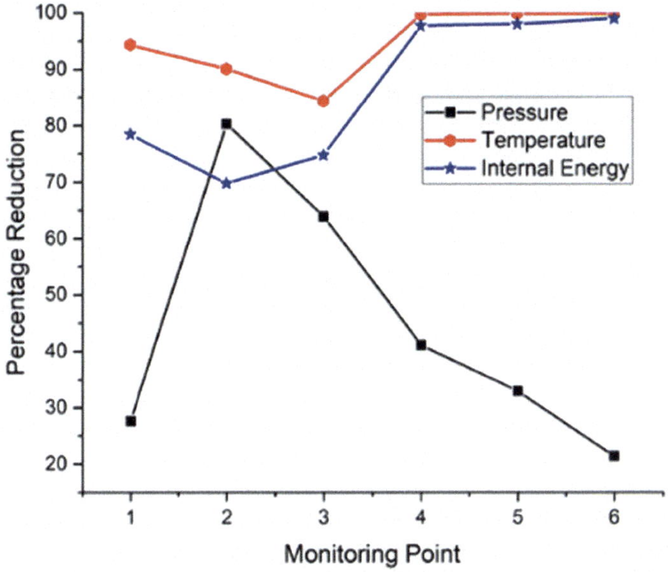

Fig. 6 Comparison of % reduction of pressure, temperature and internal energy at each monitoring point

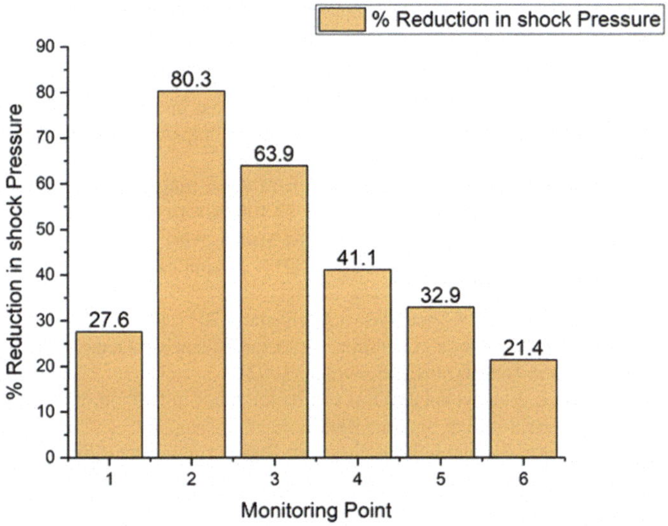

Fig. 7 Mitigation of shockwave at each monitoring point

reduction of shock temperature and internal energy, as shown in Fig. 6. Consequently, both quantitative and qualitative results demonstrate that the presence of water significantly attenuates and mitigates the shock wave.

4 Conclusion

The objective of this study is to examine the propagation and attenuation of shock waves along the interface between air and water. In the first case, the domain consists solely of air, allowing for an investigation of how the shock front propagates in the longitudinal direction and how the shock wave dissipates in the transverse directions. In the second case, the domain includes both air and water. When the shock wave encounters the air–water interface, it deforms the interface and transmits inside the water while also reflecting. The acoustic impedance and strength of water play a crucial role in reducing the intensity of the shock wave. Due to that the propagation of shock wave was significantly reduced in case 2 compared to that of case 1. Moreover, the nature of water causes it to absorb the shock's temperature and internal energy along the water's surface, leading to a significant attenuation of the shock wave. Monitoring points were placed at equal intervals to facilitate a qualitative comparison of the flow field variables in both cases. The results indicate that the presence of water along the interface has a substantial impact on mitigating the shock wave.

Acknowledgements This work was supported by the National Research Foundation of Korea (NRF) grant funded by the Korea government (MSIP) (No. NRF-2021R1I1A3044216).

References

Chen, L., Fang, Q., Zhang, L., Zhang, Y., Chen, W.: Numerical investigation of a water barrier against blast loadings. Eng. Struct. 111, 199–216 (2016). https://doi.org/10.1016/j.engstruct.2015.12.015

Cheng, M., Hung, K.C., Chong, O.Y.: Numerical study of water mitigation effects on blast wave. Shock Waves. 14, 217–223 (2005). https://doi.org/10.1007/s00193-005-0267-4

Cole, R.: Underwater Explosions. Dover Publ. Inc New York. (1948)

Jaiswal, A.: ANSYS ® Explicit Dynamics and AUTODYN ® Applications Explicit Dynamics and AUTODYN Applications. 6–12 (2011).

Jeon, H.: An Experimental Study of Shock Wave Attenuation. 2017, (2017).

Keenan, W. a, Wager, P.C.: Mitigation of confined explosion effects by placing water in proximity of explosives. 25th Dep. Def. Explos. Saf. Semin. (1992).

Malvar, L.J., Tancreto, J.E.: Analytical and test results for water mitigation of explosion effects. Twenty-Eighth DOD Explos. Saf. Semin. (1998).

Needham, C.E.: Blast Waves Second Edition, Springer, Berlin, Heidelberg. (2018)

Oshita, D., Hosseini, S.H.R., Mawatari, K., Nejad, S.M., Akiyama, H.: Two successive shock waves generated by underwater pulse electric discharge for medical applications. IEEE Trans. Plasma Sci. 42, 3209–3214 (2014). https://doi.org/10.1109/TPS.2014.2328096

Rajasekar, J., Kim, T.H., Kim, H.D.: Visualization of shock wave propagation due to underwater explosion. J. Vis. 23, 825–837 (2020). https://doi.org/10.1007/s12650-020-00664-9

Rajasekar, J., Kim, T.H., Kim, H.D.: Computational Analysis of Mitigation of Shock wave using Water Column. 20, 51–59 (2022).

Rajasekar, J., Yaga, M., Kim, H.D.: Numerical prediction on the mitigation of shock wave using geometric barriers. J. Vis. 2022 261. 26, 83–96 (2022). https://doi.org/10.1007/S12650-022-008 66-3.

Sugiyama, Y., Homae, T., Wakabayashi, K., Matsumura, T., Nakayama, Y.: Numerical study on the mitigation effect of water in the immediate vicinity of a high explosive on the blast wave. Int. J. Multiph. Flow. 99, 467–473 (2018). https://doi.org/10.1016/j.ijmultiphaseflow.2017.11.014
Sugiyama, Y., Nakayama, Y., Ohtani, K., Nishimura, K., Matsuo, A.: Propagation Behavior and Mitigation of Shock Wave Along the Water Inside a Rectangular Tube. 31st Int. Symp. Shock Waves 1. 95–101 (2019). https://doi.org/10.1007/978-3-319-91020-8_9.
Wu, Z.D., Sun, L., Zong, Z.: Computational investigation of the mitigation of an underwater explosion. Acta Mech. 224, 3159–3175 (2013). https://doi.org/10.1007/s00707-013-0909-z

Effects of Pulsed Streamwise Injections on the Mixing Performance of Supersonic Ejectors

Aswin Suresh, S. L. Anoop, and Arun Kumar Rajagopal

Abstract The current study investigates the effects of streamwise pulsed injections on the mixing performance of a supersonic ejector operating at the critical regime. Streamwise injections are alternatively pulsed at the top and bottom sides of the mixing chamber which oscillates the primary jet. This enhances the bulk mixing between the primary and secondary streams. The mixing enhancement technique resulted in the secondary stream penetrating the primary flow much upstream and the location of the onset of mixing moved upstream by 65.36%. In addition, the larger spread of the primary core flow indicated better mixing performance. The amplitude of oscillation of the primary jet decreased with an increase in the pulsation frequency in the entrance region of the mixing chamber due to the reduced time available for the jet deflection. On the other hand, the amplitude of oscillation of the primary jet remained similar towards the downstream regions of the mixing chamber mainly limited by the physical height of the mixing chamber.

Keywords Supersonic ejector · Mixing enhancement · Pulsed injections

1 Introduction

Ejectors are versatile devices used in a wide variety of applications such as thrust augmentation (Alperin and Wu 1983a, b), refrigeration systems (Besagni et al. 2016; Abdulateef et al. 2009), chemical lasers (Singhal et al. 2006) and vacuum generation (Arun Kumar and Rajesh 2018, 2016) among many other applications. Figure 1 illustrates a typical supersonic ejector with a constant area mixing chamber. The flow field of a supersonic ejector is characterized by the formation of a compressible turbulent shear layer at the interface of the primary and secondary streams in the mixing chamber. The compressibility effects stabilize the shear layer due to the hindrance

A. Suresh · S. L. Anoop · A. K. Rajagopal (✉)
Indian Institute of Technology, Jodhpur, Rajasthan, India
e-mail: arunkr@iitj.ac.in

G. Rajesh et al. (eds.), *Proceedings of the 15th International Symposium on Experimental and Computational Aerothermodynamics of Internal Flows*,
https://doi.org/10.1007/978-981-96-4082-9_22

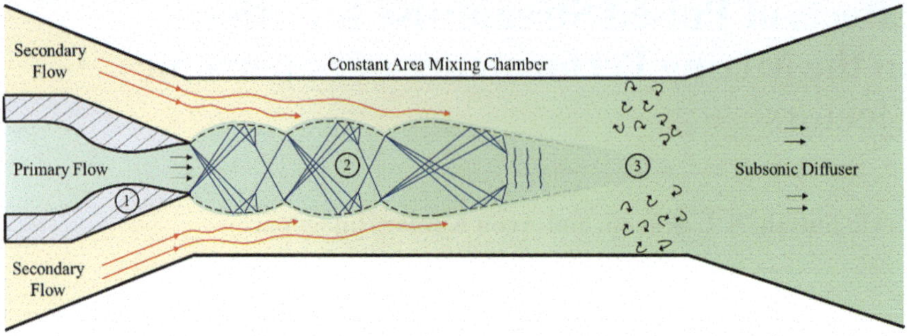

Fig. 1 The prominent flow features associated with a typical supersonic ejector with a constant area mixing chamber (1) convergent-divergent nozzle to produce the supersonic primary jet (2) shock cells created by the primary jet (under-expanded in this case) (3) turbulent mixing zone where complete mixing between the primary and the secondary stream occurs

of the cross-stream acoustic interactions. The absence of large-scale coherent structures in the compressible shear layer limits the mixing performance of supersonic ejectors. Various techniques are investigated over the years to enhance the mixing performance of supersonic ejectors and can be broadly divided into (1) passive flow control (PFC) techniques and (2) active flow control (AFC) techniques. PFC techniques introduce geometric manipulations in the flow field which lacks on-demand control such as tabs, vortex generators, and lobbed nozzles. AFC techniques, on the other hand, utilize an external energy source to manipulate the flow field enabling on-demand control such as acoustic excitation and flapping jets. Irrespective of the flow control method used, all mixing enhancement techniques destabilize the shear layer by enhancing momentum exchange between the primary and secondary streams through the formation of large-scale streamwise vortices or by introducing perturbations in the flow field. However, it is essential to ensure that the flow control methods do not cause substantial performance penalties such as stagnation pressure loss. The usage of larger control surfaces in PFC techniques can result in significant stagnation pressure loss. The operational range is also limited due to the lack of on-demand control in PFC techniques. Hence, the current study investigates the mixing enhancement effect of streamwise pulsed injections which use fluidic actuation to cause primary jet oscillation. The streamwise injections are issued at the top and bottom sides of the mixing chamber entrance alternatively causing a bulk oscillation of the flow field. Figure 2 shows the events involved in one injection cycle of the control strategy. The frequency of pulsation is varied to study the effect on mixing characteristics enabling dynamic control based on the flow conditions.

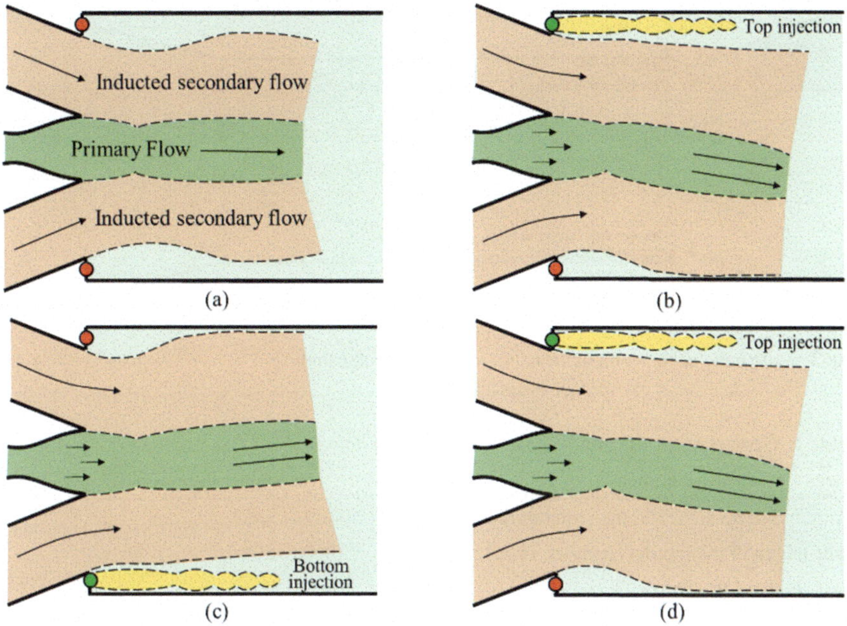

Fig. 2 The sequence of events during an injection cycle **a** no injection **b** top injection **c** bottom injection **d** top injection at the start of the next cycle

2 Computational Methodology

Figure 3 depicts the two-dimensional computational domain used in the current study. The geometric details of the domain are described in Table 1. The URANS equations are solved along with the energy and continuity equations with the ideal gas equation as the state equation. The mass distributions of the primary and secondary streams are analyzed by solving separate species transport equations for primary, secondary, and injection species. All the component species are assumed to have the properties of air. The dynamic viscosity is calculated using Sutherland's law and the eddy viscosity is modeled using the SST k-ω turbulence model (Bartosiewicz et al. 2005). The governing equations are discretized using the control volume approach. The roe-FD scheme is used for calculating fluxes at face boundaries with the second-order upwind scheme for interpolation. Second order implicit scheme is used to discretize time with a step size of 10^{-6} s.

The primary nozzle pressure ratio is selected between the second and third critical points to obtain an over-expanded primary jet. This is a preferred operating condition for the mixing performance of the supersonic ejector (Karthick et al. 2016). The primary nozzle inlet is assumed with a stagnation pressure of 5 bars with an exit Mach number of 2.19 giving a Reynolds number of 3.344×10^5 at the primary nozzle exit. The secondary fluid is assumed to be supplied from the ambient atmosphere

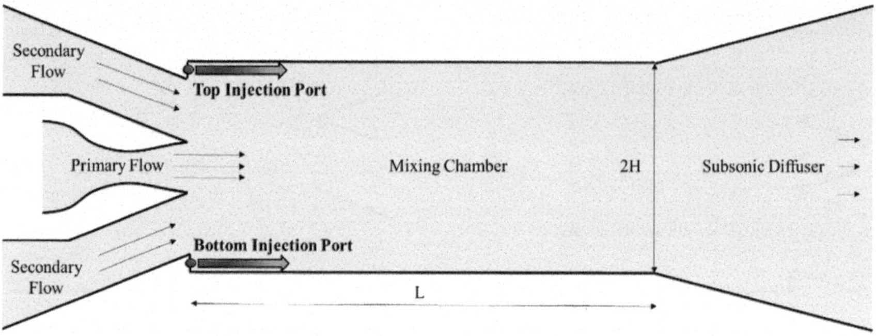

Fig. 3 Schematic of the computational domain used in the study

Table 1 Geometric specifications of the computational domain

Ejector geometric parameters	Geometric dimensions
Primary nozzle area ratio (exit area/throat area)	2
Half-height of the mixing chamber, H	0.01875 m
Exit height of the primary nozzle, h	0.32 H
Single-side secondary duct exit height	0.64 H
Total length of the mixing chamber, L	16 H
Diameter of the injection jet, d	0.053 H
Exit height of the diffuser, h_d	4.8 H
Total length of the diffuser, L_d	16 H

and the diffuser outlet is also imposed with ambient pressure conditions (1.01325 bars). The control injection jets are assumed to have a stagnation pressure of 2.5 bars. The alternate pulsation of the injections is implemented by switching between pressure inlet and wall boundary conditions of the injection positions at the required pulsation frequency. The various pulsation frequencies considered in the present study are listed in Table 2. Blowing ratio is defined as the ratio of the mass flow rate of the control injection to the primary jet.

The mesh independence study is carried out by systematically refining the grid resolution in the mixing chamber. The near-wall mesh is refined to keep the average y plus in the mixing chamber between 1 and 5. Figure 4a shows the variation of

Table 2 Details of the various cases investigated in the current study

Case	Pulsation freq. (Hz)	Strouhal no	Blowing ratio
1	0	0	0
2	250	0.098325	0.162
3	500	0.196650	0.162
4	750	0.294976	0.162

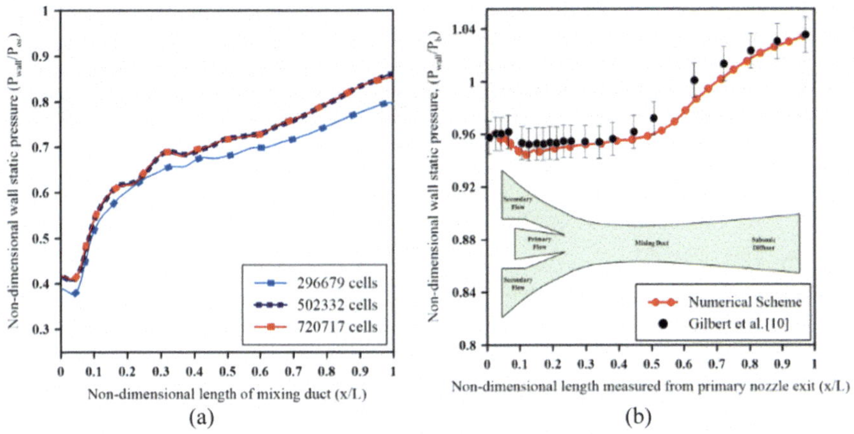

Fig. 4 **a** Variation of the wall-static pressure along the mixing chamber length for different grid resolutions **b** Comparison of the variation of the wall-static pressure along the mixing chamber length between the current numerical scheme and the experimental results (Gerald Gilbert and Hill 1973) (experimental error: ± 1.3%). (P0s: secondary stream stagnation pressure; L: mixing chamber total length; Pb: Barometric pressure)

the wall static pressure along the mixing chamber length. The numerical scheme used in the present study is validated using the experimental results from a two-dimensional variable area mixing supersonic ejector (Gerald Gilbert and Hill 1973). Figure 4b compares the variation of the wall static pressure along the mixing chamber length between the current numerical scheme and the experimental results. The inset in Fig. 4b shows the computational domain used. It can be observed that the computational results slightly under-predict the solutions as compared to the experimental results. The two-dimensionality of the simulations and Reynolds averaging can be attributed to the under-prediction. However, the variations are within reasonable limits and hence it can be ascertained that the current numerical scheme can accurately predict the supersonic ejector flows.

3 Results and Discussions

3.1 Response Characteristics

When the injection jet is issued, it pushes the secondary jet away from the mixing chamber wall and the resulting pressure perturbation deflects the primary jet away from the control injection side. This can be clearly observed from the Mach number contour shown in Fig. 5a where the top control injection pushes the primary jet downwards. The cyclic pulsation of the streamwise injections at the top and bottom positions of the mixing chamber entrance forces the primary jet to undergo oscillation

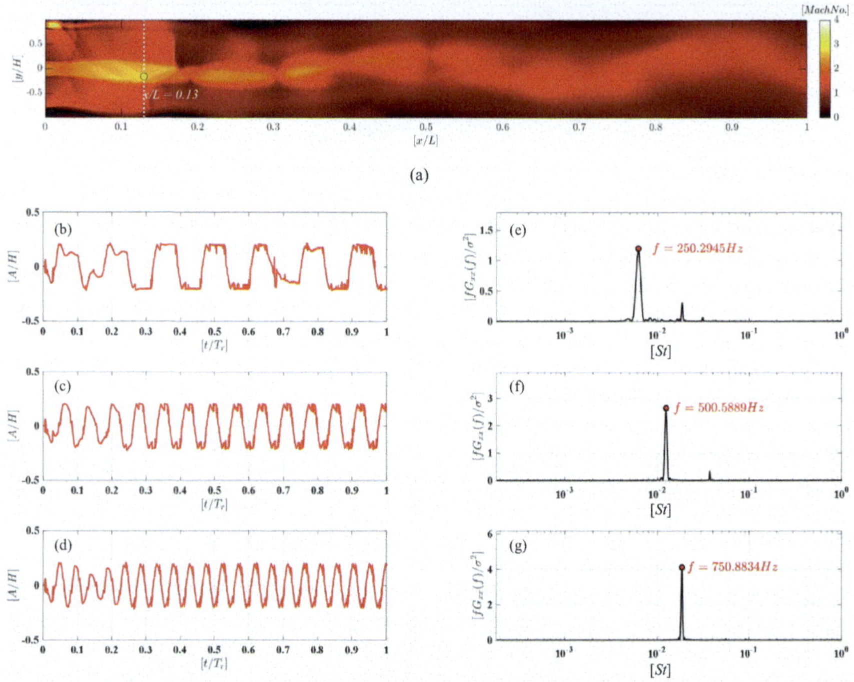

Fig. 5 **a** Mach number plot depicting the deflection of the primary jet when the top control injection is issued at $t/T_r = 0.178$. The monitoring line for the FFT analysis is shown with the yellow dot corresponding to the peak velocity point. (Left column) Temporal variation of the peak velocity point deflection for different pulsation frequencies **b** case 2 **c** case 3 and **d** case 4 (Right column) power spectra of the frequency modes for the respective cases: **e** case 2 **f** case 3 and **g** case 4. (A: y-position of the peak velocity point; H: mixing chamber half height; T_r: total run-time; f: frequency; $G_{xx}(f)$: power spectral density, σ: standard deviation of amplitude fluctuations; St: Strouhal number)

about the mixing chamber centreline enhancing the bulk mixing between the primary and secondary streams. The oscillation of the primary jet is quantified by tracking the position of the peak velocity point at a particular streamwise location ($x/L = 0.13$) in the mixing chamber of the supersonic ejector. This gives a time series of the peak velocity point deflections from the mixing chamber centreline during the injection cycle which are then subjected to the FFT analysis. Figures 5b–d show the time series of peak velocity point deflections for the pulsation frequency cases 2, 3 and 4 respectively. The corresponding power spectra (Fig. 5e–g) of the various frequency modes for the different cases of pulsation frequencies are calculated. It can be observed from Fig. 5e–g that the dominant frequency mode in the mixing chamber is corresponding to the pulsation frequency indicating a quick response time under the current control strategy. The secondary modes seen in the power spectra correspond to the low-energy higher harmonics of the dominant frequency modes and hence are neglected.

3.2 Mixing Characteristics

The mass distributions of the primary, secondary, and injection species are calculated using separate species transport equations. The primary jet oscillations create fluctuations in the mass distributions of the component species. However, the time scales associated with these fluctuations are very low. Hence, the mixing performance is analyzed after performing a time average on the mass distributions of the primary and secondary streams over the total run time. Figure 6a and b show the time-averaged secondary stream mass fraction contours for the case without control injections and for control injections pulsed at 250 Hz respectively. To quantitatively assess the improvement in mixing performance, Fig. 6c shows the time-averaged centreline mass fraction distribution of the secondary stream species under control injections compared to the baseline case without any control injections. The centreline considered is depicted in Fig. 6b. It can be seen from Fig. 6c that the centreline distribution of the secondary stream mass fraction increases due to the control injection. This indicates that the primary jet oscillation induced by the control injections increases the penetration of the secondary stream into the primary core flow. The onset of mixing is defined as the point where the mass fraction of the secondary stream becomes non-zero along the mixing chamber centreline. It can be observed that the location of the onset of mixing moves significantly upstream due to the control injections ($65.36 \pm 4.15\%$) compared to the baseline case implying an improvement in the mixing performance. The onset of mixing happens at x/L = 0.358 for the baseline case whereas it moves upstream to x/L = 0.124 under the control injections. A significant improvement in mixing performance is observed near the entrance region of the mixing chamber where the high convective Mach number limits the mixing layer growth in the baseline case. However, towards the downstream regions of the mixing chamber, the convective Mach number reduces and the mixing performance becomes closer due to the increased mixing layer growth rate for the baseline case.

The effect of control injections on the spread characteristics of the primary jet is studied by analyzing the primary stream mass fraction distribution along the mixing chamber height. Figure 7. compares the time-averaged primary stream mass fraction distribution along the mixing chamber height at different streamwise locations for the baseline without control injections (case 1) and the case with control injections (case 2). At x/L = 0.0045 (Fig. 7a), the mass distributions imply the existence of a very thin mixing layer very close to the exit of the primary nozzle and hence feeble mixing between the primary and secondary streams. At the location of the onset of mixing (x/L = 0.124), it can be seen in Fig. 7b that the primary stream mass fraction with control injections starts to spread out as compared to the baseline case. This is due to the secondary stream starting to penetrate the primary core flow. At further downstream locations (x/L = 0.445 and x/L = 0.62), the primary stream mass fraction reduces at the centreline and the primary stream spreads away from the mixing chamber centreline compared to the baseline case. This can be clearly seen in Fig. 7c and d. Hence, the current control strategy increases the spread of the primary stream due to large-scale bulk mixing induced by the primary jet oscillation.

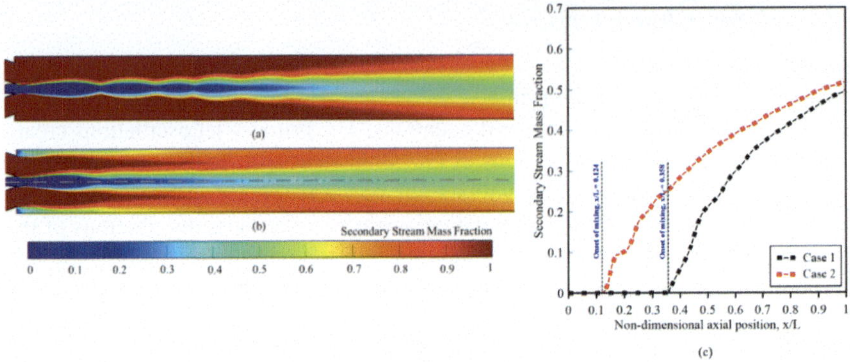

Fig. 6 Time-averaged secondary stream mass fraction contours in the mixing chamber for **a** baseline case without control injections (case 1) and for **b** control injections pulsed at 250 Hz (case 2). **c** comparison of the time-averaged centreline distribution of the secondary stream mass fraction for case 1 and case 2

Since the pulsation frequency dictates the dominant frequency mode in the mixing chamber, it is important to understand the effects of the variation in the pulsation frequency on the mixing performance of the supersonic ejector. Figure 8a shows the time-averaged centerline distribution of secondary stream mass fraction under different pulsation frequencies. It can be observed that the location of the onset of mixing remains similar irrespective of the pulsation frequency. This is because the onset of mixing is facilitated by the shock interactions between the injection jet and the secondary stream which initiates the primary jet oscillation. Since these depend only on the injection location and the blowing ratio, the change in pulsation frequency does not impact the location of the onset of mixing. A detailed explanation of the shock characteristics is given in the study published by Suresh and Rajagopal (2023).

In the entrance region of the mixing chamber ($0.124 < x/L < 0.6$), it can be seen that the centerline distribution of the secondary stream mass fraction reduces as the pulsation frequency increases. The injection jet pushes the primary stream away from the mixing chamber centerline. The time available for the primary jet to deflect farther from the mixing chamber centerline is reduced when the pulsation frequency is increased. This is because the injection jet is issued at the opposite side before the primary jet reaches its maximum deflection. Thus, near the entrance region of the mixing chamber, the amplitude of oscillation of the primary jet is reduced and so is the bulk mixing between the primary and secondary streams. This can be observed in Fig. 8b which depicts the primary jet oscillation amplitude values in the region $x/L = 0.5345$ for the different pulsation frequencies. The amplitude of oscillation of the primary jet is measured by locating the shear layer at the boundary of the primary jet. However, towards the downstream regions of the mixing chamber ($x/L > 0.6$), the amplitude of oscillation of the primary jet increases as we move away from the injection jets. The amplitude of oscillation remains almost the same between different pulsation frequencies in the downstream region as the maximum deflection is solely limited by the height of the mixing chamber. This can be observed in Fig. 8b for

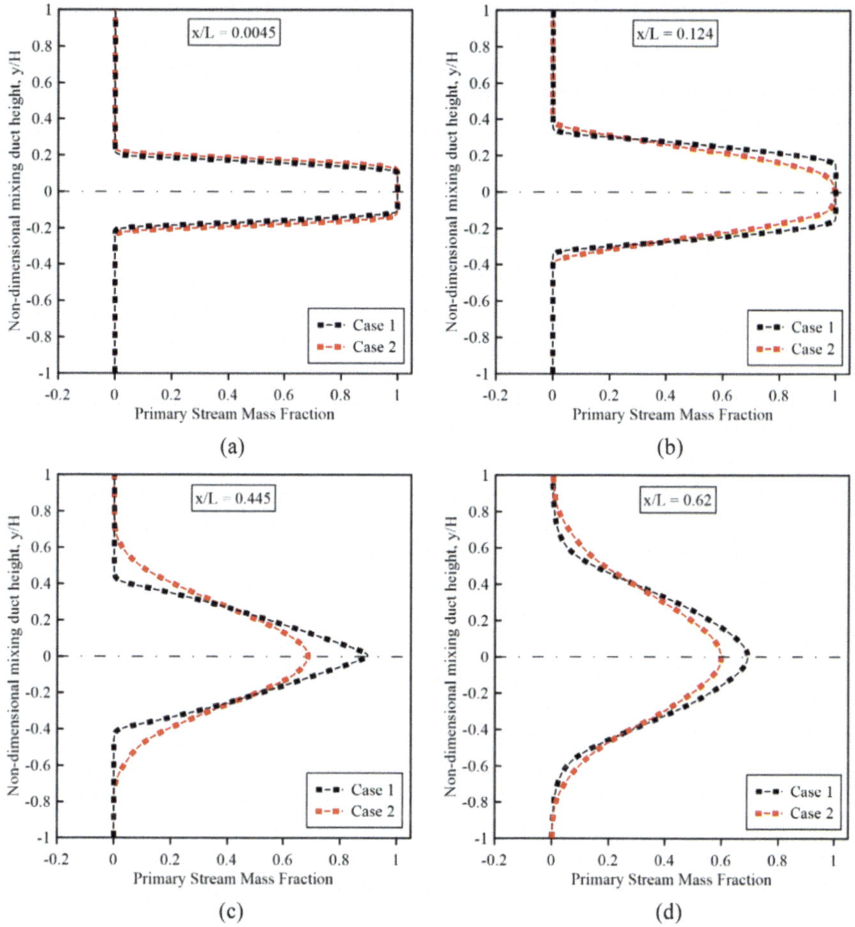

Fig. 7 Comparison of the time-averaged primary stream mass fraction along the mixing chamber height for the baseline case without control injections (case 1) and the case with control injections (case 2) at different streamwise locations **a** x/L = 0.0045 **b** x/L = 0.124 **c** x/L = 0.445 **d** x/L = 0.62 (H: mixing chamber half height; L: total mixing chamber length)

x/L = 0.945. Hence, the mixing performance becomes similar near the downstream regions of the mixing chamber for the different pulsation frequencies. However, a slight increase is observed in the centerline secondary stream mass fraction with the increase in pulsation frequency due to the increased local mixing effects.

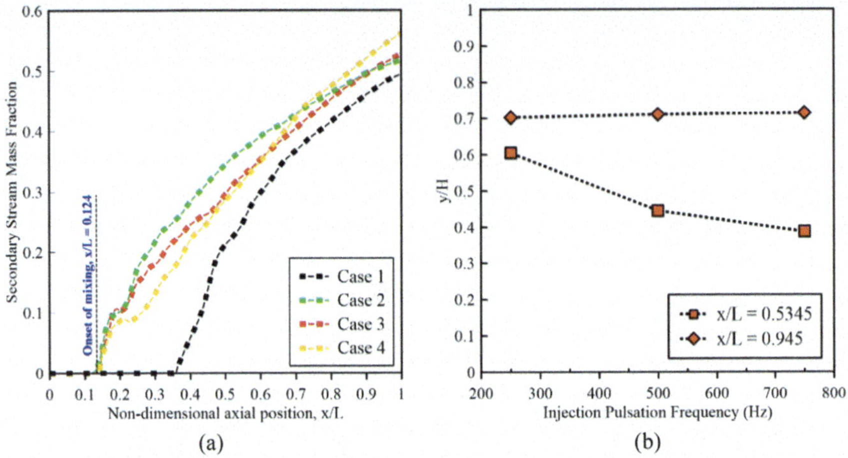

Fig. 8 **a** Time-averaged centerline distribution of the secondary stream mass fraction in the mixing chamber for various pulsation frequencies **b** non-dimensionalised amplitude of the primary jet oscillation measured at different streamwise locations (H: mixing chamber half height; L: total length of the mixing chamber)

4 Conclusions

A numerical investigation of an active mixing enhancement technique in supersonic ejectors is conducted. Control injections are issued in the streamwise direction at the top and bottom sides of the mixing chamber alternatively to oscillate the primary jet inducing a bulk mixing between the primary and secondary streams. The FFT analysis of the temporal variation of the primary jet deflection showed that the pulsation frequency of the control injections corresponds to the dominant frequency mode in the mixing chamber. Separate species transport equations are solved to obtain the mass distributions of the primary and secondary species and thus to determine the mixing characteristics of the control strategy. It is observed that the spread of the primary stream as well as the penetration of the secondary stream into the primary core flow increase due to the control injections. The amplitude of oscillation of the primary jet is primarily driving the mixing performance under the present control strategy. The enhancement in mixing performance is significant in the entrance region of the mixing chamber where the high convective Mach number reduces the mixing layer growth. As the pulsation frequency increases, the amplitude of oscillation decreases near the mixing chamber entrance due to the reduced time available for deflection. Hence, the mixing performance in the entrance region decreases as the pulsation frequency increases. However, towards the downstream regions of the mixing chamber, the amplitude of oscillation remains similar and thus the mixing performance becomes almost similar between the different pulsation frequencies.

References

Abdulateef, J. M., Sopian, K., Alghoul, M. A., Sulaiman, M. Y.: Review on solar-driven ejector refrigeration technologies. Renewable and Sustainable Energy Reviews, vol. 13, no. 6–7, 1338–1349 (2009).

Alperin, M., Wu, J.-J.: Thrust augmenting ejectors, Part I. AIAA Journal, vol. 21, no. 10, 1428–1436 (1983).

Alperin, M., Wu, J.-J.: Thrust augmenting ejectors, II. AIAA Journal, vol. 21, no. 12, 1698–1706 (1983).

Arun Kumar, R., Rajesh, G.: Flow transients in un-started and started modes of vacuum ejector operation. Physics of Fluids, vol. 28, no. 5 (2016).

Arun Kumar, R., Rajesh, G.: Physics of vacuum generation in zero-secondary flow ejectors. Physics of Fluids, vol. 30, no. 6 (2018).

Bartosiewicz, Y., Aidoun, Z., Desevaux, P., Mercadier, Y.: Numerical and experimental investigations on supersonic ejectors. Int J Heat Fluid Flow, vol. 26, no. 1, 56–70 (2005).

Besagni,G., Mereu, R., Inzoli, F.: Ejector refrigeration: A comprehensive review, Renewable and Sustainable Energy Reviews, vol. 53, 373–407 (2016).

Gerald Gilbert, A. B., Hill, P. G.: Analysis and testing of two-dimensional slot nozzle ejectors with variable area mixing sections (1973).

Karthick, S. K., Rao, S. M. V., Jagadeesh, G., Reddy, K. P. J.: Parametric experimental studies on mixing characteristics within a low area ratio rectangular supersonic gaseous ejector. Physics of Fluids, vol. 28, no. 7 (2016).

Singhal, G., Rajesh, R., Tyagi, R. K., Dawar, A. L., Subbarao, P. M. V., Endo, M.: Two-stage ejector based pressure recovery system for small scale SCOIL. Exp Therm Fluid Sci, vol. 30, no. 5, 415–426 (2006).

Suresh, A., Rajagopal, A. K.: An active mixing enhancement technique in supersonic ejectors using pulsed streamwise injections. Physics of Fluids, vol. 35, no. 7, (2023).

Relaminarization Effects on SWTBLI Due to Upstream Expansion

Kavan M. Prabtani and Rajesh Ranjan

Abstract The complex shock wave-boundary layer interactions are still a major obstacle in the accurate prediction and control of scramjet inlet/isolator flows. Moreover, the nature of interactions strongly depends on the state of the incoming boundary layer which may be relaminarized due to supersonic expansion near the ramp shoulder. While shock-turbulence interactions result in Reynolds stress anisotropy and turbulence amplification, relaminarization is characterized by the 'freezing' of Reynolds stress in outer layers. Such scenarios are typically outside the range of applicability of eddy viscosity-based models and may result in significant discrepancies in predictions of separation and aerothermal characteristics. This work focuses on a scramjet inlet/isolator flow, where the boundary layer passes through the expansion corner and shocks in close proximity. For this flow, significant discrepancies in the prediction of separation region and separation shock characteristics were observed with both the SST k-ω model and the Reynolds stress model (RSM). By investigating the flow behavior near the expansion corner, we find that the main cause of the inconsistency is the strong relaminarization of the boundary layer prior to shocks. This was confirmed by examining some characteristic behavior of relaminarization such as quick decay in turbulence levels in the relaminarizing region, stress-freezing in the relaminarized region, deviation from the law of the wall, and value of the pressure gradient parameter exceeding the threshold value for typical relaminarizing flows. The models show some ability to predict the turbulence decay in this region, however, the discrepancy may be due to incorrect estimation of the eddy viscosity level and treatment of shock-relaminarized boundary layer interactions. Current observations suggest that a calibration of the turbulence models is necessary where expansion and shock are present in close proximity.

Keywords Relaminarization · Turbulence models · Scramjet inlets · Shocks

K. M. Prabtani · R. Ranjan (✉)
Department of Aerospace Engineering, Indian Institute of Technology, Kanpur, India
e-mail: rajeshr@iitk.ac.in

G. Rajesh et al. (eds.), *Proceedings of the 15th International Symposium on Experimental and Computational Aerothermodynamics of Internal Flows*,
https://doi.org/10.1007/978-981-96-4082-9_23

1 Introduction

Hypersonic flows exhibit the interplay of several fluid-dynamical phenomena which require special attention. Figure 1 shows the general schematic of the scramjet inlet/isolator engine with various flow phenomena. Shock wave turbulent boundary layer interaction (SWTBLI) is a key feature that affects the performance of mixed compression scramjet inlet through large-scale separation regions, reduced airflow, and possibly unstart phenomenon (Emami et al. 1995). The subsequent reattachment of flow results in high heat transfer rates. Therefore, the accurate prediction of separation bubble characteristics and wall fluxes is necessary to ensure a safe design.

The behavior of interactions strongly depends on the state of the incoming boundary layer as well as the strength and location of impinging cowl shock on the ramp surface. As shown in Fig. 1 for a typical scramjet, the boundary layer of incoming flow is successively distorted by two external compression ramps, followed by an expansion corner before it goes through complex shock-shock interactions and shock-boundary layer interactions inside a constant area duct, called an isolator. The combined effects of adverse pressure gradient and favorable pressure gradient on flow structure are studied in detail by Smith and Smits (1997). It was shown that both Reynolds stress and structure parameter are greatly distorted and 'negative' turbulence production may appear near the expansion corner, indicating the inability of conventional eddy viscosity-based turbulence models for this type of flow. Furthermore, the rapid expansion of supersonic flows over the expansion corner may cause the boundary layer to return to a laminar-like or a 'relaminarized' state (Narasimha and Viswanath 1975). Some general characteristics of relaminarizing flows are thinning of the boundary layer, deviation from the well-known law of the wall, increase in skin friction as well as heat transfer coefficients before the rapid decline, and the 'freezing' of Reynolds stress along streamlines in the outer layer. Moreover, this phenomenon is gradual and not catastrophic, unlike the transition to turbulence, hence, prediction of its onset using a single parameter is challenging (Sreenivasan 1982).

The suitability of turbulence models for simulating such flows depends on their ability to capture these key features of relaminarization accurately. Rumsey and Spalart (2009) showed that widely popular SA and SST models are not able to

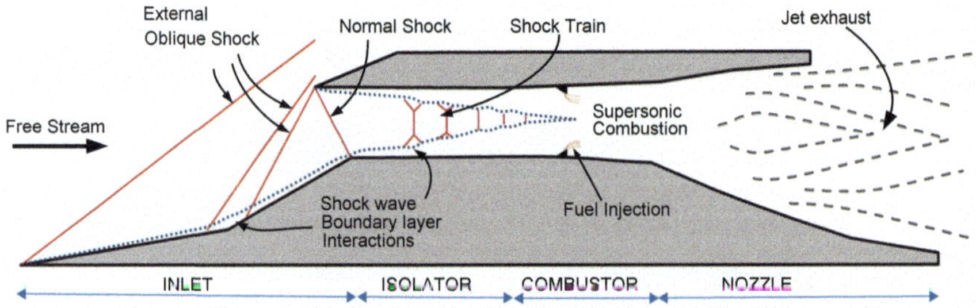

Fig. 1 General schematic of scramjet inlet/isolator (Reproduced from Sethuraman et al. (2021))

predict the relaminarization in 2-D sink flow and 3-D infinite swept-wing flow. The eddy viscosity was seen to remain turbulent even in accelerating regions with partial reversion. They concluded that these models should be applied with high precautions for relaminarizing or transitional flows as they are being applied outside of their range of applicability, and some transitional modeling should be necessary to capture the flow physics. Nguyen et al. (2009) carried out a numerical investigation of flow over the expansion corner to review the capability of turbulence models. It was shown that only the Reynolds stress model can predict mean flow features while other eddy viscosity-based models produced discrepancies in strong interaction regions. They also tested the γ-Re_θ transition model and concluded that the model needs some modifications to accurately capture the mean flow. Moreover, no Reynolds stress 'freezing' was observed in the results.

While existing turbulence models are not sufficiently accurate in predicting even canonical SWTBLI flows (Liou et al. 2000; Knight et al. 2003; Georgiadis and Yoder 2013; Raje and Sinha 2021), the presence of incoming distorted and possibly relaminarized boundary layer inside a practical scramjet make the predictions even more challenging. In a very narrow region near the ramp shoulder, the flow is successively perturbed by rapid expansion, relaminarization, retransition, and finally oblique cowl shock wave boundary layer interaction. Highest discrepancy is usually found in this region, and can be referred to as a 'critical spot'. To better understand the flow phenomena near this critical spot, simplified problems of shock impingement in the vicinity of the expansion corner are studied in the literature. Chew (1978) studied oblique shock wave boundary layer interaction near the expansion corner at Mach numbers 1.8 and 2.5. They found that the reflected shock wave was effectively neutralized by expansion waves when the shock impingement location was at the corner. Chung and Lu (1995) carried out an experimental study for Mach 8 flow past an expansion corner with shock impingement location at the corner as well as one boundary layer thickness before and after the corner. SWBLIs were affected significantly when shock impinged behind the corner with a decrease in upstream influence. White and Ault (1996) conducted a similar study for Mach 11.5 flow with a much more extensive dataset to be compared with computational codes. Sathianarayanan and Verma (2017) carried out an experimental analysis of a similar problem with Mach 3.9 with the added complexity of side walls and presented an effect of geometrical parameter variation on the three-dimensionality of SWBLI. Konopka, Meinke & Schröeder 2012 performed large eddy simulation (LES) of an expanding turbulent boundary layer at Mach 1.76 with and without shock impingement to study turbulence statistics of interactions. Tong et al. (2020) conducted direct numerical simulation (DNS) of the problem with Mach 2.9 to investigate the effect of expansion corner angle on flow phenomenon, turbulence statistics, and separation unsteadiness.

It is evident from the above discussion that the presence of an expansion corner upstream of SWTBLI changes the nature of the incoming boundary layer, and therefore makes it challenging to predict the downstream flow behavior. In this paper, we study discrepancies in flow prediction with widely used turbulence models due to such successive events in close proximity. Further, we assess the 'relaminarization' state of the boundary layer downstream of the expansion corner using the well-known relaminarization parameters available in the literature. It is expected that this study will help understand the precise limitations of the turbulence models in such flows. This knowledge can be used for improvements and calibrations of the turbulence models to make predictions better.

2 Numerical Methodology

2.1 Computational Setup

The model configuration for the scramjet inlet/isolator is taken from the experiments conducted at RWTH Aachen, Germany (Reinartz et al. 2003). It was reported in the experiment that the flowfield in the center-line symmetry plane was not affected by side walls, and therefore a two-dimensional setup was employed in the current study. The 2D computational domain is shown in Fig. 2, which encompasses the entire scramjet inlet/isolator model geometry from the beginning of the second ramp to the end. Further, a straight downstream region is added after the last divergent section to ensure smooth outflow.

The boundary conditions at the inlet were given as a pressure farfield with a total pressure of 540kPa and a total temperature of 305 K. Mach number at the inlet was specified as 2.41 which is the value after the first ramp compression. The turbulence intensity was taken as 0.5%. At the outlet, a non-reflecting boundary condition was enforced to prevent any non-physical reflections. Ramp and cowl surfaces are specified as adiabatic no-slip walls. A multi-block structured grid was created with 350×3050 grid points in wall-normal and streamwise directions, as shown in Fig. 2. The average y^+ value was kept near 1 on both ramp and cowl surfaces. Grid-sensitivity studies were performed by nearly doubling the grid size in each direction. Further details about the numerical setup are available in our earlier work (Prabtani and Ranjan 2023).

2.2 Numerical Approach

Numerical simulations for the current study are performed with Reynolds-Averaged Navier Stokes (RANS) approach using the commercial solver ANSYS®Fluent Ver-

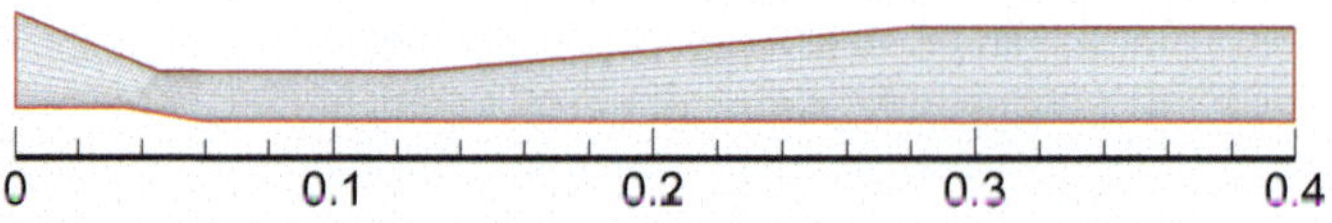

Fig. 2 Computational domain and mesh of the scramjet inlet/isolator based on experiments (Reinartz et al. 2003). Every 10th grid point displayed in the mesh

sion 21. We employ a density-based solver, which solves the governing equations of continuity, momentum, and energy in a coupled manner. Convective flux terms were computed using AUSM+ scheme (Liou 1996), while a dual-time stepping-based implicit approach is used for time-marching. MUSCL third-order approach is used for the reconstruction of fluxes. The simulations are performed using a variety of turbulence models. In this paper, we report the results of Menter (1994) SST k-ω and Reynolds stress model (RSM), which are widely used in high-speed simulations. Results from other models were either equivalent or inferior to current results.

3 Results and Discussion

Both inviscid and viscous simulations were performed for the scramjet inlet/isolator flow. For each simulation, convergence was ensured by allowing the residuals to fall below 10^{-5} value. Figure 3a, b shows the pressure distribution on the surface of the cowl and ramp, respectively due to different turbulence models as well as when the flow is assumed inviscid. The inviscid study was performed to understand the effect of shocks and expansion waves in the absence of complicated boundary layer effects. The significant difference between viscous and inviscid solutions indicates very strong shock boundary layer interactions.

The numerical results are compared with experimental data (Reinartz et al. 2003) as well as other computational studies performed with the URANS approach (Xu et al. 2016). Figure 3a shows that the pressure rise corresponding to separation shock impinging on the cowl does not match the experimental data for both the SST k-ω and Reynolds stress model (RSM), although the latter seems to provide slightly better predictions. It is to be noted that the computational time of the Reynolds stress model (RSM) is much higher than that of the SST k-ω model. The same mismatch is observed in the URANS simulation by Xu et al. (2016). The pressure plateau corresponding to separation is smaller than that of the experiment, as evident from Fig. 3b.

Figure 4 shows the qualitative comparison of the separation bubble and consequently, the location of the separation shock between the CFD simulation and experiment. The size of the separation bubble is predicted to be significantly smaller than that of the experiment, as observed from the Schlieren images. Further, the foot of the separation is predicted much downstream in the simulation. These discrepancies in the prediction of separation region and consequently in the location of separation shock are reported in most literature (FeiChao and Huang 2022; Du et al. 2020; Liu et al. 2016; Zheng et al. 2019).

To investigate the cause of these discrepancies, we probe more into the region near the expansion corner, or 'critical spot' as explained in the introduction section. The corner is present at $x = 0.0455$ m as shown in Fig. 4a. Therefore, our primary region of interest is between $x = 0.045$ m and $x = 0.05$ m. Figure 5 shows the variation of the skin friction coefficient in the vicinity of the expansion corner. Locations of the expansion corner as well as separation and reattachment points as predicted by

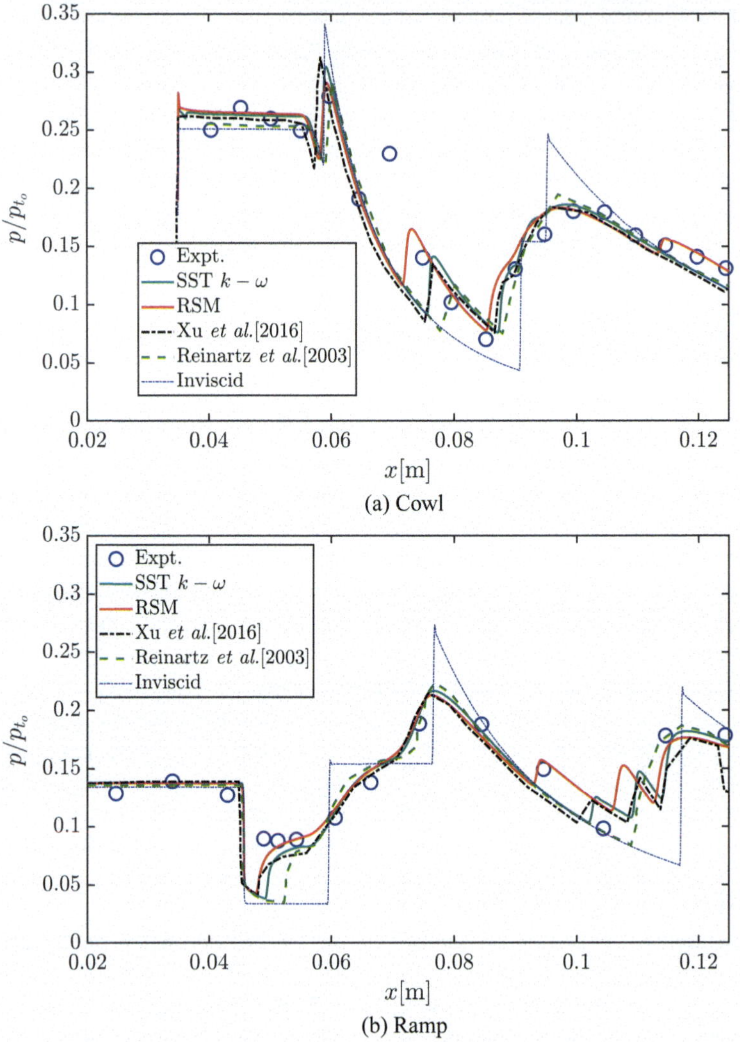

Fig. 3 Comparison of pressure distribution with experimental results as well as other simulations

both models are annotated in the figure. The skin friction, as expected, becomes negative in the separation region, but a more distinctive behavior observed in skin friction is a rapid increase and immediate decline before changing the sign. This trend is observed at the expansion corner and may indicate a relaminarized boundary layer (Sreenivasan 1982). While the onset of separation predicted by RSM is slightly earlier than that of SST, the reattachment point is predicted much upstream of that predicted by the SST model. This results in a smaller separation for RSM as compared to the SST model.

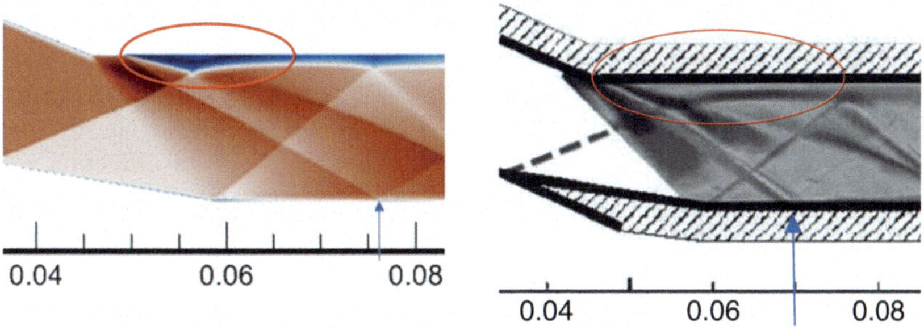

(a) Numerical prediction (b) Experimental observation (modified from [18])

Fig. 4 Comparison of separation size and separation shock location for computation and experiment using Schlieren images. Red ovals indicate the separation regions, while blue arrows represent the locations of shock impingement on the cowl. The expansion corner is around $x = 0.0455$ m. This dimension is used in further analysis for examining the nature of the boundary layer in the vicinity of the corner

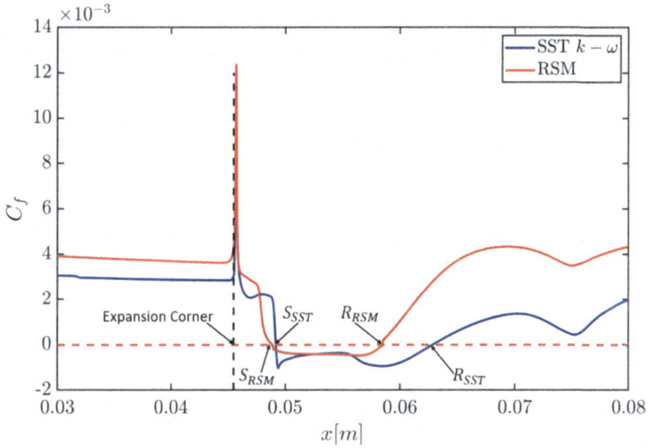

Fig. 5 Variation of skin friction coefficient in the vicinity of expansion

We further probe the nature of the boundary layer downstream of the expansion corner. As discussed earlier, it has been observed that a relaminarized boundary layer may lead to inaccurate estimations of the separation region and downstream flow features. We calculate the relaminarization parameter, $\Delta P/\tau_0$, given by Narasimha and Viswanath (1975) to predict the occurrence of relaminarization in a supersonic expansion. Authors have suggested that for relaminarizing flows, the value of $\Delta P/\tau_0$ should be greater than 70. They suggested this criterion to be valid up to $Ma_\infty < 3$. For the present case, the Mach number corresponding to the incoming flow is 2.41, hence this parameter is used for estimation of the boundary layer state. For SST k-ω model, this parameter is found to be 116, indicating the flow is indeed relaminarized near the ramp shoulder. In the RSM model also, a similar value was observed.

Since, flow is partially laminarizing, the computed turbulent viscosity ratio must decrease in the streamwise direction. For subsequent discussions, we will focus on results obtained with the SST k-ω model; the RSM model provides qualitatively similar observations. The decay in turbulent viscosity ratio is evident from Fig. 6a, where maximum values of turbulent viscosity along a wall-normal direction are plotted at different streamwise locations. Note that the location of the expansion corner is $x = 0.0455$ (see Fig. 4a), after which the viscosity ratio rapidly decays up to $x = 0.049$ before separation. The increase in ratio at $x = 0.05$ marks the onset of separation and separation shock. Fig. 6b shows the profile of μ_t/μ along a wall-normal direction at three different streamwise locations. We observe that in a very short distance from $x = 0.045$ to $x = 0.046$, the peak value decreases significantly due to expansion. Further downstream of the expansion corner, at $x = 0.048$, the profile becomes more wide compared to a very steep profile near expansion. Thus, the initial phase near expansion may be characterized as a non-equilibrium state (Sreenivasan 1982), while wide profiles indicate a relaminarized state of flow. Another interesting observation is the shifting of peak value away from the wall, towards the outer layer, as we move downstream of expansion. This makes the boundary layer less resistant towards adverse pressure gradient due to impinging shock and result in a separation. However, the separation onset predicted by the model is much downstream than that of the experimental location (see Fig. 4).

Now, we show the variation of normalized mean velocity with normalized wall normal distance, y^+ in Fig. 7. Profiles A and B are taken at $x = 0.040$ and $x = 0.045$, respectively. Both profiles are almost identical with a well-resolved viscous sublayer. The log law region shows a constant slope with a consistent dip from the log line which is characteristic behaviour in such complicated flows. As the flow passes through expansion, the profile-C ($x = 0.046$) shows highly distorted behavior with collapsed wake region and extended viscous sublayer. The existence of an extended viscous sublayer is consistent with the observation that the peak value of μ_t/μ shifts away from the wall in Fig. 6b. Profile D (at $x = 0.048$) shows an increase in mean velocity in the wake layer indicating recovery from a non-equilibrium region to a fully relaminarized state. The boundary layer, however, is separated due to impinging shock-boundary layer interaction before the relaminarized state can recover to a fully turbulent state. This is evident from profile-E at $x = 0.049$ where mean velocity changes sign. Initially, sign reversal is seen only below $y^+ = 6$ as the viscous sublayer has lower resistance toward flow separation. This is a consequence of the peak of turbulent viscosity ratio shifting away from the wall.

Finally, we look into the turbulent statistics of flow near the expansion. As proposed by Sreenivasan (1982), a characteristic behavior of a fully relaminarized boundary layer is stress freezing along a streamline. To verify this in the current flow, we calculate the stress behavior along a few streamlines in the outer layer of the boundary layer. Figure 8a shows the location of streamlines along which the stress quantities are analyzed. Figure 8b shows the variation of Reynolds stress along these streamlines with stream function evaluated in terms of mass flow rate. With the origin set on the expansion corner, the x-axis represents the downstream distance from the corner. We see that after the initial non-equilibrium region, Reynolds

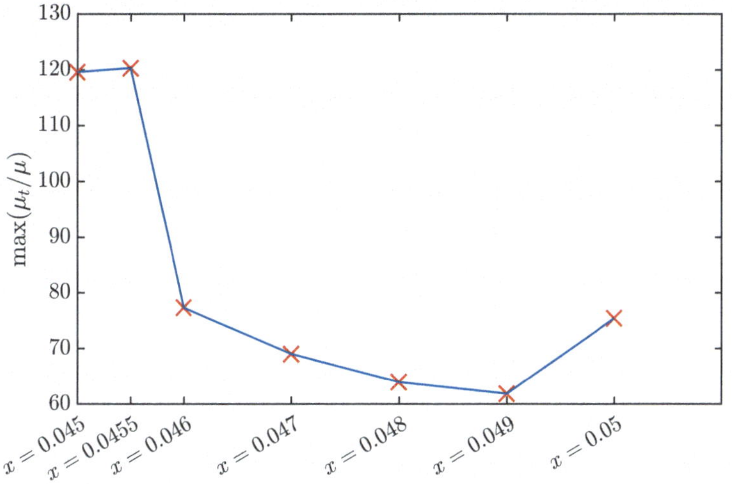

(a) Peak of μ_t/μ at various streamwise locations

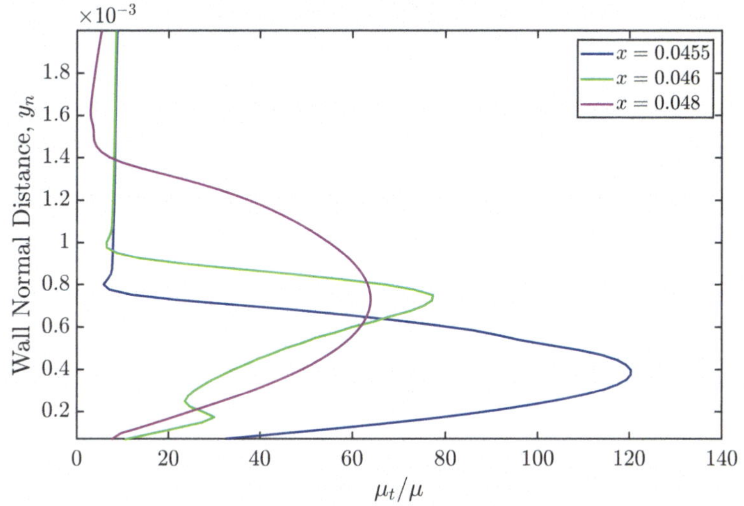

(b) Profiles of μ_t/μ along wall normal direction.

Fig. 6 Behavior of turbulent viscosity ratio near expansion corner

stress becomes nearly constant or 'frozen' along the streamlines. Figure 8c shows the behavior of streamwise normalized turbulent intensity along the same streamlines. After the initial drop in the non-equilibrium region, intensity decays along a streamline as expected for relaminarized flow.

Based on the above observations, we conclude that the boundary layer shows strongly relaminarizing behavior as it passes through the expansion corner. Consequently, the nature of SWTBLI and hence the predictive abilities of turbulence models

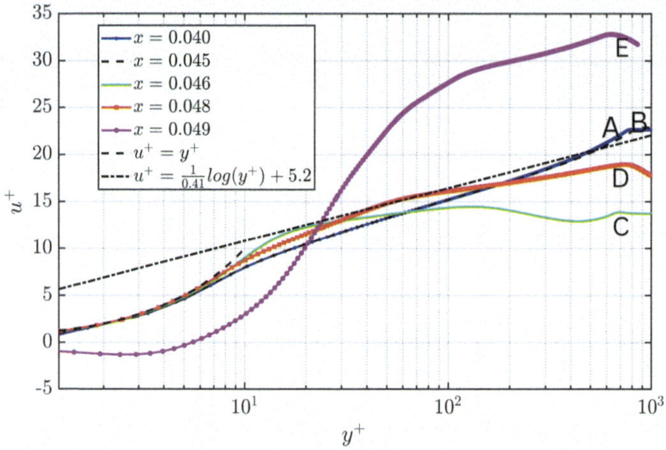

Fig. 7 Profiles of normalized mean velocity profiles along wall normal distance, y^+

are affected. One such major consequence of relaminarization is the increased susceptibility of the boundary layer to separation due to impinging shock as compared to the fully turbulent boundary layer. The employed turbulence models-SST k-ω, RSM—show some ability to predict the relaminarizing behavior qualitatively but the correct values of eddy viscosity are possibly not predicted accurately leading to discrepancies in the predictions of the separation region. This could be the primary cause of the discrepancy in the separation region.

4 Conclusions

We assess the relaminarizing behavior of flows in a scramjet inlet/isolator. RANS simulations are performed using turbulence models- Menter SST k-ω RANS model and RSM. The models exhibit significant discrepancies in the prediction of separation bubble and separation shock—this is consistent with computational studies in the literature. We analyze the flow near the discrepancy region to better understand the cause of the discrepancy and the nature of the boundary layer. The following observations are made based on analysis of the boundary layer:

- The skin friction behavior near expansion as well as the criterion of (Narasimha and Viswanath 1975) indicates the relaminarized boundary layer before it goes through complex interactions with the shock wave, thus affecting the predictive capability of the model.
- The peak value of μ_t/μ decreases consistently with the streamwise location near the expansion corner. The profile of μ_t/μ along the wall-normal direction shows that the peak value shifts away from the wall with the profile becoming flatter as

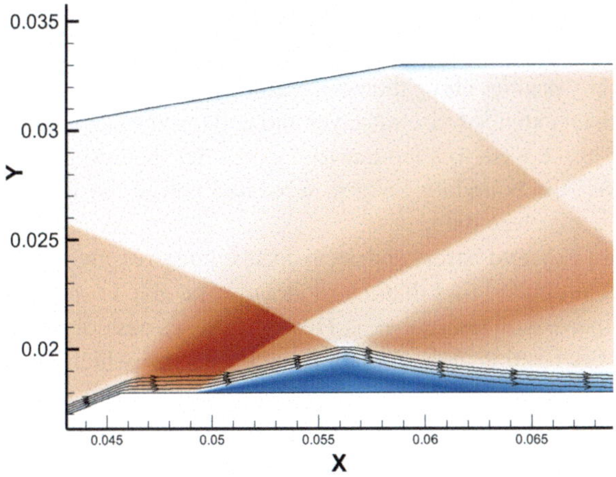

(a) Streamline locations for analysis of stress-freezing. Streamline-1 is closest to the wall

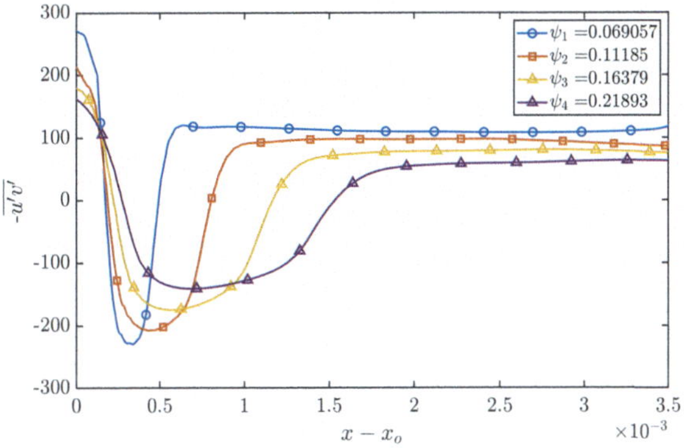

(b) Reynolds stress

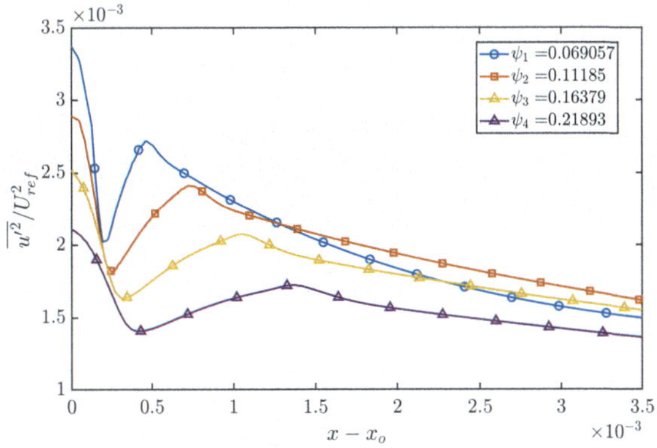

(c) Normalized streamwise turbulent intensity

Fig. 8 Turbulent statistics along streamlines in the outer layer, as shown in **a**. The region shown is between $x = 0.0455$ (from the expansion corner) to $x = 0.0490$ (before the onset of separation). **b** suggests the stress freezing in this region, while **c** shows the decay in turbulence intensity compared to the steep profile near the corner. This indicates the existence of a non-equilibrium region before the flow becomes laminarized.

- The mean velocity profiles along the wall-normal direction indicate the highly distorted flow with an extended viscous layer and collapsed wake layer. The collapse in the wake layer recovers to a laminarized state after the initial non-equilibrium region. However, the boundary layer is separated before the flow goes through retransition to a turbulent state.
- The 'freezing' of Reynolds stress and decay in absolute streamwise turbulent intensity along a streamlines shows that the model may be able to show the characteristics of relaminarization qualitatively. However, the current discrepancy may be due to incorrect estimation in the drop of turbulence level, which can be calibrated based on prior knowledge.

Acknowledgements The authors would like to acknowledge the ISRO for support of this project via Space Technology Cell, IIT Kanpur. IIT Kanpur's computational facility HPC2013 was employed for the simulations.

References

A Sathianarayanan and SB Verma. Experimental investigation of an incident shock-induced interaction near an expansion corner. *Journal of Spacecraft and Rockets*, 54(3):769–773, 2017.

Birgit U Reinartz, Carsten D Herrmann, Josef Ballmann, and Wolfgang W Koschel. Aerodynamic performance analysis of a hypersonic inlet isolator using computation and experiment. *Journal of Propulsion and Power*, 19(5):868–875, 2003.

Christopher L Rumsey and Philippe R Spalart. Turbulence model behavior in low reynolds number regions of aerodynamic flowfields. *AIAA journal*, 47(4):982–993, 2009.

Douglas R Smith and Alexander J Smits. The effects of successive distortions on a turbulent boundary layer in a supersonic flow. *Journal of Fluid Mechanics*, 351:253–288, 1997.

Doyle Knight, Hong Yan, Argyris G Panaras, and Alexander Zheltovodov. Advances in cfd prediction of shock wave turbulent boundary layer interactions. *Progress in Aerospace Sciences*, 39(2-3):121–184, 2003.

FeiChao Cai and Xing Huang. Study on self-excited oscillation suppression of supersonic inlet based on parallel cavity. *Frontiers in Energy Research*, page 418, 2022.

Florian R Menter. Two-equation eddy-viscosity turbulence models for engineering applications. *AIAA journal*, 32(8):1598–1605, 1994.

Fulin Tong, Xinliang Li, Xianxu Yuan, and Changping Yu. Incident shock wave and supersonic turbulent boundarylayer interactions near an expansion corner. *Computers & Fluids*, 198:104385, 2020.

K Liu, Y Zhu, W Gao, J Yang, Y Jin, and Y Wu. Influence of combustion-preheating vitiation on operability of a hypersonic inlet. *Shock Waves*, 26(6):815–824, 2016.

Kavan M Prabtani and Rajesh Ranjan. Turbulence modeling for swtbli with expansion wave interference inside scramjet inlets. *International Journal of Computational Fluid Dynamics*, 2023.

Kejing Xu, Juntao Chang, Weixing Zhou, and Daren Yu. Mechanism and prediction for occurrence of shock-train sharp forward movement. *AIAA Journal*, 54(4):1403–1412, 2016.

KR Sreenivasan. Laminarescent, relaminarizing and retransitional flows. *Acta Mechanica*, 44:1–48, 1982.

Kuan Zheng, Wei Tian, Jiang Qin, Silong Zhang, and Hui Hu. Effect of film cooling injection on aerodynamic performances of scramjet isolator. *Aerospace Science and Technology*, 94:105383, 2019.

Kung-Ming Chung and Frank K Lu. Hypersonic turbulent expansion-corner flow with shock impingement. *Journal of Propulsion and Power*, 11(3):441–447, 1995.

Martin Konopka, Matthias Meinke, and Wolfgang Schröeder. Large-eddy simulation of relaminarization in supersonic flow. In *42nd AIAA Fluid Dynamics Conference and Exhibit*, page 2978, 2012.

Meng-Sing Liou. A sequel to ausm: Ausm+. *Journal of computational Physics*, 129(2):364–382, 1996.

Michael E White and David A Ault. Expansion corner effects on hypersonic shock wave/turbulent boundary-layer interactions. *Journal of propulsion and power*, 12(6):1169–1173, 1996.

Nicholas Georgiadis and Dennis Yoder. Recalibration of the shear stress transport model to improve calculation of shock separated flows. In *51st AIAA Aerospace Sciences Meeting Including the New Horizons Forum and Aerospace Exposition*, page 685, 2013.

Pratikkumar Raje and Krishnendu Sinha. Anisotropic sst turbulence model for shock-boundary layer interaction. *Computers & Fluids*, 228:105072, 2021.

R Narasimha and PR Viswanath. Reverse transition at an expansion corner in supersonic flow. *AIAA Journal*, 13(5):693–695, 1975.

Saied Emami, Carl A Trexler, Aaron H Auslender, and John P Weidner. Experimental investigation of inlet-combustor isolators for a dual-mode scramjet at a mach number of 4. Technical report, 1995.

Tue Nguyen, Marek Behr, and Birgit Reinartz. Numerical investigation of compressible turbulent boundary layer over expansion corner. In *16th AIAA/DLR/DGLR International Space Planes and Hypersonic Systems and Technologies Conference*, page 7371, 2009.

Vignesh Ram Petha Sethuraman, Tae Ho Kim, and Heuy Dong Kim. Shock train control by boundary layer suction in a scramjet isolator. *Proceedings of the Institution of Mechanical Engineers, Part G: Journal of Aerospace Engineering*, 235(10):1211–1224, 2021.

William W Liou, George Huang, and Tsan-Hsing Shih. Turbulence model assessment for shock wave/turbulent boundary-layer interaction in transonic and supersonic flows. *Computers & fluids*, 29(3):275–299, 2000.

YT Chew. Shockwave and boundary layer interaction in the presence of an expansion corner. *Aeronautical Quarterly*, 30(3):506–527, 1979.

Zhao-bo Du, Wei Huang, and Li Yan. Investigation on the supplementary combustion scheme for the divergent section of a solid rocket engine nozzle. *Energy*, 190:116295, 2020.

Numerical Investigation of Film Cooling on Second Throat Ejector-Diffuser Used in High Altitude Testing Facilities

M. V. V. Srinivas, R. Arun Kumar, B. Lebonah, M. Jegesh David, and A. Narayanan

Abstract The present study aims to investigate the flow and cooling characteristics in second throat ejector diffuser used in the High-Altitude Testing (HAT) Facility using gaseous film cooling technique. A detailed computational analysis with three different coolant injection configurations (normal, inclined, and tangential) and with varying mass flow rate ratios (ratio of coolant to ejector-diffuser mass flow rate) were performed. It was found that the injection of secondary fluid into the mainstream produced additional secondary shock structure apart from the primary shock structure produced by the supersonic nozzle flow impingement with the diffuser wall. The secondary shock wave reflection was clearly observed for the normal and inclined injection configurations, whereas for the tangential injection configuration the secondary shock structure was not observed. It was found that the cooling effectiveness of the normal injection was higher than the inclined and the tangential injection for the same mass flow rate ratio. The film cooling effectiveness increases with an increase in mass flow rate ratio for all the injection configurations. It was found that, with the highest injection mass flow rate considered in the present study, the average diffuser wall temperature drops by nearly 40% for all the injection cases. It was also observed that the coverage length is higher for normal injection followed by inclined and the tangential injection. The coverage length has been considered to be the distance at which the diffuser wall attains 900 K (60% temperature of that of the without injection case (1500 K)).

Keywords Supersonic ejector · Film cooling · High altitude testing facility · Shock waves

M. V. V. Srinivas · R. Arun Kumar (✉)
Department of Mechanical Engineering, Indian Institute of Technology, Jodhpur 342030, India
e-mail: arunkr@iitj.ac.in

B. Lebonah · M. Jegesh David · A. Narayanan
ISRO Propulsion Complex, Mahendragiri 627109, Tamil Nadu, India

G. Rajesh et al. (eds.), *Proceedings of the 15th International Symposium on Experimental and Computational Aerothermodynamics of Internal Flows*,
https://doi.org/10.1007/978-981-96-4082-9_24

1 Introduction

Rocket engines used in space vehicles are subjected to a continuously decreasing ambient pressure condition as the altitude increases. The ambient pressure levels in the upper atmosphere are extremely low, resulting in a highly under expanded nozzle jet. As a result, it is extremely crucial to test the performance characteristics of upper-stage rocket engines under high vacuum conditions and the testing facilities for such purposes are generally referred to as the High-Altitude Testing (HAT) facilities. In the HAT facility, the high momentum of the rocket plume at the nozzle exit entrains the surrounding fluid from the chamber upstream to the nozzle exit maintaining the required vacuum pressure in the chamber. Ejectors are also connected with a properly designed diffuser system for pressure recovery. Various types of diffuser configurations for supersonic ejector flows, such as straight ejector diffuser (SED), second throat ejector diffuser (STED), and center body diffuser can be found in the literature (Amatucci et al. 1982; Park et al. 2012; Ferri and Louis 1954). The second throat ejector diffuser requires a lesser pressure ratio for the operation of the HAT tunnel and hence is the most preferred diffuser configuration. The schematic of the second throat ejector diffuser used in high altitude testing facility can be seen in Fig. 1.

Many experimental and numerical studies have been done in the past to analyze and optimize the performance of the ejector diffuser system for high altitude testing facility applications (Massier and Roschke 1960; Bauer and German 1961; German et al. 1963; German and Panesci 1965). One of the initial experimental studies on various ejector diffuser systems for high altitude testing facilities was done by Massier and Roschke (1960). They studied the effect of geometrical variations like the length of the second throat (L_{st}, second throat length), throat diameter of the nozzle (D_1, C-D nozzle diameter), and the expansion ratio of the nozzle on the performance of the rocket motor in straight ejector diffuser and second throat ejector diffuser. Their study found that the second-throat diffuser achieved the best performance when the second throat's length-to-diameter ratio (L_{st}/D_3, where D_3 is second throat diameter) was greater than 8. Bauer and German (1961) conducted experimental studies on second throat ejector diffuser without induced flow using twenty-seven ejector configurations by varying several geometric parameters like second throat location (L_d, length before converging portion), nozzle exit area, and second throat length (L_{st})

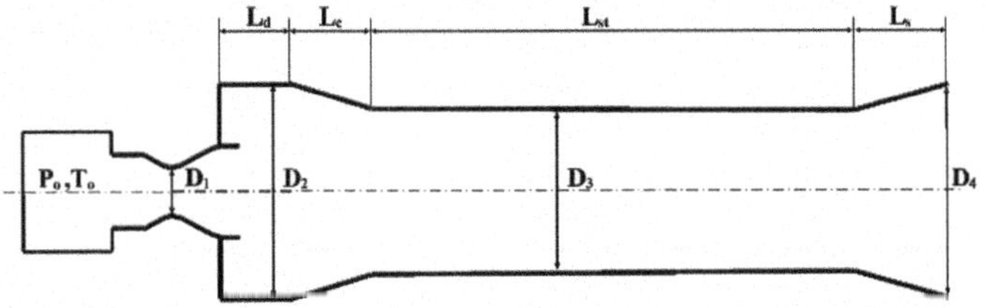

Fig. 1 Schematic of high-altitude testing facility with second throat ejector diffuser

(Massier and Roschke 1960; Bauer and German 1961; German et al. 1963; German and Panesci 1965; Panesci and German 1963). They concluded that the optimum location of the second throat is very critical and depends on the total pressure at the nozzle inlet. Arun Kumar and Rajesh (2016, 2018) performed both experimental and computational studies on zero secondary flow ejectors using rectangular and round ejector geometries to investigate the vacuum generation process. They reported that the recirculation bubbles in the annular gap between the nozzle and the diffuser wall play a major role in determining the vacuum pressure generation process. The past studies show that the flow and performance characteristics of the ejector diffuser system used in high-altitude testing tunnels have been well understood.

However, a key issue with the ejector diffuser system that has not been addressed well in the past is in improving the cooling effectiveness of the diffuser inner wall. It should be noted that the walls of the converging part and the second throat of the diffuser are subjected to enormous thermal load due to the impingement of rocket plume gas at extremely high temperatures. As a result, proper cooling should be ensured to restrict the temperature rise within the limits of safe operation. At present, most of the working tunnels employ a jacket cooling technique and the major drawback of such a cooling system is the enormous mass flow requirement of the coolant and high power required for operating heavy pumps/pressurized systems. Hence it is essential to consider alternate cooling techniques.

Film cooling is one of the potential alternatives for diffuser wall cooling in HAT facilities. In film cooling, a secondary fluid (in gaseous or in liquid state) is injected into the inner wall side of the diffuser, which produces a thin fluid layer attached to the diffuser inner wall. This fluid layer acts like an insulating shield between the high-temperature core flow and the inner wall surface. Goldstein et al. (1966) conducted one of the first experimental studies on tangential film cooling on a flat plate with a supersonic main flow. They reported that the film cooling effectiveness was nearly equal to one close to the injection location for both the subsonic and supersonic cases. However, for the supersonic case, the film cooling effectiveness of unity can be observed for a longer downstream distance compared to the subsonic case. Zakkay et al. (1970) performed an experimental study on tangential injection and concluded that injecting a secondary supersonic stream resulted in superior film cooling than injecting a secondary subsonic stream into the supersonic mainstream. Shan et al. (2018) numerically compared the film cooling effectiveness for both subsonic and supersonic secondary fluid flow for different mass flux. They observed that the mass flux ratio was the dominating factor for both subsonic and supersonic flow. However, for the same blowing ratio, the supersonic secondary flow proved to be more effective when compared with the subsonic secondary flow. Laganelli (1970) analytically calculated the film cooling effectiveness and found that to enhance the film cooling effectiveness the film cooling technique can be employed locally and can be injected through multiple injection ports rather than a single injection port. Heufer and Olivier (2008) investigated film cooling on a flat plate inclined at 30° in a supersonic flow both experimentally and numerically and observed that increasing the mass flux ratio above a critical value result in transition of the boundary layer from laminar to turbulent decreasing the film cooling effectiveness.

The coolant injection in high altitude testing facility will be more complicated compared to conventional film cooling because the film injection is subjected to an adverse pressure gradient prevailing in the diffuser. Moreover, the film characteristics will be more complicated in a supersonic ejector diffuser due to the shock wave interaction with the film. Hence, the film flow characteristics and cooling effectiveness is expected to show deviations compared to the well-investigated film cooling in uniform supersonic flows. The present study, thus, aims to investigate the cooling effectiveness and flow characteristics of the second throat ejector diffuser with the film cooling technique using three different injection configurations. The film cooling effectiveness with various mass flow rate ratios was also investigated.

2 Computational Details

The computational domain of the axisymmetric second throat ejector diffuser model consists of a converging–diverging nozzle, a flow straightening duct, a convergent supersonic diffuser, a second throat, and a diverging subsonic diffuser. In the present study, the effectiveness of film cooling using three distinct coolant injection geometries (Normal, 45° Inclined, and Tangential) were investigated as shown in Fig. 2. The geometric details of the second throat ejector diffuser are mentioned in Table 1. The computational domain is discretized into quadrilateral cells for performing the numerical simulation. The flow field is simulated by solving the axisymmetric Reynolds averaged Navier–Stokes equation using the commercial CFD solver ANSYS Fluent 2019 R2. The governing equations in the conservative form are considered for the present simulation and are solved in a coupled manner. The ideal gas equation is used to represent the state equation. The governing equations are discretized using the finite volume method. The cell centre values are extrapolated to the face centres using the second-order scheme. The fluxes crossing the boundaries are computed using the Roe scheme. The time integration was carried out using a second-order implicit scheme. The turbulence is modelled using the shear stress transport (SST) k-ω turbulence model. Separate transport equations for the turbulent kinetic energy and dissipation rate are solved along with the conservation equations of mass, momentum, and energy for predicting the flow field.

A pressure inlet boundary condition has been used at the inlet of the converging–diverging nozzle with the total pressure fixed to be 6 bar absolute which is the started mode pressure required for the present ejector diffuser configuration. The total temperature at the nozzle inlet was fixed to 1500 K. A pressure outlet boundary condition has been used at the exit of the subsonic diffuser with a pressure value fixed to 1 bar absolute. For secondary flow or coolant injection, pressure inlet boundary condition has been employed. The stagnation pressure and the static pressure have been varied to achieve the required coolant mass flow rates. Various cases with different mass flow rate ratios have been simulated for the present study as shown in Table 2. In the current study, the mass flow rate ratio has been determined using the below Eq. 1,

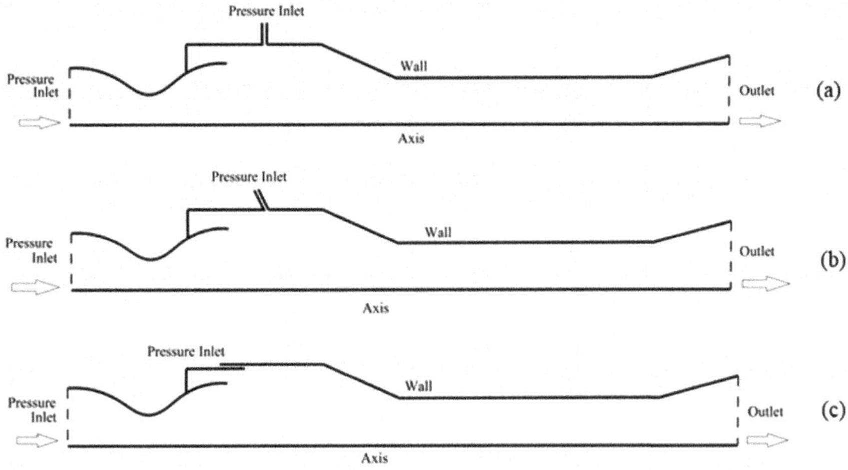

Fig. 2 Boundary conditions used in the computation study of second throat ejector diffuser **a** normal injection **b** inclined injection **c** tangential injection

Table 1 Geometric details of the second throat ejector diffuser

Parameter	Location	Value
D_1	First throat (C-D Nozzle throat)	33.34 mm
D_2	Before second throat	66.68 mm
D_3	Second throat	60 mm
D_4	Subsonic diffuser at exit	73.63 mm
L_{st}/D_3	Second throat	8
L_d	Before second throat	60 mm
L_c	Second throat	63.73 mm
L_s	Subsonic diffuser	300 mm
L_{inj}	Coolant injection	2 mm

$$\text{Mass Flow Rate Ratio (MFR)} = \frac{\dot{m}_c}{\dot{m}_p} \tag{1}$$

$\dot{m}_c$ represents the mass flow rate of the coolant flow.
$\dot{m}_p$ represents the mass flow rate of the primary flow.

The mass flow rate ratio in the present study has been varied by changing the secondary injection mass flow rate (0.02, 0.05, 0.07, and 0.1 kg/s) with a constant primary jet mass flow rate (0.54 kg/s) and total temperature of 1500 K.

Table 2 Computational details of coolant injection

Type of injection	Coolant mass flow rate (kg/s)	Mass flow rate ratio ($\dot{m}_c/\dot{m}_p$)
Normal injection	0.02	0.04
	0.05	0.09
	0.07	0.13
	0.1	0.19
Inclined (45°) injection	0.02	0.04
	0.05	0.09
	0.07	0.13
	0.1	0.19
Tangential injection	0.02	0.04
	0.05	0.09
	0.07	0.13
	0.1	0.19

3 Mesh Independence Study and Validation

A mesh independence analysis was carried out with various grid sizes to determine an ideal mesh size for carrying out further computational studies. Three different mesh systems having 60,000, 150,000 and 240,000 element count have been considered for the mesh independence analysis. Figure 3 shows the wall temperature variation along the diffuser wall for the three different mesh systems. Figure 3 reveals that the wall temperature is unaffected by the mesh refinement beyond a mesh size of 150,000. As a result, a mesh with 150,000 elements is used for further studies. The current computational model has been validated with an experimental study done by Aupoix et al. (1998). The experimental study investigated the wall temperature variation with tangential film injection in a supersonic primary jet.

Figure 4 shows the validation plot in which the temperature variation along the wall predicted from the simulation has been compared with the experimentally measured values. For the numerical simulation, the wall temperature is seen to be constant up to some distance indicating that the film is attached to the wall surface before the breaking of the film, whereas for the experimental study, temperature initially reduces and thereafter gradually increases. The deviation in computational prediction might be due to the turbulence modelling approach used in the numerical model that might lead to slight inaccuracy in the prediction of thermal and momentum diffusion characteristics. This deviation can also be attributed to the fact that the numerical model used for the validation case is a two-dimensional model which will not be able to address the three-dimensional effects present in the actual experiment. Nevertheless, the wall temperature prediction from the numerical simulation fairly agrees well with the experimental measurement.

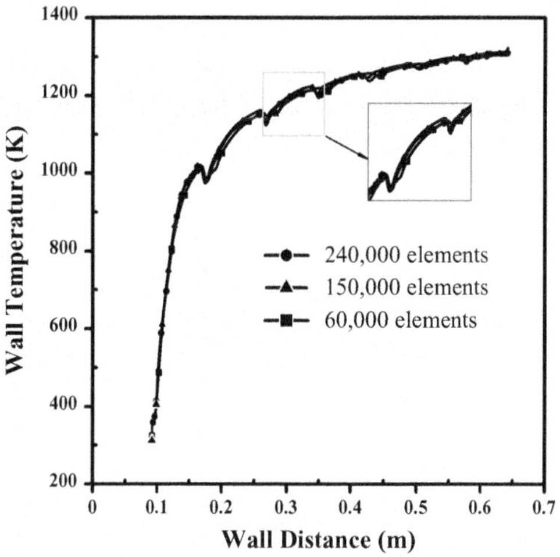

Fig. 3 Wall temperature variation along the diffuser wall for various mesh sizes

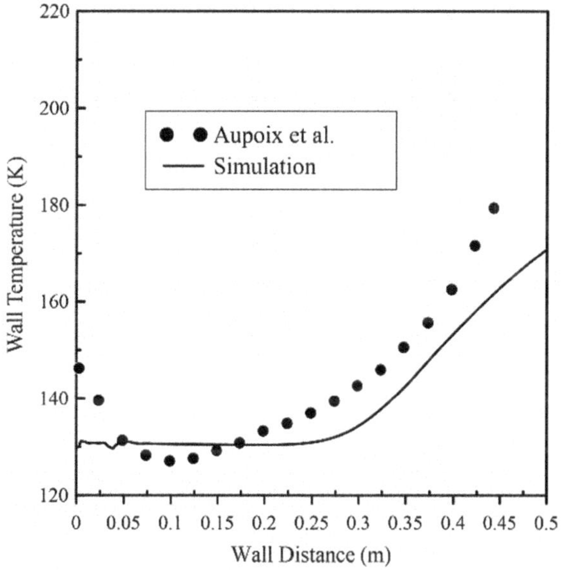

Fig. 4 Wall temperature comparison of CFD study and the experimental study by Aupoix et al. (1998)

4 Results and Discussion

4.1 Normal Injection with Various Mass Flow Rate Ratios

The effect of the coolant injection mass flow rate (varying from 0.02 to 0.1 kg/s) on the flow physics has been investigated with a primary mass flow rate, $\dot{m}_p = 0.54$ kg/s and a Mach number of 1.8 at the nozzle exit. Figure 5 shows the

Mach number contours with various coolant injection mass flow rates (Fig. 5b–e). It is clearly seen from Fig. 5a that the diffuser is started with the supersonic flow attaching to the diffuser wall. The impingement of supersonic jet with the diffuser wall results in the formation of shock wave which in turn reflects from the symmetry line and results in a shock reflection structure which is referred as the "primary shock wave reflection" (PSWR). This shock wave reflection from the symmetry line is found to be a Mach reflection structure with three distinct shock discontinuities, the incident shock wave, the reflected shock wave, and the Mach stem marked as i, r and m respectively in Fig. 5a. It is seen from Fig. 5b–e that the coolant flow injection to the core supersonic flow results in the formation of additional secondary shock front at the injection location. The strength of the secondary shock front increases with an increase in coolant mass flow rate as clearly evident from Fig. 5b–e. This is obvious since the increase in coolant mass flow rate results in more penetration of the secondary jet into the core supersonic jet and produce a more prominent shock structure. The oblique shock wave produced by the coolant jet injection (marked as 'i_s' in Fig. 5e) interacts with the reflected shock wave from the primary jet impingement shock structure (marked as 'r' in Fig. 5e) and results in further shock reflection patterns. The secondary jet injection thus results in the formation of 'secondary shock wave reflections' (SSWR) downstream to the primary shock wave reflection produced by the primary jet interaction with diffuser wall. This can be clearly seen in the numerical schlieren images shown in Fig. 6, where the shock reflections in the diffuser with and without the secondary injections are shown.

Figure 7 shows the comparison of temperature variation along the diffuser wall for normal injection with different secondary coolant injection mass flow rates. It is clearly seen from Fig. 7 that without any coolant injection (MFR = 0), the wall temperature in the straight portion of the diffuser reaches to an average value of 1500 K. It is found that with the coolant injection, the average wall temperature in the straight diffuser portion reduces and reduces further with increase in secondary coolant injection mass flow rate. The average straight diffuser wall temperature is found to reduce nearly 30% with an injection mass flow rate of 0.1 kg/s (MFR = 0.19) when compared to the no injection case. This is clearly evident from Fig. 7.

It is obvious that the wall temperature reduces with the secondary fluid injection due to the shielding effect produced by the coolant gas. It is also observed from Fig. 7 that there is a localized decrease and increase in wall temperature near the merging portion of the convergent diffuser with the straight diffuser. The localized decrease can be attributed to the fact that the reflected shock wave in the core jet interacts with the diffuser wall and cause an adverse pressure gradient. This adverse pressure gradient results in an incipient point for flow separation near the second throat region. A qualitative representation of this is shown in Fig. 8, which depicts the temperature contours with various coolant injection mass flow rates. From Fig. 8, it is seen that with increase in coolant injection mass flow rate, the downstream extent to which the cooling can be effectively felt also increases. This is due to the formation of a thicker film with increase in mass flow rate and thereby enhancing the shielding effect, as seen in Fig. 7.

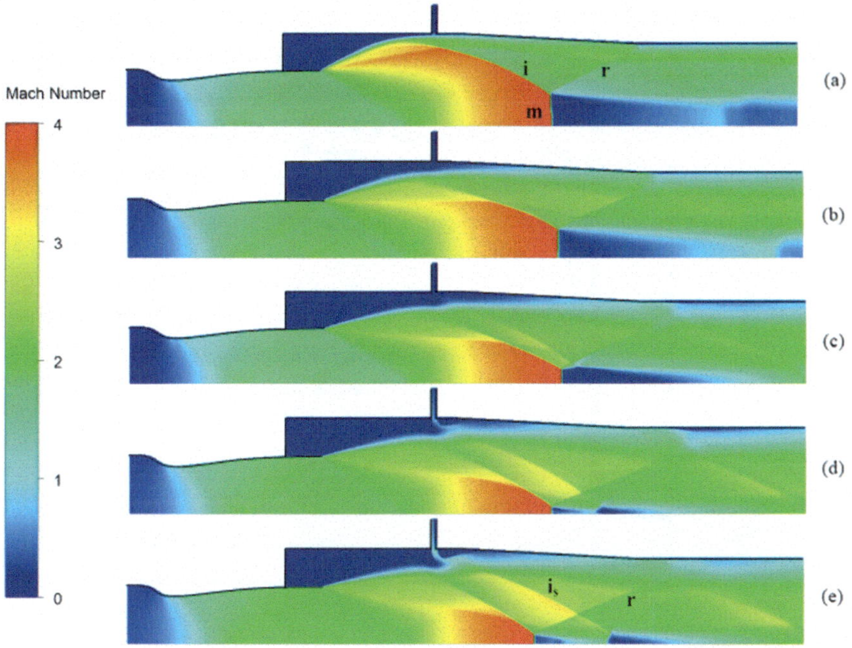

Fig. 5 Mach contour comparison for different coolant mass flow rates in normal injection configuration. **a** $\dot{m}_p = 0$ kg/s **b** $\dot{m}_p = 0.02$ kg/s **c** $\dot{m}_p = 0.05$ kg/s **d** $\dot{m}_p = 0.07$ kg/s **e** $\dot{m}_p = 0.1$ kg/s

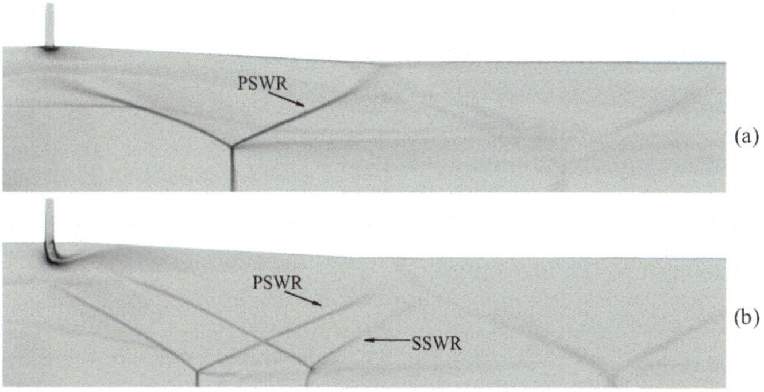

Fig. 6 Numerical schlieren comparison for with and without coolant injection in normal injection configuration **a** $\dot{m}_c = 0$ kg/s **b** $\dot{m}_c = 0.1$ kg/s

The film cooling effectiveness (η) is one of the most important parameters in any film cooling investigation. The effectiveness of film cooling has been determined using the equation below,

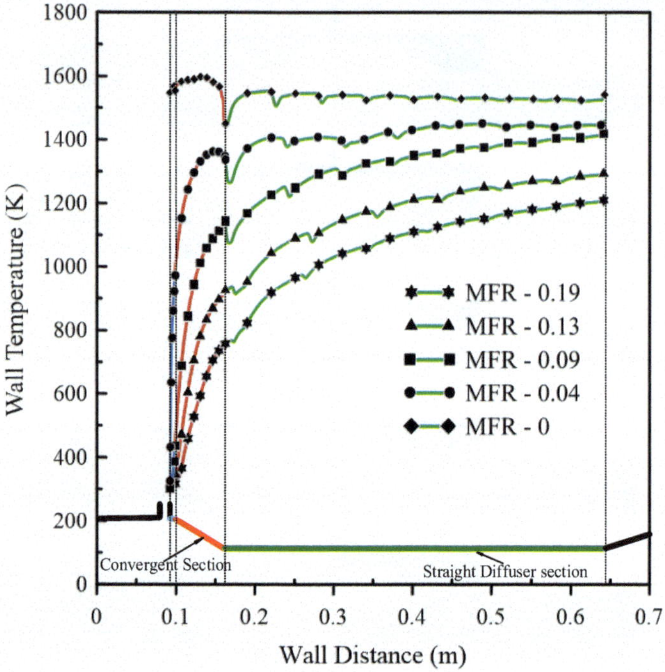

Fig. 7 Temperature variation along the diffuser wall for various mass flow rate ratios in normal coolant injection

$$\eta = \frac{(T_{o,p} - T_{wall})}{(T_{o,p} - T_{o,c})} \tag{2}$$

where, $T_{o,p}$ is the total temperature at the primary inlet

$T_{o,c}$ is the total temperature at the coolant inlet.

T_{wall} is the adiabatic wall temperature.

Figure 9 shows the variation of film cooling effectiveness with different mass flow rate ratio cases along the diffuser length. The effectiveness at the location just after the coolant injection is nearly equal to one as the temperature of the wall is nearly equal to the coolant temperature. As the axial distance increases, the coolant fluid absorbs heat from the high-temperature core flow. As a result of this, the coolant temperature increases which in turn increases the wall temperature and this results in the decrease in effectiveness as the axial distance increases. The drop in effectiveness is more rapid for the coolant injection mass flow rate of 0.02 kg/s when compared with the coolant injection mass flow rate of 0.1 kg/s.

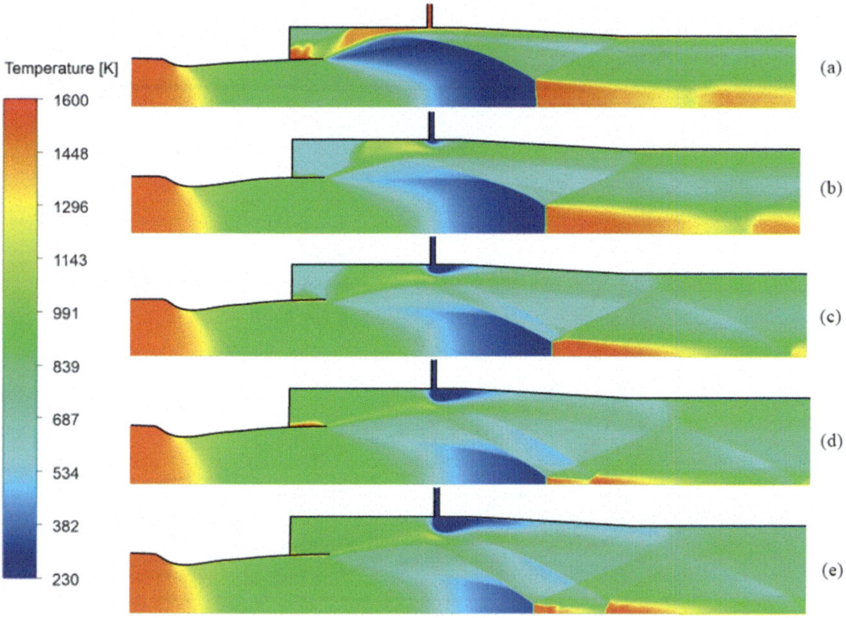

Fig. 8 Temperature contour comparison for different coolant mass flow rates in normal injection configuration. **a** $\dot{m}_c = 0$ kg/s **b** $\dot{m}_c = 0.02$ kg/s **c** $\dot{m}_c = 0.05$ kg/s **d** $\dot{m}_c = 0.07$ kg/s **e** $\dot{m}_c = 0.1$ kg/s

4.2 Inclined Injection with Various Mass Flow Rate Ratio

In this section, the flow characteristics and cooling effectiveness with a 45° inclined injection configuration has been investigated. Figure 10 shows the Mach number contours for various cases with different coolant mass flow rates. Similar to the normal injection, it is observed that a secondary shock structure occurs due to the penetration of secondary jet into the supersonic core flow. As the coolant mass flow rate increases, the strength of the secondary shock structure increases and results in additional shock reflection pattern other than that of the primary shock reflection produced by the underexpanded core jet interaction with the diffuser wall. The strength of the additional shock structures depends on the penetration of coolant flow which is in turn dependent on injection configuration. The inclined injection results in a lower penetration compared to the normal injection for the same coolant mass flow rate. This is due to the fact that, for the inclined injection the total momentum is split between the tangential and normal direction, whereas, for the normal injection the entire momentum is only in the normal direction. Figure 11 shows the effectiveness of different cases with various coolant mass flow rate along the diffuser wall. From Fig. 11, it can be inferred that, as the coolant mass flow rate increases the effectiveness also increases.

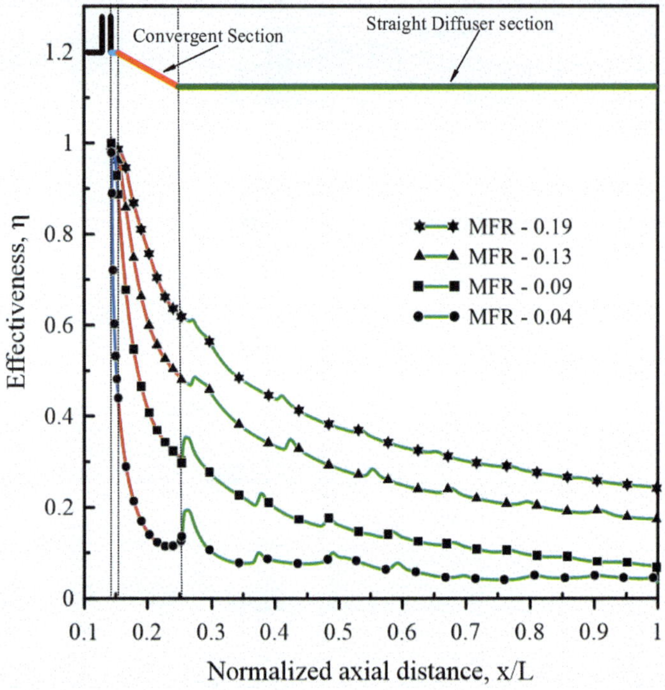

Fig. 9 Effectiveness plot for normal coolant injection

4.3 Tangential Injection with Various Mass Flow Rate Ratio

In this section, the effect of the coolant injection mass flow rate with tangential injection configuration on the core flow physics and cooling effectiveness have been studied. Figure 12 depicts the Mach number contours for different coolant injection mass flow rates. For the tangential injection case, the entire injection momentum is in the tangential direction and hence the injected jet is not penetrating into the core jet due to the absence of normal momentum. As a result, the secondary shock structure due to the coolant injection seen in the normal and inclined injection is not prominent for the case of tangential injection and this is clearly evident from Fig. 12.

Figure 13 shows the cooling effectiveness along the diffuser length for various tangential injection cases with different injection mass flow rate. From Fig. 13, it can be seen that the cooling effectiveness increases with increase in coolant injection mass flow rate.

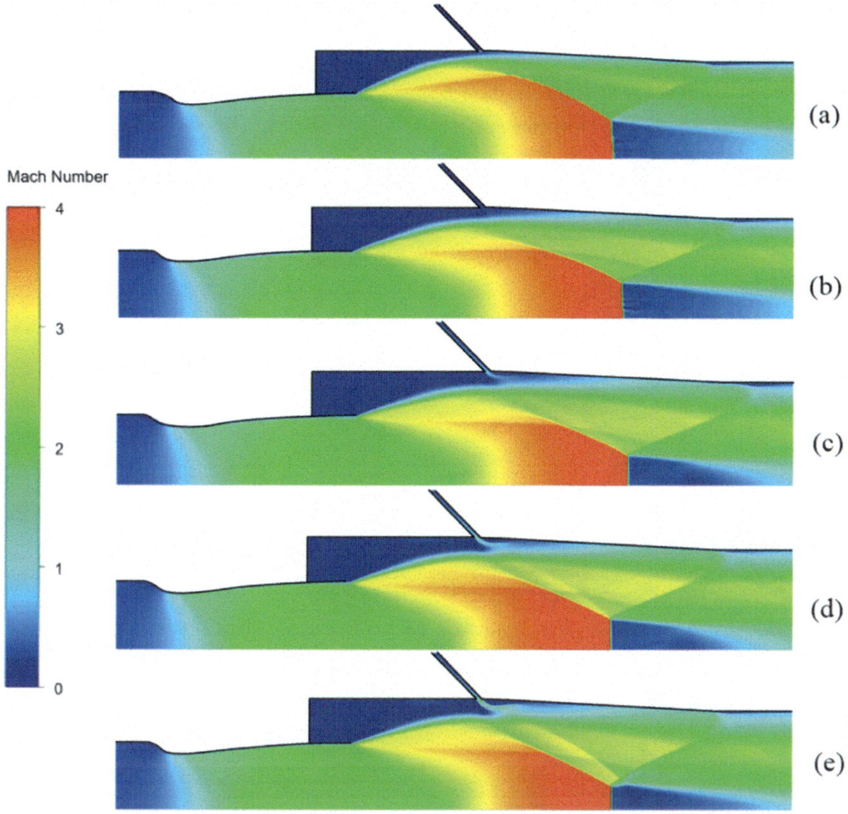

Fig. 10 Mach contour comparison for different coolant mass flow rates in inclined injection configuration. **a** $\dot{m}_c = 0$ kg/s **b** $\dot{m}_c = 0.02$ kg/s **c** $\dot{m}_c = 0.05$ kg/s **d** $\dot{m}_c = 0.07$ kg/s **e** $\dot{m}_c = 0.1$ kg/s

4.4 Comparative Study of Different Coolant Injection Configurations

In this section, a comparative study of the flow and cooling characteristics produced by various injection configurations have been reported. Figure 14 shows the numerical schlieren for three different injection configurations (normal, inclined, and tangential) with same injection mass flow rate (0.1 kg/s). It is seen from Fig. 14 that the additional shock structure due to coolant injection is clearly visible for the normal and inclined injection cases, whereas, for the tangential injection the additional shock structure is not observed. In order to compare the cooling characteristics with various injection configurations, the diffuser wall temperature data is extracted to calculate the film cooling effectiveness and has been compared for different injection configurations with coolant injection mass flow rate of 0.1 kg/s and is shown in Fig. 15. It is seen from Fig. 15 that the effectiveness at any section in the diffuser

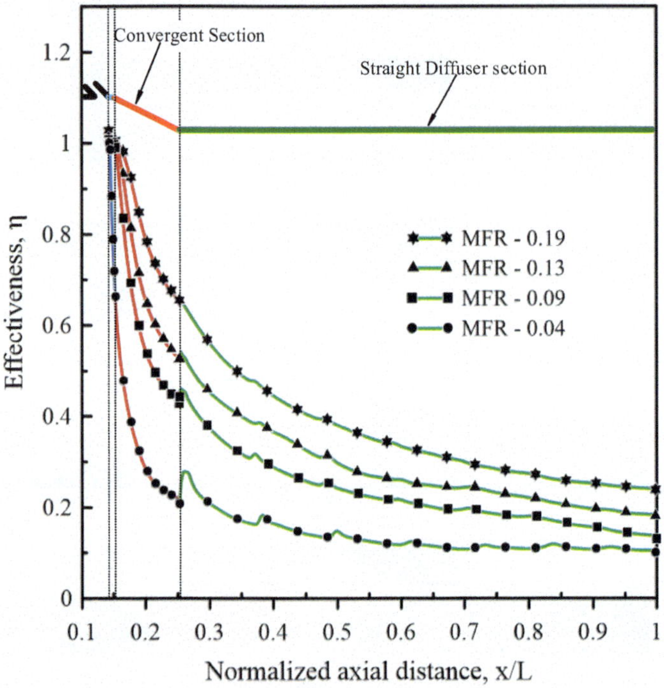

Fig. 11 Effectiveness plot for inclined coolant injection

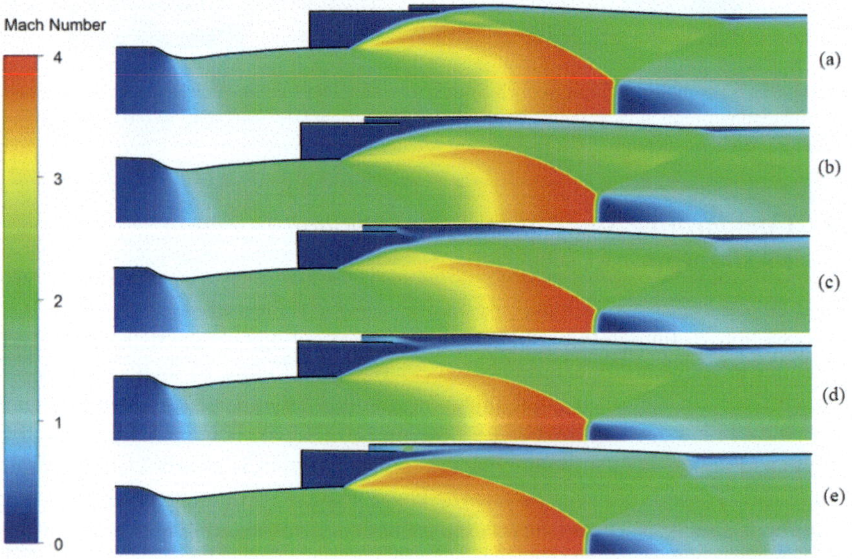

Fig. 12 Mach contour comparison for different coolant mass flow rates in tangential injection configuration. **a** $\dot{m}_c = 0$ kg/s **b** $\dot{m}_c = 0.02$ kg/s **c** $\dot{m}_c = 0.05$ kg/s **d** $\dot{m}_c = 0.07$ kg/s **e** $\dot{m}_c = 0.1$ kg/s

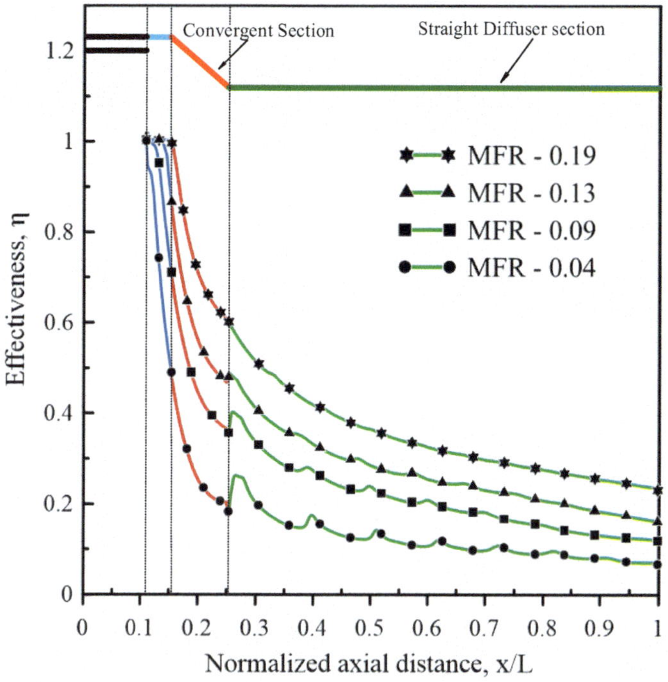

Fig. 13 Effectiveness plot for tangential coolant injection

is almost the same for the normal and inclined injection configurations. However, the effectiveness at any section in the diffuser is lower for the tangential injection compared to the other two configurations. In the far downstream of the diffuser all injection configurations show the same effectiveness distribution. The higher effectiveness with normal and inclined injection compared to tangential injection can be attributed to the larger penetration and the subsequent formation of a thicker film.

5 Conclusions

Computational investigation on the film cooling characteristics in second throat ejector-diffuser have been carried out in the present study. The flow characteristics and cooling effectiveness with various injection configurations, such as, normal, 45° inclined and tangential injection and with varying coolant mass flow rates have been studied in detail. It was observed that the coolant injection results in the formation of additional secondary shock reflection structures for normal and inclined injection cases. However, the secondary shock wave reflection structures are not seen for the tangential injection case. The formation of the secondary shock wave in the normal and inclined injection cases can be attributed to the penetration of the coolant jet into

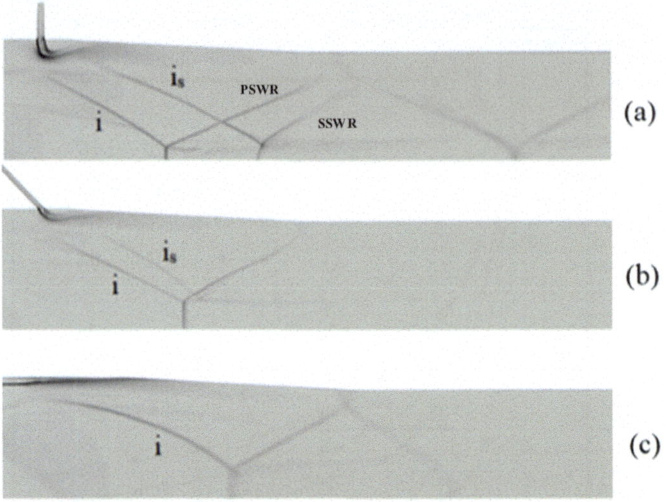

Fig. 14 Numerical schlieren comparison for different coolant injection configurations for $\dot{m}_c$= 0.1 kg/s. **a** normal injection **b** inclined injection **c** tangential Injection

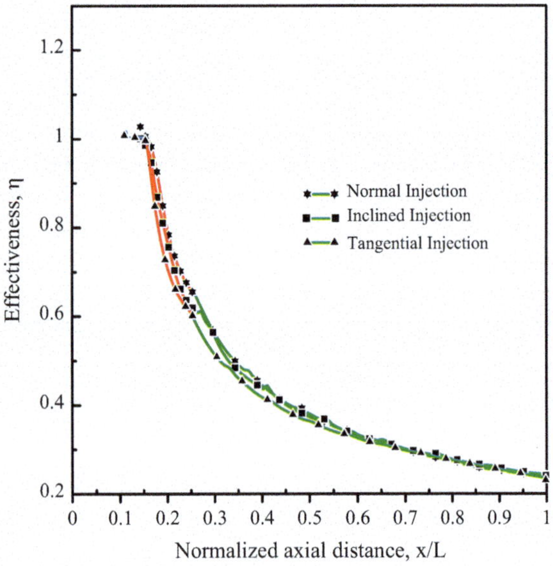

Fig. 15 Comparison of effectiveness for different injection configurations for coolant mass flow rate, $\dot{m}_c$= 0.1 kg/s

the primary supersonic ejector-diffuser flow. Further, the film cooling effectiveness has been evaluated for all the injection configurations and with change in coolant mass flow rates. It was observed that the film cooling effectiveness increases with increase in mass flow rate ratio for all the cases and beyond an optimum value of mass flow rate, the percentage increment in cooling effectiveness decreases. The film cooling effectiveness was found to be maximum for the normal coolant injection configuration followed by inclined injection and film cooling effectiveness was minimum for the tangential injection. It is also seen that the diffuser wall attains a particular temperature at a larger axial distance for the normal injection followed by inclined injection and tangential injection. This implies that the effective cooling length is higher for normal injection followed by inclined injection and tangential injection.

References

Amatucci VA, Addy AL, Dutton JC. Pressure recovery in a constant-area, two-stream supersonic diffuser. AIAA J 1982; 20(9):1308-9.

Arun Kumar R, Rajesh G., Flow Transients in un-started and started modes of vacuum ejector operation, Phys. Fluids, 2016 26; 28(5):056105.

Arun Kumar R, Rajesh G., Physics of vacuum generation in zero-secondary flow ejectors, Phys. Fluids 2018 25;30(6):066102.

Aupoix B., Mignosi A., Viala S., et al., Experimental and Numerical Study of Supersonic Film Cooling, AIAA J 1998; Vol 36, No.6.

Bauer RC, German RC., "The effect of second throat geometry on the performance of ejectors without induced flow", ARNOLD ENGINEERING DEVELOPMENT CENTER ARNOLD AFB TN; 1961 Nov 1.

Ferri, A, and Louis MN, "Theoretical and Experimental Analysis of Low-Drag Supersonic Inlets Having a Circular Cross Section and a Central Body at Mach Numbers 3.30, 2.75, and 2.45", 1954, No. NACA-TR-1189.

German RC, Panesci JH., "Improved Methods for Determining Second-Throat Diffuser Performance of Zero-Secondary-Flow Ejector Systems", ARNOLD ENGINEERING DEVELOPMENT CENTER ARNOLD AFB TN; Jul 1, 1965.

German RC, Panesci JH, Clark HK.,"Zero secondary flow ejector-diffuser performance using annular nozzles", ARO INC ARNOLD AFS TN; Jan 1, 1963.

Goldstein RJ, Eckert ERG, Tsou FK, et al., Film Cooling with Air and Helium Injection through a Rearward-Facing Slot into a Supersonic Air Flow, AIAA J 1966; Vol. 4, No. 6, pp. 981-985.

Heufer, KA, and Olivier H., Experimental and numerical study of cooling gas injection in laminar supersonic flow, AIAA J, 2008; 46, no. 11: 2741-2751.

A. Laganelli, A comparison between film cooling and transpiration cooling systems in high-speed flow, AIAA J 1970, No. 70–153.

Paul F. Massier and E. John Roschke, Experimental Investigation of Exhaust Diffusers for Rocket Engines, AIAA J.1960.

J. H. Panesci and R.C. German, "An Analysis of Second-Throat Diffuser Performance for Zero-Secondary-Flow Ejector Systems", AEDC TDR-63–249, Arnold Air Force Station, Tenn, October 1963.

Park BH, Lim J, Park S, et al., Design and analysis of a second-throat exhaust diffuser for altitude simulation, J.Propuls. Power. 2012;28(5):1091-104.

Yong Shan, Xiao-ming Tan, Jing-zhou Zhang et al., Numerical Study on Flow and Cooling Characteristics for Supersonic Film Cooling, Heat Transf. Eng. 2018, 39:13-14, 1318-1330.

Zakkay V, Sakell L, and Parthasarathy K, An Experimental Investigation of Supersonic Slot Cooling, Proceedings of 1970 Heat Transfer and Fluid Mechanics Institute, Ca. pp. 88–103.

Strouhal Number Variation in Multi-phase Flow Past a Cylinder

Shesh Narayan Dhurandhar, Annie Rose Elizabeth, and T. Jayachandran

Abstract Air-water two-phase flow over a cylinder is investigated numerically. The two-dimensional simulation using a six-equation formulation is performed. The Reynolds number (Re_∞) of the flow is considered to be 100 and 150 for Mach number (Ma_∞) of 0.1. The droplet size is assumed as 100μm in diameter, and the droplet volume fraction is varied from 0 to 1%. The results show that increasing the droplet fraction significantly influences the flow regimes. The Strouhal number, lift coefficient, and drag coefficient are affected by the droplet fraction present in the flow.

Keywords Multiphase flow · Compressible flow · Wake

1 Introduction

Flow past a cylinder is a canonical setup to study many bluff body dynamics in fluid mechanics. Flow past a cylinder replicate a moving maritime platform, cross-flow analysis in heat exchangers (Kumar et al. 2016), airflow around the buildings, chimneys, cooling towers, etc. There has been extensive studies on the variation aspects of flow past cylinder for a large range of Reynolds number (Re_∞) (Taneda 1956; Coutanceau and Bouard 1977; Fornberg 1980). The effect of compressibility also took a lot of attention because of its applications in aircraft design, high-speed vehicles, wind tunnel testing, etc. (Karagiozis et al. 2010; Canuto and Taira 2015). The study of flow past a heated cylinder holds significant relevance for various applications, including heat exchangers, chimney stacks, and hot-wire anemometers (Kumar et al. 2016).

The present study examines the frequency variation of shedding due to the multi-phase composition of the oncoming flow over the cylinder. The fluid is assumed to be a mixture of gas and water droplets. The compressible multifluid formulation is used

S. N. Dhurandhar (✉) · A. R. Elizabeth · T. Jayachandran
Department of Aerospace Engineering, Indian Institute of Technology Madras, Chennai, Tamil Nadu, India
e-mail: sheshdhurandhar@gmail.com

G. Rajesh et al. (eds.), *Proceedings of the 15th International Symposium on Experimental and Computational Aerothermodynamics of Internal Flows*,
https://doi.org/10.1007/978-981-96-4082-9_25

to simulate multi-phase flow, wherein both phases are assumed compressible with the suitable equations of state for both phases. The two-phase fluid is used, where one phase is gas, and the other is water droplets. The analysis requires two additional parameters to study the two-phase flow over a bluff body (Yeom et al. 2012). The flow depends on the volume fraction (ϕ_p) and the size of the water droplets in the mixture.

2 Governing Equations

A compressible flow past a cylinder is characterized by Re_∞ and Ma_∞ as:

$$Re_\infty = \frac{\rho_{\infty,g} U_{\infty,g} D}{\mu}, \tag{1}$$

$$Ma_\infty = \frac{U_{\infty,g}}{\sqrt{\gamma_g \mathbb{R} T_{\infty,g}}}, \tag{2}$$

where ρ_∞, U_∞, T_∞, D, μ, γ, and $\mathbb{R}$ are free stream density, free stream velocity, free stream temperature, the diameter of the cylinder, dynamic viscosity, the ratio of specific heats, and gas constant, respectively. Subscript g is used for the gas phase. The Strouhal number (St) of the vortex shedding is defined as

$$St = \frac{fD}{U_{\infty,g}}, \tag{3}$$

where f is the shedding frequency. The lift coefficient, C_d, and drag coefficient, C_l, calculated over the surface of the cylinder using

$$C_d = \frac{F_d}{\frac{1}{2}\rho_{\infty,g} {U_{\infty,g}}^2 D}, \tag{4}$$

$$C_l = \frac{F_l}{\frac{1}{2}\rho_{\infty,g} {U_{\infty,g}}^2 D}. \tag{5}$$

Where F_d is the mean drag force, and F_l is the maximum value of lift force achieved over the surface of the cylinder.

In the six-equation model, the pressures in both phases are assumed to be the same. The model is thermodynamically consistent and conditionally hyperbolic (Yeom et al. 2012). The formulation in conservation form is shown below.

$$\frac{\partial U}{\partial t} + \frac{\partial F}{\partial x} + \frac{\partial G}{\partial y} = S; \tag{6}$$

$$
\text{Where, } U = \begin{bmatrix} \phi_g\rho_g \\ \phi_g\rho_g u_g \\ \phi_g\rho_g v_g \\ \phi_g\rho_g E_g \\ \phi_p\rho_p \\ \phi_p\rho_p u_p \\ \phi_p\rho_p v_p \\ \phi_p\rho_p E_p \end{bmatrix}, F = \begin{bmatrix} \phi_g\rho_g u_g \\ \phi_g\rho_g {u_g}^2 + \phi_g P - \phi_g\tau_{xx,g} \\ \phi_g\rho_g u_g v_g - \phi_g\tau_{xy,g} \\ \phi_g u_g(\rho_g E_g + P) - \phi_g(u_g\tau_{xx,g} + v_g\tau_{xy,g}) \\ \phi_p\rho_p u_p \\ \phi_p\rho_p {u_p}^2 + \phi_p P - \phi_p\tau_{xx,p} \\ \phi_p\rho_p u_p v_p - \phi_p\tau_{xy,p} \\ \phi_p u_p(\rho_p E_p + P) - \phi_p(u_p\tau_{xx,p} + v_p\tau_{xy,p}) \end{bmatrix},
$$

$$
G = \begin{bmatrix} \phi_g\rho_g v_g \\ \phi_g\rho_g u_g v_g - \phi_g\tau_{xy,g} \\ \phi_g\rho_g {v_g}^2 + \phi_g P - \phi_g\tau_{yy,g} \\ \phi_g v_g(\rho_g E_g + P) - \phi_g(u_g\tau_{xy,g} + v_g\tau_{yy,g}) \\ \phi_p\rho_p v_p \\ \phi_p\rho_p u_p v_p - \phi_p\tau_{xy,p} \\ \phi_p\rho_p {v_p}^2 + \phi_p P - \phi_p\tau_{yy,p} \\ \phi_p v_p(\rho_p E_p + P) - \phi_p(u_p\tau_{xy,p} + v_p\tau_{yy,p}) \end{bmatrix}, \text{ and}
$$

$$
S = \begin{bmatrix} 0 \\ D_x + P_I\frac{\partial\phi_g}{\partial x} \\ D_y + P_I\frac{\partial\phi_g}{\partial y} \\ U_I D_x + V_I D_y + P_I U_I\frac{\partial\phi_g}{\partial x} + P_I V_I\frac{\partial\phi_g}{\partial y} \\ 0 \\ -D_x - P_I\frac{\partial\phi_g}{\partial x} \\ -D_y - P_I\frac{\partial\phi_g}{\partial y} \\ -U_I D_x - V_I D_y - P_I U_I\frac{\partial\phi_g}{\partial x} - P_I V_I\frac{\partial\phi_g}{\partial y} \end{bmatrix}.
$$

where U, F, G, and S denote the conserved variables, flux in the x direction, flux in the y direction, and the source terms, respectively. The subscripts $g,\ p,\ x$, and y denote the gas phase, liquid/water phase, x-direction, and y-direction. The variables $\phi,\ \rho,\ u,\ v,\ E,\ U_I,\ V_I,\ P_I$ and D represent volume fraction, density, velocity in x direction, velocity in y direction, total energy, interface velocity in x direction, interface velocity in y direction, interface pressure and drag between the two phases, respectively. The interface pressure is defined as (Yeom and Chang 2013)

$$
P_I = P - \delta\frac{\phi_g\phi_p\rho_g\rho_p}{\phi_g\rho_p + \phi_p\rho_g}\left[(u_g - u_p)^2 + (v_g - v_p)^2\right], \tag{7}
$$

where δ is the damping coefficient, and its value 2 for the hyperbolic equation system. The volume fraction of gas and water droplets satisfies $\phi_p + \phi_g = 1$. The unstructured finite volume method with second-order HLLC scheme (Batten et al. 1997) for spatial discretization and eSSPRK(4,3) (Isherwood et al. 2018) for time integration is used as the numerical model.

3 Validation

The computational domain of size $(x, y) \epsilon [-20D, 40D] \times [-20D, 20D]$ is used for the simulation which is shown in Fig 1. A total of 0.15 million tetrahedral computational cells have been used. The O-grid mesh has been used near the cylinder surface. The boundary condition used for the simulation is given in Table 1.

The wave transmissive boundary conditions for velocity and temperature at the outlet are applied as follows:

$$T_{g,f}{}^{n+1} = wT_{g,f}{}^{n} + (1-w)(T_{g,c} + \nabla(T_{g,ref})), \tag{8}$$

$$u_{g,f}{}^{n+1} = wu_{gf}{}^{n} + (1-w)(u_{g,c} + \nabla(u_{g,ref})), \tag{9}$$

$$v_{g,f}{}^{n+1} = wv_{g,f}{}^{n} + (1-w)(v_{g,c} + \nabla(v_{g,ref})), \tag{10}$$

subscript f, c, and ref are the boundary face value, boundary cell value, and reference value, respectively. Superscript n and $n+1$ specify the old-time step and new-time step values, respectively. The fraction value, w and α is calculated from

$$w = \frac{1}{1+\alpha}; \qquad \alpha = \frac{U\Delta t}{\delta_x}. \tag{11}$$

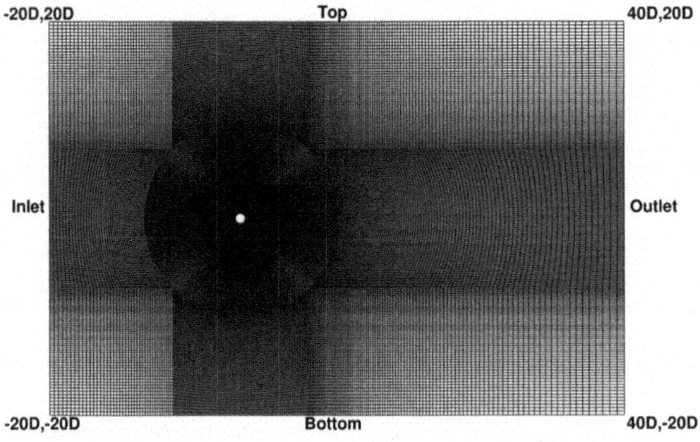

Fig. 1 Computational domain

Table 1 Boundary Conditions for Gas and Droplets

Boundary	Gas	Droplet
Inlet	P_g = zero gradient	P_p = zero gradient
	$T_g = T_{\infty,g}$	$T_p = T_{\infty,g}$
	$u_g = u_{\infty,g}$	$u_p = u_{\infty,g}$
	$v_g = 0$	$v_p = 0$
		$\phi_p = \phi_{\infty,p}$
Outlet	$P_g = P_{\text{amb}}$	$P_p = P_{\text{amb}}$
	$T_g = T_{g,f}{}^{n+1}$	T_p = zero gradient
	$u_g = u_{g,f}{}^{n+1}$	u_p = zero gradient
	$v_g = v_{g,f}{}^{n+1}$	v_p = zero gradient
		ϕ_p = zero gradient
Top/bottom	P_g = zero gradient	P_p = zero gradient
	T_g = zero gradient	T_p = zero gradient
	u_g = zero gradient	u_p = zero gradient
	$v_g = 0$	$v_p = 0$
		ϕ_p = zero gradient
Cylinder wall	P_g = zero gradient	P_p = zero gradient
	T_g = zero gradient	T_p = zero gradient
	$u_g = 0$	$u_p = 0$
	$v_g = 0$	$v_p = 0$
		ϕ_p = zero gradient

where δ_x is the distance between the cell center and the face. The value of advection speed U for the velocity vector perpendicular to the outlet boundary is

$$U = v_{g,f} + \sqrt{\frac{\gamma_g P_{g,f}}{\rho_{g,f}}}, \tag{12}$$

and for temperature and velocity vectors in parallel to the outlet boundary

$$U = v_{g,f}. \tag{13}$$

The solver has been validated by comparing the St with the experimental data and numerical results for $Re_\infty = 100$ and 150 for $Ma_\infty = 0.1$. The cut-off water droplet fraction of 10^{-8} is initially used. The St calculated from the present simulation is 0.164 for $Re = 100$ and $Ma_\infty = 0.1$. The St comparison is present in Table 2. The result shows that the present solver can calculate the St for $Re_\infty = 100$, and 150 accurately.

Table 2 Strouhal number comparison

Re_∞	Ma_∞	St			
		Present simulation	Mittal and Raghuvanshi (2001)	Karagiozis et al. (2010)	Williamson (1996)
100	0.1	0.1640	0.1680	0.1680	0.1634
150	0.1	0.1847	–	–	0.1843

4 Results and Discussion

In the present work, we studied the effect of ϕ_p for $Re_\infty = 100$ and 150 for $Ma_\infty = 0.1$. The droplet diameter (d_p) is fixed at $d_p = 100 \times 10^{-6}$ m, and the droplet volume fraction is varied from 0 to 1%. The flow is simulated for the given set of parameters. To study the dynamic behavior of flow, the velocity magnitude signal is plotted at probe location (x^*, y^*):($1.25D$, $-1.25D$) as shown in Fig. 2. The small amount of the droplet fraction $\phi_p = 0.01\%$ does not influence the velocity signal significantly. However, there is a slight change in the amplitude of the oscillations when ϕ_p increases to 0.1%. A similar observation is made for $Re_\infty = 150$. The flow becomes steady at $\phi_p = 1\%$, and the velocity signal becomes linear and is therefore not shown here.

The effect of droplet droplets on the wake dynamics is analyzed by examining the mean gas velocity along the stagnation line, as illustrated in Fig. 3a. The solid line is used for $Re_\infty = 100$, and the dashed line is used for $Re_\infty = 150$. The figure shows that for $Re_\infty = 100$ and $\phi_p = 0\%$ and 0.01%, the mean velocities of the flows overlap, even after the flow crosses the cylinder. This shows that at lower Re_∞, the

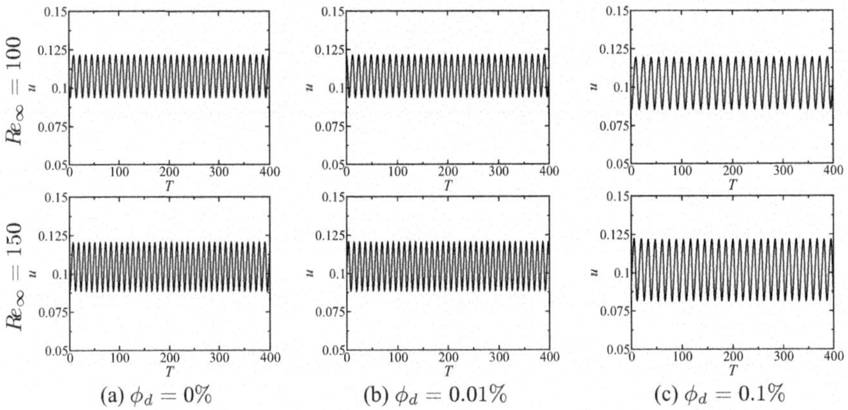

Fig. 2 Velocity magnitude (u) signal at probe location (x^*, y^*):($1.25D$, $-1.25D$) from the center of the cylinder

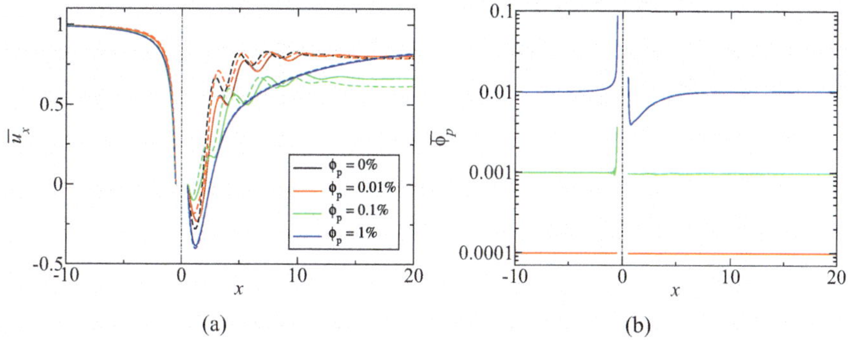

Fig. 3 **a** Mean velocity $\overline{u_x}$ and **b** mean droplet fraction $\overline{\phi_p}$ variation along the center (stagnation) line of the cylinder. Solid line for $Re_\infty = 100$ and dash line for $Re_\infty = 150$. Location $x = 0$ represents the center of the cylinder

flow is not influenced by the less amount of water droplet fraction. However, when the droplet fraction is increased, the mean velocity starts to reduce, and for 1% of the droplet fraction, it shows the trend of the steady flow. A similar observation was made for the $Re_\infty = 150$. However, at this Re_∞, the small amount of the droplet fraction $\phi_p = 0.01\%$ influences the mean velocity. This is due to the increased inertia force of the flow at high Re_∞.

Figure 3b shows the variation of the droplet fraction along the streamwise direction at the center (stagnation) line. It is observed that for $\phi_p = 0.1\%$, the maximum droplet fraction in front of the cylinder is ≈ 4 of the free stream droplets. This is due to the droplets accumulating in front of the cylinder. The droplets increase the pressure difference between the front and back of the cylinder. For the $\phi_p = 1\%$, the maximum droplet concentration in front of the cylinder is around ≈ 9 of the free stream value. Fig. 4 shows the instantaneous shedding from the cylinder. The presence of droplets in the flow reduces the separation from the cylinder. The droplets slow the flow, which influences the separation.

Figure 5 shows the variation of C_d and C_l over the surface of the cylinder. The C_d is found to be increasing with an increase in the ϕ_p and Re_∞. As the ϕ_p increases in the flow, more droplets are accumulating to form a layer over the front surface of the cylinder. That increases the pressure difference between the front and back of the cylinder. Due to this, the pressure gradient increases, which in turn increases the pressure drag. There is a slight increase in the C_d with an increase in the Re_∞ of the flow. The C_l is found to reduce with the increase in the ϕ_p. This reduction in C_l is due to the weakening of vortex shedding with an increase in ϕ_p, which is caused by the damping of flow instabilities as ϕ_p increases. The C_l is found to be reducing with ϕ_p, and its value is zero for $\phi_p = 1\%$.

Figure 6 shows the variation of St with the droplet fraction. The St for the case of $Re_\infty = 100$ is 0.164, and for $Re_\infty = 150$ is 0.1847. It is seen that even a small amount of droplet $\phi_p = 0.01\%$ affects the St of the flow, and it reduces. Further

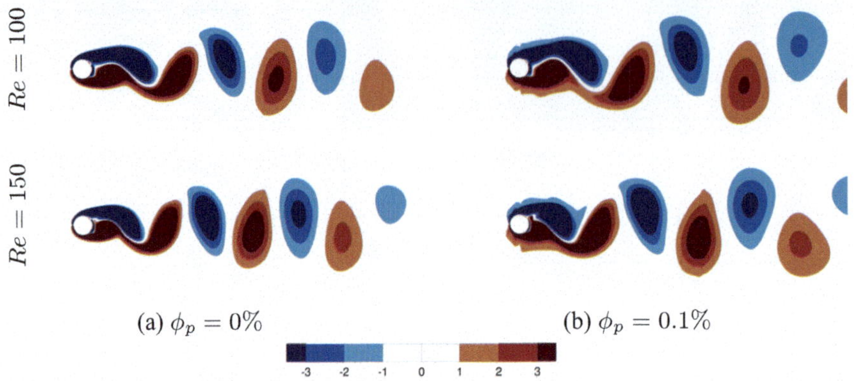

Fig. 4 Comparison of vorticity fields for different droplet fractions

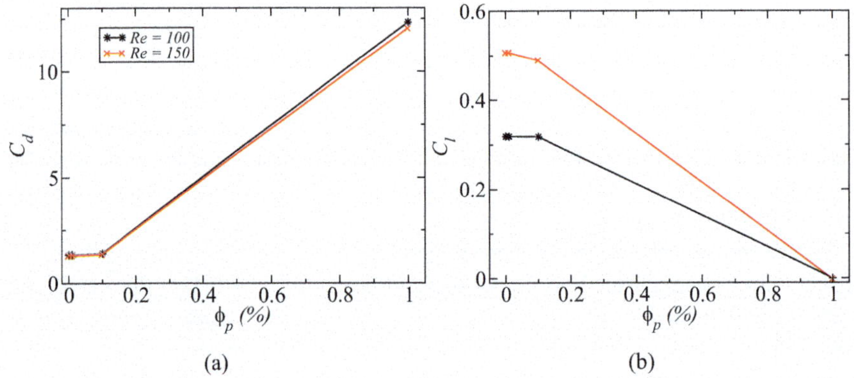

Fig. 5 **a** Drag force coefficient (C_d) and **b** lift force coefficient (C_l) variation with the droplet fraction

Fig. 6 Strouhal number variation with droplet fraction

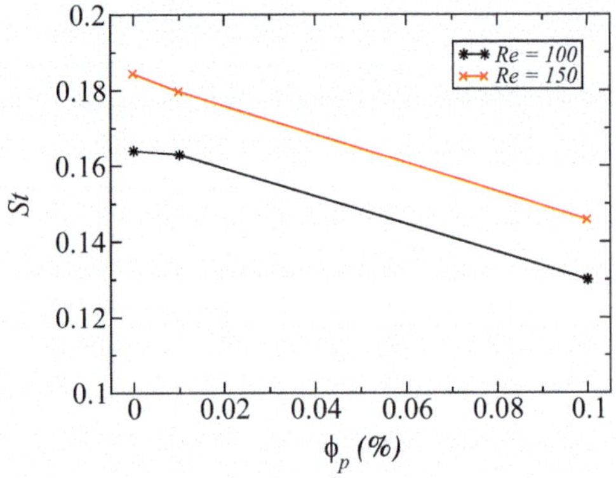

increasing the to $\phi_p = 0.1\%$, significantly reduces the St of the flow, and its value is reduced to 0.13 and 0.146 for $Re_\infty = 100$ and 150, respectively. The reduction in St is attributed to the increase in drag due to slower droplet motion.

5 Conclusions

In this work, air-water two-phase flow past a cylinder is simulated for $Re_\infty = 100$ and 150 for $Ma_\infty = 0.1$. The six-equation model in the compressible multi-phase is used. The ϕ_p is varied from 0 to 1% for droplet diameter of $d_p = 100\mu m$. The cylinder wakes start to disperse when the droplet fraction is increased, and the flow becomes steady when the droplet fraction increases to 1%. Increasing the droplet fraction increases the accumulation of the droplets in the front surface of the cylinder, which increases the drag coefficient. The lift coefficient decreases with the increase in the droplet fraction. The strouhal number of the flow reduces with the increase in the droplet fraction value and increase in the Reynolds number of the flow.

Acknowledgements The authors would like to express their sincere gratitude to Prof. A. Sameen for the assistance with data analysis, simulations, and reviewing the work.

References

Batten, P., Clarke, N., Lambert, C. and Causon, D.M.: On the choice of wavespeeds for the HLLC Riemann solver. SIAM Journal on Scientific Computing, 8(6), 1553–70 (1997). https://doi.org/10.1137/S1064827593260140

Canuto, D., Taira, K.: Two-dimensional compressible viscous flow around a circular cylinder. J. Fluid Mech. 785(1956), 349–371 (2015). https://doi.org/10.1017/jfm.2015.635

Coutanceau, M., Bouard, R.: Experimental determination of the main features of the viscous flow in the wake of a circular cylinder in uniform translation. Part 1. Steady flow. J. Fluid Mech. 79(2), 231–256 (1977). https://doi.org/10.1017/S0022112077000135

Fornberg, B.: A numerical study of steady viscous flow past a circular cylinder. J. Fluid Mech. 98(4), 819–855 (1980). https://doi.org/10.1017/S0022112080000419

Isherwood, L., Grant, Z.J. and Gottlieb, S.: Strong stability preserving integrating factor Runge–Kutta methods. SIAM Journal on Numerical Analysis, 56(6), 3276–307 (2018). https://doi.org/10.1137/17M1143290

Karagiozis, K., Kamakoti, R., Pantano, C.: A low numerical dissipation immersed interface method for the compressible Navier-Stokes equations. Comput. Phys. 229(3), 701–727 (2010). https://doi.org/10.1016/j.jcp.2009.10.005

Kumar, S. A., Mathur, M., Sameen, A., Lal, S. A.: Effects of Prandtl number on the laminar cross flow past a heated cylinder. Phys. Fluids 28(11), 2016. https://doi.org/10.1063/1.4966937. https://doi.org/10.1063/1.4966937

Mittal, S., Raghuvanshi, A.: Control of vortex shedding behind circular cylinder for flows at low Reynolds numbers. Int. J. Numer. Methods Fluids 35, 421–447 (2001). https://doi.org/10.1002/1097-0363(20010228)35

Taneda, S.: Experimental Investigation of the Wakes behind Cylinders and Plates at Low Reynolds Numbers. J. Phys. Soc. Jpn. 11(3), 302–307 (1956). https://doi.org/10.1143/JPSJ.11.302

Williamson, C. H. K.: Three-dimensional wake transition. J. Fluid Mech. 328, 345–407 (1996). https://doi.org/10.1017/S0022112096008750

Yeom, M., Chang, K., Baek, S.: Numerical prediction of airfoil characteristics in a transonic droplet-laden air flow. Int. J. Heat Mass Transfer 55(1-3), 453–469 (2012). https://doi.org/10.1016/j.ijheatmasstransfer.2011.09.045

Yeom G. and Chang K.: A modified HLLC-type Riemann solver for the compressible six-equation two-fluid model. Comput. Fluids 76, 86–104 (2013). https://doi.org/10.1016/j.compfluid.2013.01.021

Numerical Simulation on Coaxial Injector Using a Coupled Volume of Fluid and Discrete Phase Model

P. Shyam Raj, P. Parthasarathy, and Ganesh Paramasivan

Abstract This research paper uses the method of the coupled volume of fluid and discrete phase model (coupled VOF-DPM) to analyse liquid-gas coaxial injector. The gas-liquid interface is resolved using a local solution-adaptive mesh refinement technique. The study is to evaluate the effect of the mass flow ratio of gas and liquid on the droplet velocity and Sauter mean diameter (SMD) at different injection times. Simulations were conducted using a coaxial injector with an airflow rate of 9.852×10^{-4} kg/s and three different water flow rates: 1.254×10^{-3} kg/s, 9.408×10^{-3} kg/s, and 3.136×10^{-2} kg/s. The results show that droplet size generally increases over time at all the mass flow rates. The size of droplets also increases with an increase in water flow rate. However, the overall size of droplets decreases as they move from the injection plane to the outlet. The number of droplets formed also increases with a higher water flow rate. Droplet velocity increases with a decrease in the mass flow rate of water.

Keywords Coaxial injector · VOF to DPM · Atomization · Sauter mean diameter

1 Introduction

An effective injector design for a liquid rocket engine has been the subject of extensive research. The coaxial injector is one of the most commonly used injectors for introducing liquid oxidizer and gaseous fuel into the combustion chamber. Coaxial injectors are used in liquid rocket engines such as the Space Shuttle Main Engine (SSME), the Vulcain 2.1, the Vinci, the RS-25, the J-2, the LE-9, the LE-5, the RD-0146, etc. Atomization is a primary factor determining how efficiently combustion

P. S. Raj · P. Parthasarathy (✉)
Aerospace Research Laboratory, Department of Mechanical Engineering, National Institute of Technology Karnataka, Surathkal, Manglore, India
e-mail: parthasarathy@nitk.edu.in

G. Paramasivan
ISRO Propulsion Complex, Mahendragiri, Tirunelveli District, Tamil Nadu, India

G. Rajesh et al. (eds.), *Proceedings of the 15th International Symposium on Experimental and Computational Aerothermodynamics of Internal Flows*,
https://doi.org/10.1007/978-981-96-4082-9_26

occurs in rocket engines. It is crucial to have a better grasp of these factors since the size, velocity, and distribution of the droplets substantially impact how the engine operates. The relative shear force between the rapidly flowing coaxial gas stream and the slowly moving liquid jet causes the coaxial injector to atomize.

Numerous experimental and numerical studies have been conducted on coaxial injection under various operating conditions as well as on coaxial injectors with simple modifications. The experiment, carried out by Zallar (1990) at the NASA Lewis Research Centre, provided a preliminary understanding of the drop size and velocity distributions in Liquid Oxygen/Hydrogen coaxial injectors at supercritical pressure. The impact of chamber pressure, which varies from subcritical to supercritical, on the coaxial injection has been the subject of extensive experimental study by numerous researchers, including Sankar et al. (1991), Beisler et al. (1994), Mayer and Krulle (1995), Mayer and Tamura (1996), and Chehroudi et al. (2002). Most investigations found that higher chamber pressure led to improved atomization.

The experimental investigation by Kendrick et al. (1999) examines the influence of a recess on cryogenic flame stabilisation in the case of a coaxial injector with no swirl. It is demonstrated that when the LOX tube is recessed with respect to the injection plane, the flame inside the injector is stabilised, the flame expansion angle is increased, the thickness of the flame brush is increased, and the size of the volume where the reaction occurs is increased. The experimental study by Liu et al. (2006) reveals the effect of liquid jet diameter on the performance of liquid/gas airblast atomizers. The size of the droplets highly depends on the liquid jet diameter at a high liquid/gas mass flux ratio and has negligible effect at a lower liquid/gas mass flux ratio. Woodward et al. (2006) experimented on LOX/GH_2 shear coaxial injectors at higher momentum flux values. The flow field characteristics were found to be the same for both high and low momentum flux ratio injections. The findings demonstrated that the flow field is characterised by a long, sinuous LOX core region that breaks into large LOX structures even at high injection momentum flux ratios. The experimental study by Gautam and Gupta (2007) shows that the coaxial gaseous jet of the LN_2/He coaxial rocket injector has a heat-shielding effect. The existence of the gaseous jet considerably reduces the heat transfer from the surroundings to the cold LN_2 jet, which has a considerable impact on the potential core length of the liquid nitrogen jet. Woodward et al. (2007) experimentally studied the effects of non-concentricity and recess. Non-concentric injectors show lower performance and greater core length variability than concentric injectors. The results also show that the recess has a negligible effect on c* efficiency and oxygen dense-core length.

The data provided by the experiment in an injection is often restricted to the far-field region from the injection plane. Access to the primary atomization region near the injector plane in an experimental study is limited. It is not easy to get detailed information on atomization and droplet characteristics in the near-field region of an injector through an experiment. The numerical modelling of injection is challenging as it includes complicated processes. The numerical study of Liang et al. (1986) is one of the primary efforts to simulate such flows. In this study, the solution for the gas and liquid phases is obtained with the combination of the implicit continuous Eulerian technique and the fractional volume of the fluid-free surface description method. The discrete Lagrangian approach is used to solve the particulate phase. O'Rourke and Amsden (1987) developed another numerical method called TAB (Tayler Analogy Breakup) to calculate the droplet dynamics in sprays using the stochastic particle method.

Przekwas et al. (1989) presented a numerical model for tackling primary atomization. In this model, the wave growth and droplet separation processes are computed using the Jet-Embedding method in conjunction with surface wave dispersion. This method is unable to anticipate jet breakup due to turbulence-generated perturbations. Chuech et al. (1991) devised a numerical model for primary atomization. They calculated the mean flow equations of a liquid jet by solving the surface-wave dispersion equations in conjunction with the jet-embedding technique. This approach is not suitable for high-speed jets. It cannot forecast jet breakups caused by turbulence perturbations. This method was developed for the one-dimensional mean flow equations of axisymmetric jets. Mayer et al. (1992) calculated droplet flow through the coaxial injector by the method of Lagrangian tracking together with the stochastic turbulence model based on the Lagrange Integral Time Scale. The model does not calculate the secondary break-up for situations where the Lagrange integral time scale is less than the characteristic break-up time.

A numerical study on mixing and injecting cryogenic fluid jets under supercritical conditions was done by Zong et al. (2004). The turbulence is simulated using the large-eddy-simulation technique. The analysis enables a thorough investigation of the temporal and spatial evolution of a cryogenic jet. The spatial growth rate of the surface instability wave is found to increase with increasing pressure. Additionally, Zong and Yang (2005) performed a numerical analysis on the High-Pressure Oxygen/Methane shear coaxial injector using the large-eddy simulation (LES) method. They observed the formation of two mixing layers close to the injection. They found that turbulent mixing gets improved as the injection velocity of the methane stream increases and the cores of the inner and outer streams get reduced.

Masquelet and Menon (2010) used a large-eddy simulation technique to numerically simulate nonpremixed turbulent combustion and its effects on the wall heat flow in a shear coaxial injector. This large-eddy simulation uses a hybrid central-upwind approach with a thermally ideal formulation to solve the compressible multispecies Navier-Stokes equations. The current 3-D LES formulation, with its hybrid solver, is able to forecast the fundamental physics of the flow. Tani et al. (2017) simulated the breakup and atomization of the LOX core in LOX and GH_2 coaxial combustion at subcritical pressures using the CIP-LSM along with the CIP-CUP method with the hybrid Level-set and MARS method for volume tracking. The findings suggested that the current numerical technique could give extensive information on the LOX core's ignition, breakup and atomization. Shinjo and Umemura (2019) used the method of hybrid atomization LES to investigate the initial spray formation and subsequent autoignition. The method is effective in investigating the initial development stages of evaporative and reactive sprays.

Nowadays, the advancement in technology makes simulations computationally effective. New approaches are developed to simulate complex problems. Li et al. (1999) used a combined method, combining the discrete particle method (DPM) and the volume-of-fluid (VOF) method, to simulate the gas-liquid-solid flow in a fluidized bed. The computation uses the volume-tracking approach for the bubble flow, the Lagrangian simulation for the particle flow and the volume-averaging method for the liquid phase flow. Xu et al. (2013) used the coupled discrete particle model and volume of fluid method to simulate the effect of suspended particles on bubbling behaviour in gas-liquid-solid flow systems. The outcomes are consistent with those of the experiment. The bubble rising and coalescence phenomena in low hold-up particle-liquid suspension systems are simulated using the combined CFD-VOF-DPM method by Liu and Luo (2018). In this case, the particles are thought of as the discrete phase and the liquid as the primary phase. Bubbles or bubble parcels are modelled using the VOF method.

This paper intends to utilize the coupled volume of fluid and discrete phase model (coupled VOF-DPM) to analyze the flow behaviour in the coaxial injector. This research work computationally studies the consequences of the mass flow ratio of gas and liquid on the Sauter mean diameter (SMD) and droplet velocity for coaxial injectors at different injection times. The study provides detailed insight into the primary breakup of the injection and subsequent flow path in the chamber for the coaxial injector.

2 Methodology

The model of the present study includes the injector part and part of the combustion chamber. The Shear Stress Tensor (SST) k-ω model proposed by Menter (1993) and Wilcox (1994) is used for the turbulence modeling in the present numerical study. The liquid core and primary breakup are modeled using the Volume of Fluid (VOF) model, and the droplets formed from the primary breakup are tracked further using the Discrete Phase Model (DPM). The VOF model will solve the single set of momentum equations and track the volume fraction of each fluid throughout the domain. The momentum equation can be written as follows:

$$\frac{\partial(\rho\mathbf{u})}{\partial t}+\nabla\cdot(\rho\mathbf{u}\mathbf{u})=-\nabla p+\nabla\cdot\left[\mu\left(\nabla\mathbf{u}+(\nabla\mathbf{u})^{T}-\frac{2}{3}\nabla\cdot\mathbf{u}I\right)\right]-\nabla\cdot(\rho\overline{\mathbf{u}'\mathbf{u}'})+F_{s}+\rho\mathbf{g} \tag{1}$$

Here $\mathbf{u}$ represents the mean flow velocity vector, $\mathbf{g}$ represents the gravitational acceleration vector, ρ represents the fluid density, μ represents viscosity, p is the static pressure, the I is unit tensor, $\mathbf{u}'$ is the velocity fluctuation due to turbulence and F_s represents the momentum transfer term exerted by discrete phase.

The interface between the phases is obtained by solving the continuity equation for the volume fraction of phases. The equation for the volume fraction ith phase is

$$\frac{1}{\rho_i}\left[\frac{\partial}{\partial t}(\alpha_i\rho_i)+\nabla\cdot(\alpha_i\rho_i\mathbf{u}_\mathbf{i})=S_{\alpha_i}+\sum_{i=1}^{n}\left(m_{ij}-m_{ji}\right)\right] \tag{2}$$

where ρ_i is the density of phase i, α_i is the volume fraction of phase i, $\mathbf{u_i}$ is the velocity of phase i, m_{ij} is the mass transfer from phase i to j phase and m_{ji} is the mass transfer from phase j to i phase. The volume fraction for the primary phase is computed from the following equation:

$$\sum_{j=1}^{n} \alpha_j = 1 \tag{3}$$

The DPM model tracks the droplets formed from the primary breakup. The equation for force balance on the particle, which equates the inertia of the particles with the forces acting on the particles, is as follows:

$$\frac{dx_p}{dt} = \mathbf{u_p} \tag{4}$$

$$\frac{d\mathbf{u_p}}{dt} = F_D(\mathbf{u} - \mathbf{u_p}) + \frac{\mathbf{g}\left(\rho_p - \rho\right)}{\rho_p} \tag{5}$$

Here x_p is particle displacement, $\mathbf{u}$ is the fluid phase velocity, $\mathbf{u_p}$ is the particle velocity, $\mathbf{g}$ is the gravitational acceleration vector, ρ is the fluid density, ρ_p is the density of the particle, and $F_D(\mathbf{u} - \mathbf{u_p})$ is the drag force per unit particle mass and :

$$F_D = \frac{18\mu C_D Re}{\rho_p d_p^2 24} \tag{6}$$

Where μ is the molecular viscosity of the fluid, d_p is the diameter of the particle, Re is the Reynolds number and C_D is the drag force coefficient.

$$Re = \frac{\rho d_p |\mathbf{u_p} - \mathbf{u}|}{\mu} \tag{7}$$

The drag coefficient, C_D can be obtained from the equation given by Morsi and Alexander (1972) for the smooth spherical particles. It is in the form

$$C_D = a_1 + \frac{a_2}{Re} + \frac{a_3}{Re^2} \tag{8}$$

Where, a_1, a_2,and a_3 are constants for different ranges of Re.

The term F_s in the Eq. (1) is defined in the form given in Eq. (9).

$$F_s = \sum \left[F_D(\mathbf{u} - \mathbf{u_p}) + \frac{\mathbf{g}\left(\rho_p - \rho\right)}{\rho_p} \right] \dot{m}_p \Delta t \tag{9}$$

where $\dot{m}_p$ is the mass flow rate of the particulate phase and Δt is the time step.

The VOF-to-DPM model transition algorithm automatically finds compact liquid masses detached from the liquid core in the VOF solution. If the liquid lump satisfies user-specified criteria, it is removed from the resolved liquid in the VOF simulation and converted to a particle parcel in the Lagrangian formulation. The size of the liquid mass is the first criterion. There is a minimum and a maximum diameter for the volume equivalent. The liquid lump's asphericity is measured as the next criterion. The lump's asphericity reveals how much it resembles a perfect sphere. Two methods are used to determine asphericity. The normalised radius standard deviation (or variance) is the first one. The distance between a facet centroid and the lump centre of gravity divided by the average radius is known as the normalised radius. The second is the average radius-surface orthogonality. It is described as the degree of relative orthogonality of the vector joining the lump centroid to the facet centroid and facet normal. In the scenario of an ideal sphere, these two vectors overlap. The conversion of droplets from the VOF framework to DPM is done with the help of the VOF-DPM algorithm employed in ANSYS fluent 2021 R2.

Local solution-adaptive mesh refinement is used to resolve the gas-liquid interface. When coupled with an adaptive mesh, the solver enables effective utilisation of computational resources by coarsening the mesh after lumps are transformed into the DPM.

2.1 Numerical Simulation

The numerical validation of the simulation models is carried out based on the experiment by Krulle et al. (1990). The geometry of the coaxial injector used for the study is shown in Fig. 1. The numerical simulation is conducted with water and air fluids under ambient conditions. One-fourth of the actual domain is selected for simulation, and the symmetry boundary condition is applied.

ANSYS 2021 R2 Design Modeller was used to construct the 3D geometry. The fluent meshing module of ANSYS 2021 R2 was utilized to generate the mesh. The domain's outer region comprises polyhedra cells, while the interior comprises hexa

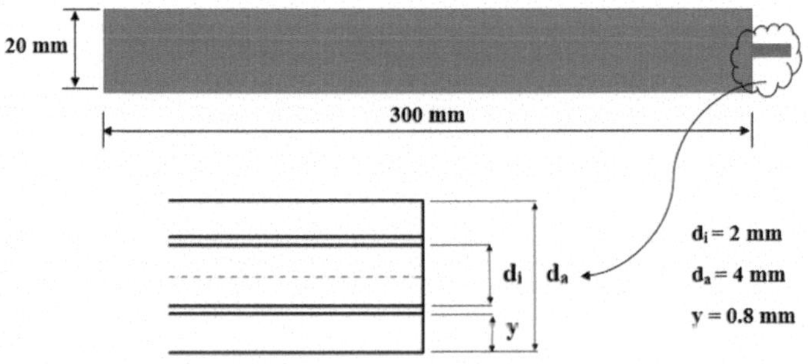

Fig. 1 Schematic representation of the coaxial injector

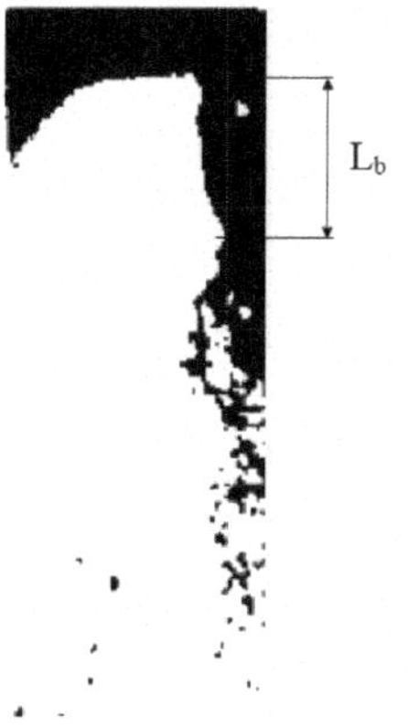

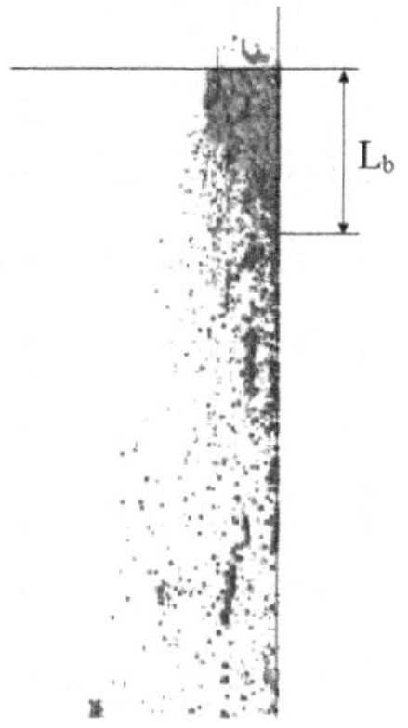

Fig. 2 Coaxial injector with a water velocity of 0.4 m/s and air velocity of 100 m/s

cores. All domain walls have undergone mesh refinement (inflation layers). The liquid gas interface mesh can be refined further up to a maximum of six levels via dynamic mesh adaptation. The basic mesh has 1214411 nodes and 456635 cells. The water inlet and the air inlet use the mass flow inlet boundary condition. At the outlet, a pressure outlet boundary condition is applied. Perpendicular surfaces along the length of the chamber are subject to the symmetry boundary condition. All other surfaces are subject to the wall boundary condition. A pressure-based transient simulation was performed. Multiphase modelling is done using the VOF-DPM transition tool. The pressure velocity coupling was resolved using the coupled scheme. Pressure is discretized using the PRESTO! Scheme. Volume fraction discretization is accomplished using the geo-Reconstruct scheme. For maintaining interface sharpness, it is the best scheme. The second order upwind scheme discretizes kinetic energy, specific dissipation rate, and momentum. For transient formulation, the first-order implicit method is utilized.

In the experiment (Krulle et al., 1990), water and air are passed through the coaxial injector with velocities of 0.4 m/s and 100 m/s, respectively. A comparison of the results obtained from the work of Krulle et al. (1990) and the present work is shown in Fig. 2. It is observed that both have identical breakup lengths.

The simulations of flow through the coaxial injector are carried out for one air flow rate of 9.852×10^{-4} kg/s and three water flow rates of 1.254×10^{-3}, 9.408×10^{-3}, and 3.136×10^{-2} kg/s. The properties of the water and air considered for this study are shown in Table 1.

Table 1 The properties of the water and air

Properties	Water	Air
$Density(kg/m^3)$	998.2	1.225
Dynamic viscosity (kg/ms)	0.001003	1.7894×10^{-5}
Surface tension coefficient (N/m)	0.0708	

3 Results and Discussion

The simulation results of flow through the coaxial injector with the water and air under ambient conditions are discussed below. Figure 3 shows the isosurface of the liquid phase volume fraction and the droplets converted to DPM particles at three different water mass flow rates. It is taken after 0.5 ms of time of injection. When the mass flow rate of water increases, the distance traveled by the droplets at a given time increases due to an increase in flow velocity. The breakup length increases with the increase in water mass flow rate. The liquid-gas momentum ratio and liquid inertia are high at a high mass flow rate. The liquid jet becomes less susceptible to the effects of air flow as the liquid mass flow rate increases. So liquid can move long distances without breakup. At a low mass flow rate, the liquid surface is unstable. Then it gets dispersed quickly.

Figure 4 shows the velocity contours through coaxial injector at time of injection = 0.5 ms. Velocity contours show that the velocity increases slightly near the injection plane in comparison to the inlet velocity. A constant mass flow rate of water is applied at the inlet. The velocity decreases due to the formation of the boundary layer, and the

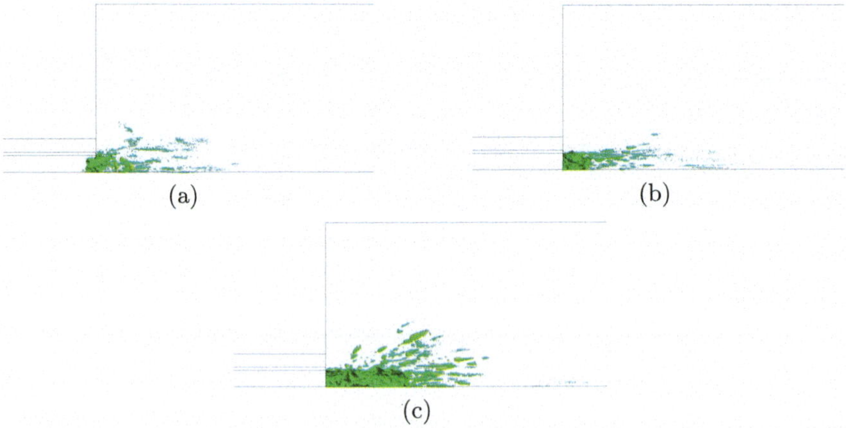

Fig. 3 Flow through coaxial injector for different water mass flow rates of (**a**) $\dot{m} = 1.254 \times 10^{-3}$kg/s (**b**) $\dot{m} = 9.408 \times 10^{-3}$kg/s (**c**) $\dot{m} = 3.136 \times 10^{-2}$kg/s

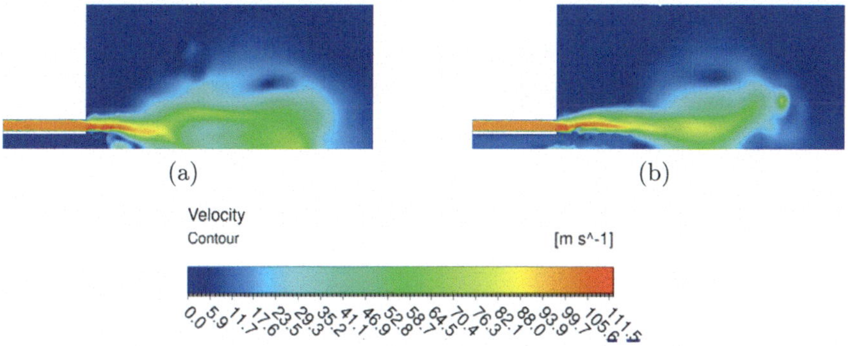

Fig. 4 Velocity contours through coaxial injector for different water mass flow rates of **a** $\dot{m} = 9.408 \times 10^{-3}$kg/s **b** $\dot{m} = 3.136 \times 10^{-2}$kg/s

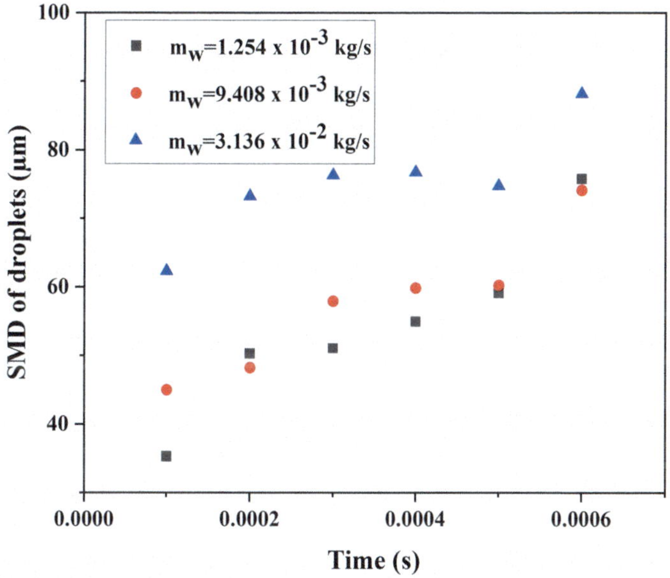

Fig. 5 Effect of water mass flow rate on Sauter Mean Diameter (SMD)

flow area remains constant, so to maintain a constant mass flow rate, the centerline velocity increases.

Figure 5 shows the variations of Sauter Mean Diameter (SMD) for different water mass flow rates. It is found that SMD in the complete chamber increases with time at all mass flow rates of water. It is also found that SMD increases with an increase in the mass flow rate of water at any particular time after injection.

The effect of the mass flow rate of water on the number of droplets formed is shown in Fig. 6. The number of droplets formed is higher at a higher mass flow rate of water. The number of droplets increases with the time of injection.

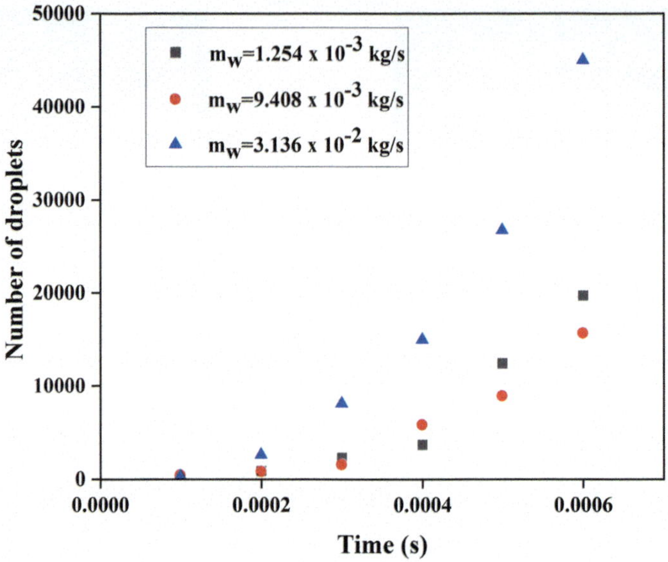

Fig. 6 Effect of water mass flow rate on number of droplets formed

Variation of SMD along the axis for different water mass flow rates is shown in Fig. 7. An overall decrease in SMD is observed when it moves from the injection plane to the outlet along the axial direction at all water mass flow rates. The size of droplets in the spray varied depending on their distance from the injection plane. The largest droplets were found closest to the injection plane, where there was the greatest amount of liquid. As the distance from the injection plane increased, the droplets became smaller. The airflow caused some of the larger droplets to break apart, creating smaller droplets near the edge of the spray. Additionally, small droplets were formed when the liquid cluster broke apart. These small droplets dispersed more quickly than the larger ones.

Variation of droplet velocity along the axis for different mass flow rates of water is shown in Fig. 8. It has been noticed that droplets move slower near the injection plane. However, their velocity increases as they receive momentum from the high-velocity air. The air velocity is lower at a greater distance from the injection plane, which affects the droplets in that region, causing them to also move at a slower pace. Droplet velocity increases with a decrease in the mass flow rate of water. The higher mass flow rate of water produces large droplets with low velocity.

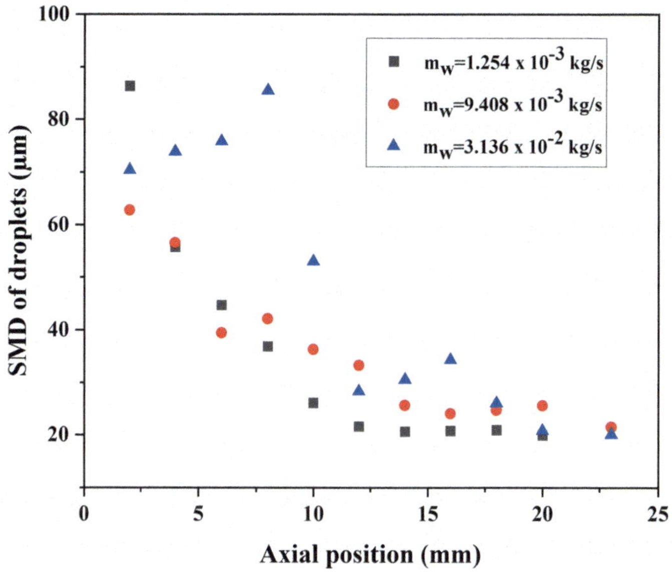

Fig. 7 Variation of Sauter Mean Diameter along the axis for different mass flow rate of water

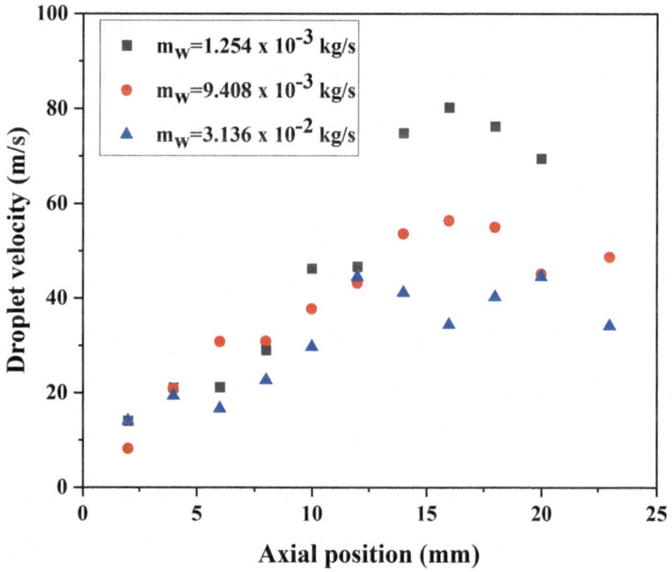

Fig. 8 Variation of droplet velocity along the axis for different mass flow rate of water

4 Conclusion

The numerical simulation of the flow through the coaxial injector is done. It is found that SMD shows an overall increase with time at all mass flow rates. The size of the droplet increases with increase in the mass flow rate of water. An overall decrease in SMD is observed when it moves from the injection plane to the outlet. The number of droplets formed increases with an increase in the mass flow rate of water. Droplet velocity increases with a decrease in the mass flow rate of water.

References

Beisler, M.A., Pal, S., Moser, M.D., Santoro, R. J.: Shear coaxial injector atomization in a LOX/GH^2 propellant rocket. In: 30th AIAA/ASME/SAE/ASEE Jt. Propuls. Conf., Indianapolis, IN, (1994)

Chehroudi, B., Cohn, R., Talley, D.: Cryogenic shear layers: experiments and phenomenological modelling of the initial growth rate under subcritical and supercritical conditions. Int. J. Heat Fluid Flow 23, 554–563 (2002)

Chuech, S. G., Przekwas, A. J., Singhal, A. K.: Numerical modelling for primary atomization of liquid jets. J. Propuls, 7(6), 879–86 (1991)

Gautam, V., Gupta, A. K.: Characterization of cryogenic flow and atomization from a coaxial rocket injector. In: 43rd AIAA/ASME/SAE/ASEE Jt. Propuls. Conf. Exhib., Cincinnati, OH, AIAA 2007–5567 (2007)

Kendrick, D., Herding, G., Scouflaire, P., Rolon, C., Candel, S.: Effects of a recess on cryogenic flame stabilization. Combust. Flame, 118, 327–339 (1999)

Krulle., G., Mayer, W., Schley, C.-A.: Recent advances in H^2/O^2 high pressure coaxial injector performance analysis. In: 26th AIAA/SAE/ASME/ ASEE Jt. Propuls. Conf., Orlando, Florida, AIAA 90-1959 (1990)

Li, Y., Zhang, J., Fan, L. S.: Numerical simulation of gas-liquid-solid fluidization systems using a combined CFD-VOF-DPM method: bubble wake behaviour. Chem. Eng. Sci. 54, 5101–5107 (1999)

Liang, P. Y., Fisher, S., Chang, Y. M.: Comprehensive modelling of a liquid rocket combustion chamber. J. Propuls., 2(2), 97–104 (1986)

Liu, Q., Luo, Z.H: CFD-VOF-DPM simulations of bubble rising and coalescence in low holdup particle-liquid suspension systems. Powder Technol. 339, 459–469 (2018)

Liu, H. F., Li, W. F., Gong, X., Cao, X. K., Xu, J. L., Chen, X. L., Wang, Y. F., Yu, G. S., Wang, F. C., Yu, Z. H.: Effect of liquid jet diameter on performance of coaxial two-fluid airblast atomizers. Chem. Eng. Process. 45, 240–245 (2006)

Masquelet, M., Menon, S.: Large-Eddy Simulation of flame-turbulence interactions in a shear coaxial injector. J. Propuls. Power 26 (5), 924–935 (2010)

Mayer, W., Labani, R., Krulle, G.: Theoretical Investigations of Droplet Flow under Typical Coaxial Injector Flow Conditions in Cryogenic Rocket Engines. In: 28th AIAA/ASME/SAE/ASEE Jt. Propuls. Conf. Exhib., Nashville, TN, AIAA 92-3121 (1992)

Mayer, W., Krulle, G.: Rocket engine coaxial injector Liquid/Gas interface flow phenomena. J. Propuls. Power 11 (3), 513–518 (1995)

Mayer, W., Tamura, H.: Propellant injection in a Liquid Oxygen/Gaseous Hydrogen rocket engine. J. Propuls. Power 12(6), 1137–1147 (1996)

Menter, F. R.: Zonal two equation k-omega turbulence models for aerodynamic flows. In: 24th Fluid Dyn. Conf., Orlando, Florida, AIAA 93-2906 (1993)

Morsi, S.A., Alexander, A.J.: An investigation of particle trajectories in two-phase flow systems, J. Fluid Mech. 55 193–208 (1972)

O'Rourke, P. J., Amsden, A. A.: The TAB method for numerical calculations of spray droplet breakup. In: Int. Fuels Lubr. Meet. Expo., Toronto, Ontario. (1987)

Przekwas, A., Chuech, S., Singhal, A.: Numerical modelling for primary atomization of liquid jets. In: 27th Aerosp. Sci. Meet., Reno, Nevada, AIAA 89-0163 (1989)

Sankar, S. V, Rosa, A. B. D. la, Isakovic, A., Bachalo, W. D.: Liquid atomization by coaxial rocket injectors. In: 29th Aerosp. Sci. Meet., Reno, Nevada, pp. 9100691, (1991)

Shinjo, J., Umemura, A.: Fluid dynamic and auto ignition characteristics of early fuel sprays using hybrid atomization LES. Combust. Flame 203, 313–333 (2019)

Tani, H., Umemura, Y., Daimon, Y.: Interface-Tracking simulations of liquid Oxygen/gaseous Hydrogen coaxial combustion at subcritical pressures. In: 55th AIAA Aerosp. Sci. Meet., Grapevine, Texas, AIAA SciTech Forum, AIAA 2017–0144 (2017)

Wilcox, D. C.: Turbulence modelling for CFD, DCW Industries, Inc., La Canada, California (1994)

Woodward, R. D., Pal, S., Farhangi, S., Jensen, G. E., Santoro, R. J.: LOX/GH^2 shear coaxial injector atomization studies: Effect of recess and non-concentricity. In: 45th AIAA Aerosp. Sci. Meet. Exhib., Reno, Nevada, AIAA 2007–571 (2007)

Woodward, R. D., Pal, S., Farhangi, S., Santoro, R. J.: LOX/GH2 Shear coaxial injector atomization studies at large momentum flux ratios, In: 42nd AIAA/ASME/SAE/ASEE Jt. Propuls. Conf. Exhib., Sacramento, California, AIAA 2006–5203 (2006)

Xu, Y., Liu, M., Tang, C.: Three-dimensional CFD-VOF-DPM simulations of effects of low-holdup particles on single-nozzle bubbling behavior in gas-liquid-solid systems. Chem. Eng. J. 222, 292–306 (2013)

Zaller, M.: LOX / Hydrogen coaxial injector atomization test program. NASA Contract. Report, 187037, (1990)

Zong,N., and Yang, V.: A numerical study of high-pressure Oxygen/Methane mixing and combustion of a shear coaxial injector. In: 43rd AIAA Aerosp. Sci. Meet. Exhib., Reno, Nevada, AIAA-2005-0152 (2005)

Zong,N., Meng, H., Hsieh, S.Y., Yang, V.: A numerical study of cryogenic fluid injection and mixing under supercritical Conditions. Phys. Fluids 16(12), 4248–4261 (2004)

The manufacturer's authorised representative in the EU is Springer
Nature Customer Service Centre GmbH, Europaplatz 3, 69115 Heidelberg,
Germany. If you have any concerns regarding our products, please
contact ProductSafety@springernature.com

Printed and bound by CPI Group (UK) Ltd, Croydon, CR0 4YY
01/10/2026
02225926-0011